Statistical Methods for Food Science

To Jonathan and Cassandra

and to Pushka-Latitia with love

"Everything is on a scale . . ."

Statistical Methods for Food Science

Introductory procedures for the food practitioner

Second Edition

John A. Bower
Former Lecturer and Course Leader (BSc Food Studies)
Queen Margaret University, Edinburgh, UK

Library of Congress Cataloging-in-Publication Data

Bower, John A. (Lecturer in food science)
    Statistical methods for food science : introductory procedures for the food practitioner / John A. Bower, former lecturer and Course Leader (BSc Food Studies) Queen Margaret University, Edinburgh, UK. – Second edition.
        pages cm
    Includes bibliographical references and index.
    ISBN 978-1-118-54164-7 (softback : alk. paper) – ISBN 978-1-118-54159-3 – ISBN 978-1-118-54160-9 (epdf) – ISBN 978-1-118-54161-6 (emobi) – ISBN 978-1-118-54162-3 (epub) 1. Food–Research–Statistical methods. I. Title.
    TX367.B688 2013
    664.0072–dc23
                                                                            2013001794

A catalogue record for this book is available from the British Library.

Contents

Preface

The recording and analysis of food data are becoming increasingly sophisticated. Consequently, the practicing food scientist in industry or at study faces the task of using and understanding statistical methods. Unfortunately, statistics is often viewed as a difficult subject and tends to be avoided because of complexity and lack of specific application to the food field. While this situation is changing and there is much material on multivariate applications for the more advanced reader, a case exists for a univariate approach for the non-statistician. That is the intent of this book. It provides food scientists, technologists and other food practitioners with a source text on accessible statistical procedures. Material for students and workers in the food laboratory is included, covering the food analyst, the sensory scientist and the product developer. Others who work in food-related disciplines involving consumer survey investigations will also find many sections of use. Emphasis is on a 'hands-on' approach with worked examples using computer software with the minimum of mathematical formulae.

For the second edition, the content has been revised, some errors corrected and additional information and detail given at various points. The main thrust of the analyses, Excel use, has been updated for Excel 2010 format, whilst retaining the instructions for Excel 2003. This update includes some of the amended formulae as well as the new style of menu interaction when using the calculation and charting facilities.

About the companion website

This book is accompanied by a companion website:

www.wiley.com/go/bower/statistical

Excel spreadsheet files with data are available on the website in both formats (2010 and 2003), so that readers can perform the calculations in the text or use new data for their own examples and exercises.

Acknowledgements

I thank all students of food at Queen Margaret University, Edinburgh (1985–2007), for their enthusiasm and contribution to experimentation in the study of food science and consumer studies. Also, thanks are due to my colleagues in the Department of Consumer Studies, in particular Dr. Monika Schröder for many stimulating conversations and for her help with reading and editing of early drafts.

The publishers and author thank the following for permission to include output material:

SPSS Software (IBM®/SPSS®)[1] graphs and tables by International Business Machines. Corporation (pp. 193, 295, 299, 300, 301, 303, 307 and 308).
Excel spreadsheet tables and graphs by Microsoft® Corp.
Design-Expert(R) software output by Stat-Ease, Inc.

Examples of Minitab® Statistical Software output by Minitab Inc.: Portions of information contained in this publication/book are printed with permission of Minitab Inc. All material remains the exclusive property and copyright of Minitab Inc. All rights reserved.

Trademark notice

The following are trademarks or registered trademarks of their respective companies:

Excel is a trademark of Microsoft® Corporation.
Minitab (Minitab® Statistical Software™) is a trademark of Minitab Inc.
Design-Expert(R) software is a trademark of Stat-Ease Inc., Minneapolis, MN.
SPSS is a trademark or registered trademark of International Business Machines Corp., registered in many jurisdictions worldwide.
MegaStat is a trademark of McGraw-Hill/Irwin, McGraw-Hill Companies Inc.

Minitab® and all other trademarks and logos for the Company's products and services are the exclusive property of Minitab Inc. All other marks referenced remain the property of their respective owners. See minitab.com for more information.

---

[1] SPSS Inc. was acquired by IBM in October 2009.

Part I
Introduction and basics

Chapter 1
Basics and terminology

1.1 Introduction

Food issues are becoming increasingly important to consumers, most of whom depend on the food industry and other food workers to provide safe, nutritious and palatable products. These people are the modern-day scientists and other prac titioners who work in a wide variety of food-related situations. Many will have a background of science and are engaged in laboratory, production and research activities. Others may work in more integrated areas such as marketing, consumer science and managerial positions in food companies. These food practitioners encounter data interpretation and dissemination tasks on a daily basis. Data come not only from laboratory experiments but also via surveys on consumers, as the users and receivers of the end products. Understanding such diverse information demands an ability to be, at least, aware of the process of analysing data and inter-preting results. In this way, communicating information is valid. This knowledge and ability gives undeniable advantages in the increasingly numerate world of food science, but it requires that the practitioners have some experience with statistical methods.

Unfortunately, statistics is a subject that intimidates many. One need only con-sider some of the terminology used in statistic text titles (e.g. 'fear' and 'hate'; Salkind 2004) to realise this. Even the classical sciences can have problems. Pro-fessional food scientists may have received statistical instruction, but application may be limited because of 'hang-ups' over emphasis on the mathematical side. Most undergraduate science students and final-year school pupils may also find it difficult to be motivated with this subject; others with a non-mathematical back-ground may have limited numeracy skills presenting another hurdle in the task.

These issues have been identified in general teaching of statistics, but like other disciplines, application of statistical methods in food science is continually progressing and developing. Statistical analysis was identified, two decades ago, as one subject in a set of 'minimum standards' for training of food scientists at

Statistical Methods for Food Science: Introductory Procedures for the Food Practitioner,
Second Edition. John A. Bower.
© 2013 John Wiley & Sons, Ltd. Published 2013 by John Wiley & Sons, Ltd.

undergraduate level (Iwaoka *et al.* 1996). Hartel and Adem (2004) identified the lack of preparedness for the mathematical side of food degrees, and they describe the use of a quantitative skills exercise for food engineering, a route that merits attention for other undergraduate food science courses.

Unfortunately, for the novice, the subject is becoming more sophisticated and complex. Recent years have seen this expansion in the world of food science, in particular in sensory science, with new journals dealing almost exclusively with statistical applications. Research scientists in the food field may be cognizant with such publications and be able to keep abreast of developments. The food scientist in industry may have a problem in this respect and would want to look for an easier route, with a clear guide on the procedures and interpretation, etc. Students and pupils studying food-related science would also be in this situation. Kravchuk *et al.* (2005) stress the importance of application of statistical knowledge in the teaching of food science disciplines, so as to ensure an ongoing familiarity by continual use.

Some advantages of being conversant with statistics are obvious. An appreciation of the basis of statistical methods will aid making of conclusions and decisions on future work. Other benefits include the increased efficiency achieved by taking a statistical approach to experimentation. Guiding the reader on the path to such knowledge and skills begins with a perusal of the book contents.

What will this book give the reader?

The book will provide the reader with two main aspects of statistical knowledge. One is a workbook of common univariate methods (Part I) with short explanations and implementation with readily available software. Secondly (Part II), the book covers an introduction to more specific applications in a selection of specialised areas of food studies.

1.2   What the book will cover

Chapter 1 introduces the book and gives a summary of how the chapter contents will deal with the various aspects. Accounts of the scope of data analysis in the food field, its importance and the focus of the text lead on to a terminology outline and advice on software and bibliography.

Chapter 2 begins with consideration of data types and defines levels of measurement and other descriptions of data. Sampling, data sources and population distributions are covered.

Chapter 3 introduces the style of the analysis system used with the software and begins with simple analysis for summarising data in graph and table format. Measures including mean, median, mode, standard deviation and standard error are covered, along with various types of graphs. Definitions and application of some of these methods to measures of error, uncertainty and sample character are also given.

Chapters 4–6 cover various aspects of analysis of effects. Firstly, Chapter 4 gives a detailed account of significance testing. Analysis of significant differences, probability and hypothesis testing and its format are described and discussed. The chapter concludes with consideration of types of comparison and factors deciding selection of a test, including assumptions for use of parametric methods. Chapter 5 continues with significance tests themselves, with tests for parametric and non-parametric data, two or more groups, and related and independent groups. Chapter 6 describes effects in the form of relationships as association (cross tabulation) and correlation (coefficients) and their significance. The topic of correlation is then applied in simple regression and prediction.

Chapter 7 concludes cover of basic material by detailing the nature and terminology of experimental design for simple experiments. Stages in the procedure, such as identification of factors and levels, and sources of experimental error and their elimination are explained. Details of design types for different sample, factor, treatment and replication levels are then described.

Chapters 8 and 9 start the applications part of the book. In Chapter 8, sensory and consumer data are described in terms of level of measurement, sources, sampling via surveys, sensory panels and consumer panels. Summary methods and evaluation of error, reliability and validity in these data sources are considered along with checking on assumptions for parametric nature. Specific methods of analysis are then illustrated for a range of consumer tests and survey data, and for specific sensory tests and monitoring of sensory panels.

Chapter 9 uses a similar approach to instrumental data. They are described in terms of level of measurement, sources and sampling via chemical and physical methodologies in food science. Analytical error, repeatability and accuracy are defined followed by use of calibration and reference materials. An account is then given of specific significance analysis methods for laboratory work results and experiments.

Chapter 10 applies experimental design to formulation procedures in food product development. Identification of factors and levels as ingredients for simple designs is given viewed from the formulation aspect. Decisions on the response and its measurement are described along with the issues in objective versus hedonic responses. Examples of some formulation experiments are used to illustrate the analysis methods and their interpretation.

Chapter 11 deals with the application of the basic methods and experimental design to the case of quality control procedures. Key features of sampling quality measurement are outlined and then application of statistical methods to production is explained. The data forms generated and their analysis are displayed with reference to control charts and acceptance sampling.

Chapter 12 provides some indication of how to take the univariate methods a stage further and apply them to multivariate situations. A selection of commonly used more advanced and multivariate methods are briefly described, with more detail on principal component analysis. The analysis of sensory and instrumental data and how to combine them in the multivariate context concludes.

Despite the above examples and the description of content, many scientists and workers in the food field may still ask why they are required to have such knowledge. This question can be answered by considering the importance and application of the subject in the food field.

1.3   The importance of statistics

Why are statistical methods required?

It is possible to evaluate scientific data without involving statistical analysis. This can be done by experienced practitioners who develop a 'feel' for what the data are 'telling them', or when dealing with small amounts of data. Once data accumulate and time is limited, such judgement can suffer from errors. In these cases, simple statistical summaries can reduce large data blocks to a single value. Now, both the enlightened novice and the experienced analyst can judge what the statistics reveal. Consequent decisions and actions will now proceed with improved confidence and commitment. Additionally, considerable savings in terms of time and finance are possible.

In some instances, decision-making based on the results of a statistical analysis may have serious consequences. Quantification of toxins in food and nutrient content determination rely on dependable methods of chemical analysis. Statistical techniques play a part in monitoring and reporting of such results. This gives confidence that results are valid and consumers benefit in the knowledge that certain foods are safe and that diet regimes can be planned with surety. Other instrumental and sensory measures on food also receive statistical scrutiny with regard to their trustworthiness. These aspects are also important for food manufacturers who require assurance that product characteristics lie within the required limits for legal chemical content, microbiological levels and consumer acceptability. Similarly, statistical quality control methods monitor online production of food to ensure that manufacturing conditions are maintained and that consumer rights are protected in terms of net weights, etc. Food research uses statistical experimental design to improve the precision of experiments on food.

Thus, manufacturers and consumers both benefit from the application of these statistical methods. Generally, statistics provides higher levels of confidence and uncertainty is reduced. Food practitioners apply statistical methods, but ultimately, the consumer benefits.

1.4   Applications of statistical procedures in food science

There are many applications of statistics in the field of food studies. One of the earliest was in agriculture where Fisher used experimental design to partition variation and to enable more precise estimation of effects in crop plot experiments. There was even an early sensory experiment on tea tasting (Fisher 1966), and since then, statistical applications have increased as food science emerged as a distinct

Table 1.1   Some applications of statistics in the food field.

Method	Application
Summaries of results	Tables, graphs and descriptive statistics of instrumental, sensory and consumer measures of food characteristics
Analysis of differences and relationships	Research applications on differences in food properties due to processing and storage; correlation studies of instrumental and sensory properties
Monitoring of results	Statistical control of food quality and parameters such as net filled weight
Measurement system integrity	Uncertainty of estimates for pesticides and additives levels in food
Experimental design	Development and applications of balanced order designs in sensory research

applied science subject. Some examples of the form of statistical applications in food are given in Table 1.1.

Preparation of data summaries is one general application of statistics that can be applied across the board. It is one of the simplest applications and can be done manually if necessary, depending on the requirements. A variety of simple graphs and table methods are possible, which allow rapid illustration of results. These summaries are taken further in statistical quality control where measures such as the mean value are plotted 'live', as a process is ongoing. The graphs (control charts) used include limit lines that are set by using other statistical methods, which allow detection of out-of-limit material, e.g. food product packs that are below statutory minimum net weight. Statistical methods can also be applied to evaluate the trustworthiness of data obtained by any method of measurement. This application has been used extensively in evaluation of chemical data generated by analytical laboratories. The statistical analysis provides an evaluation of how dependable the analytical results are. This can range from within-laboratory to between-laboratory comparisons, globally. Enforcement agencies rely on such checks so that they can monitor adherence to legal requirements with confidence.

Food research application brings in analysis of differences and relationships. Here, hypotheses are put forward on the basis of previous work or new ideas and then magnitudes of effects in sample statistics can be assessed for significance, for instance, examination of the change in colour pigment content during frozen storage of vegetables.

Examination of relationships requires that different measurement systems are applied and then compared. There are many examples of this in studies of food where data from instrumental, sensory and consumer sources are analysed for interrelationships.

The process of sampling of items, including food material and consumer respondents, can be controlled using statistical methods, and here, a statistical appreciation of variability is important. Experimental design takes this further, where sources of such variation are partitioned to improve precision or controlled and minimised if extraneous. A common example is the unwanted effect of order of samples in the sensory assessment of foods – design procedures can minimise this.

In fact, *all* the above examples rely on design procedures if the result is to be valid and adequately interpreted. Experimental design is dealt with fully in a later chapter, but an introduction to some aspects of experimentation is important at this point to provide a foundation.

1.4.1   The approach to experimentation

Why are experiments and research necessary?

Progress in food science and all its associated disciplines is underpinned by research activity. New information is gathered by investigations and experiments, and in this way knowledge is advanced. The scientific approach to research and exploration follows an established paradigm called the ***positivism method***. This postulates that events and phenomena are objective and concrete, able to be measured and can be explained in terms of chemical and physical reactions. All scientists are familiar with this viewpoint, which is described as the ***scientific deductive approach*** (Collis and Hussey 2003). It is largely based on ***empirical methods***, i.e. observations from experiments. The scientific style of approach can be used for any type of investigation in any subject.

The procedure uses deduction from theory based on current knowledge. To advance knowledge, experiments can be designed to test advances on existing or new theory, using a ***hypothesis process***. The findings can then be disseminated and knowledge increased. Results are generalised and can be used to establish new theories and to model processes and event reactions, which in turn allows prediction in the formation of new hypotheses. The term ***quantitative research*** is also used in reference to the scientific approach. This strictly refers to the nature of the data generated, but it implies the deductive positivistic viewpoint.

In this process, the researcher is assumed to be objective and detached. Ultimately, the deductive method searches for an explanation on the basis of *cause–effect* relationships. Without such procedures, there would be no progress and they form the foundation of the scientific approach in many food disciplines.

A more recent approach is that of ***phenomenology*** where an ***inductive approach*** can be used to examine phenomena on the basis that they are socially constructed. Theories and explanations are generated and built up from data gathered by methods and techniques such as interviews (Blumberg *et al.* 2005). These methods are often described as ***qualitative***, which again refers to the data that are in the form of words rather than numbers. The modern food practitioner needs to be aware of such data as there are several qualitative methods (e.g. interviews and focus groups) used in sensory and consumer work. Analysis of data from ***qualitative methods*** can be summarised by numerical techniques such as counting the incidence of certain words and phrases, but usually statistical analysis as such is not involved.

Typical use of the scientific approach in food studies entails identifying a topic for research or investigation then posing a ***research question(s)***. Deductive reasoning from existing knowledge is examined to develop a ***research hypothesis***.

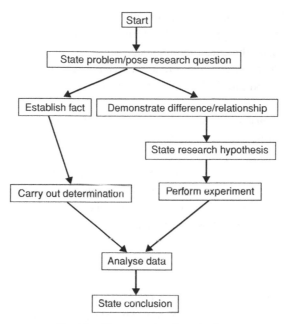

Fig. 1.1   The approach to investigation.

A plan can then be drawn up with an experimental design and specification of measurement system, etc. Data are gathered and then statistical analysis is used to test the hypothesis (quantitative). The scope of the procedure can be from a simple investigation of the 'fact-finding' type, e.g. determination of chemical content values, to a complex experimental design, e.g. a study on the effect of temperature, pressure and humidity levels on the drying properties of a food. In this latter case, the objective would be to identify any significant differences or relationships. Experimental control means that results can be verified and scrutinised for validity and other aspects.

Simple experiments do not usually require stating of hypotheses, etc. In circumstances where differences or relationships are being examined, e.g. 'Does process temperature affect yield of product', a more formal procedure is used or, at least, assumed (Fig. 1.1).

The conclusion of one investigation is not the end of the process as each piece of work leads to new ideas and further studies, etc.

1.5   Focus and terminology

It is important at this point to give some indication of what this text will concentrate on in terms of applications. Data from three main areas are drawn on:

- Instrumental methods of analysis
- Sensory methods
- Consumer tests and measures

These divisions are broad enough in scope to provide examples of a range of statistical applications. The terms above are used throughout this book in reference to data (thus, instrumental data, sensory data, etc.).

Instrumental measures itself can cover any measurement system from chemical and physical analysis to specific food instrumentation methods and process measures, e.g. protein content, Lovibond colour measures and air speed setting. *Sensory measures* include all sensory tests used by trained assessors such as discrimination tests and descriptive analysis methods. *Consumer tests* include some sensory methods, which are affective or hedonic in nature, e.g. preference ranking. The above systems cover mostly laboratory measurements.

Consumer measures refer to questionnaire measures in surveys, such as consumers' views and opinions on irradiated foods. These consumer applications are usually non-laboratory in nature.

1.5.1   Audience

As stated above, this book is aimed at the food scientist or food practitioner who has to undertake some aspect of data analysis and interpretation. The intention is not to include formulae and mathematical content wherever possible. All calculations are done using appropriate statistical software. In this respect, the text cannot be viewed as a statistical source, but numerous references for those readers who wish more on formulae are given. The emphasis is on providing a basic account of how methods and tests are selected, how to avoid inappropriate use, how to perform the tests and how to interpret the results.

Software packages

All examples of statistical tests and methods are selected from the appropriate menus of a software package usable with *Microsoft*® *Windows*®. Readers will require some familiarity with an analysis package. *Microsoft*® *Excel*® is probably the most accessible and available package that includes statistical analysis and functions. Applications employ the *Excel 2003* and *Excel 2010*[1] versions, as used in Windows for PC. Explanation of the differences between the functions for these versions is detailed in some sections, otherwise the 2010 function is stated followed by the 2003 form in brackets. As far as possible, Excel functions and data analysis add-ins are used, followed by those of *Minitab* (Minitab® Statistical Software™; any basic version). Some guidance is given on use of functions and commands for these packages, but the instructions cannot be viewed as comprehensive and some readers may wish to consult texts such as Middleton (2004), Carlberg (2011), and McKenzie *et al.* (1995) and Wakefield and McLaughlin (2005) for advice on Excel and Minitab, respectively. The Excel **Analysis ToolPak** and **Solver Add-ins** are required for some examples. Another useful addition is the *MegaStat*

---

[1] *Excel 2007* has a similar interface and commands to *Excel 2010*.

(McGraw-Hill/Irwin, McGraw-Hill Companies Inc.) add-in for Excel (Bowerman *et al.* 2008), which provides other facilities including some non-parametric tests.

Some of the more advanced design examples and the multivariate applications (Chapter 12) require Excel with more sophisticated facilities, fuller versions of Minitab, or packages such as SPSS (International Business Machines Corporation) and specialist software (Design-Expert(R), Stat-Ease, Inc.). It must be stressed that these are not required for the majority of examples in this book, and that it is appreciated that many readers will not have access to all these packages. Several are available as student versions and this is indicated. Additionally, there is a wealth of statistical tools and facilities on the Internet.

1.5.2   Conventions and terminology

Already in this chapter, new terminology is accumulating, which may cause confusion for some readers. Additionally, certain terms can be used in more than one sense. For example, 'analysis' can refer to chemical analysis or statistical analysis. 'Sample' can be used in more than one way. 'Measure' is a general term, but it signifies the act of carrying out of a determination, etc. Most key terms are defined or explained where they occur and usually the context will aid understanding, but a brief list of common terms and their meaning is presented in Table 1.2.

In addition to software and terminology, etc., indicated earlier, much reference is made to certain key texts and research publications in food science and statistics. Most chapters include at least one of the following sources, which are introduced at this point.

Advice on bibliography

Textbooks on specific basic applications of statistics to food science are not numerous. Some 'standard' texts include Gacula and Singh (1984) on general material for food research and O'Mahony (1986) on sensory applications. There are several texts on multivariate sensory applications. Also, sensory evaluation data analysis is dealt with in other general sensory texts, but only as an add-on section or a chapter in the main contents. For the chemical analyst, there are some general texts of interest such as Miller and Miller (1999), but these are not specific to food. As seen, the publication dates on several of these are more than a decade ago, although they are still used extensively by practitioners. More recent texts in general biosciences are appearing, but they often deal with advanced newer methods. Other useful texts take a more technical view (e.g. Chatfield 1992) or a gentler approach (e.g. Rowntree 2000), but again these are not specific to food science. Ultimately, the food scientist may need to go to journal and article publications to get a particular method, from research journals and the work of organisations such as the Laboratory of the Government Chemist series on Valid Analytical Measurement and the Food Standards Agency. Older texts have the advantage that in some cases the descriptions and statistical analyses are simpler and easier to understand.

Table 1.2   Terminology and conventions.

Term	Meaning
Food practitioner	The person carrying out the food investigation and the statistical analysis (scientist, technologist, researcher, student, pupil, etc.)
Food study	Any investigation on food by the above practitioner
Analysis	Statistical analysis or instrumental analysis
Analyst	Chemical analysis practitioner
Analytical	Chemical analysis
Analyte	Constituent being determined by chemical analysis
End determination	A single measure on a sampling unit or subsample
Observation	As determination
Sample	A selection of sampling units
Sampling unit	An individual object taken from the population
Subsample	A portion of a sampling unit used for an end determination
Instrumental	Instrumental method (covers, chemical, physical, biological, etc.)
Sensory panel	Group of selected, trained assessors
Sensory test	Sensory analysis tests using trained assessors or consumers
Consumer panel	Group of untrained consumers for sensory testing
Consumer test	Sensory measure using consumer respondents
Consumer measure	A measure used in a consumer survey
Consumer	A member of a consumer population
Data	The values as measured
Results	Data that have been summarised and analysed
Assessor	A member of a sensory or consumer panel
Panellist	As assessor
Judge	As assessor
Ranking	Specific use of a ranking test
Rating/scoring	Allocating a measure from an ordinal, interval or ratio scale
Scaling	Allocating a measure from any scale
Variable	A measure that can take any value in a range of possible values
Object	An item upon which measures are made (food material, food machinery, methods, consumer, etc.)
Item	As object

References

Blumberg, B., Cooper, D. R. and Schindler, P. S. (2005) *Business Research Methods.* McGraw-Hill Education, Maidenhead, pp. 18–25.

Bowerman, B. L., O'Connell, R. T., Orris, J. B. and Porter, D. C. (2008) *Essentials of Business Statistics*, 2nd edn. McGraw Hill International Edition, McGraw-Hill/Irwin, McGraw-Hill Companies, New York.

Carlberg, C. (2011) *Statistical Analysis: Microsoft® Excel 2010*. Pearson Education, QUE Publishing, Indianapolis, IN.

Chatfield, C. (1992) *Statistics for Technology*, 3rd edn. Chapman & Hall, London.

Collis, J. and Hussey, R. (2003) *Business Research*. Palgrave MacMillan, Basingstoke, pp. 46–79.

Fisher, R. A. (1966) *The Design of Experiments*, 8th edn. Hafner, New York.

Gacula, M. C. and Singh, J. (1984) *Statistical Methods in Food and Consumer Research*. Academic Press, Orlando, IL.

Hartel, R. W. and Adem, M. (2004) Math skills assessment. *Journal of Food Science Education*, **3**, 26–32.

Iwaoka, W. T., Britten, P. and Dong, F. M. (1996) The changing face of food science education. *Trends in Food Science and Technology*, **7**, 105–112.

Kravchuk, O., Elliott, A. and Bhandari, B. (2005) A laboratory experiment, based on the maillard reaction, conducted as a project in introductory statistics. *Journal of Food Science Education*, **4**, 70–75.

Malhotra, N. K. and Peterson, M. (2006) *Basic Marketing Research*, 2nd edn. International Edition, Pearson Education, Upper Saddle River, NJ.

Mckenzie, J., Schaefer, R. L. and Farber, E. (1995) *The Student Edition of Minitab for Windows*. Addison-Wesley Publishing Company, New York.

Middleton, R. M. (2004) *Data Analysis Using Microsoft Excel*. Thomson Brooks/Cole Learning, Belmont, CA.

Miller, J. C. and Miller, J. N. (1999) *Statistics and Chemometrics for Analytical Chemistry*, 4th edn. Ellis Horwood, Chichester.

O'Mahony, M. (1986) *Sensory Evaluation of Food – Statistical Methods and Procedures*. Marcel Dekker, New York.

Rowntree, D. (2000) *Statistics without Tears – A Primer for Non-mathematicians*. Pelican Books, London.

Salkind (2004) *Statistics for Those Who (Think They) Hate Statistics*. Sage publications, London.

Wakefield, D. and McLaughlin, K. (2005) *An Introduction to Data Analysis Using Minitab® for Windows*, 3rd edn. Pearson Education, Pearson Prentice Hall, Upper Saddle River, NJ.

Software sources and links

Microsoft© Excel for Windows. Available at www.microsoft.com (accessed 28 February 2013).

Excel add-in: MegaStat® for Microsoft® Excel, McGraw-Hill Higher Education. Available at www.mhhe.com (accessed 28 February 2013).

(also available with Bowerman *et al.* 2008)

Minitab: Minitab® Statistical Software, Minitab Inc. Available at www.minitab.com (accessed 28 February 2013).

Design-Expert(R) software: Stat-Ease, Inc., Minneapolis, MN. Available at www.statease.com (accessed 28 February 2013).

SPSS for Windows Student Edition 13.0, 2006, International Business Machines Corporation, New York, (formerly SPSS Inc.). (Available with Malhotra and Peterson (2006)).

SPSS for Windows, Rel. 15.0.0 2006. International Business Machines Corporation, New York (formerly SPSS Inc.).

Laboratory of the Government Chemist. Available at www.lgc.co.uk (accessed 28 February 2013).

Department of Trade and Industry's VAM. Available at www.vam.org.uk (accessed 28 February 2013).

Food Standards Agency. Available at www.food.gov.uk (accessed 28 February 2013).

Chapter 2
The nature of data and their collection

2.1   Introduction

This chapter describes the nature of data, covers methods for data gathering by sampling and details the characteristics of populations. Data originate from a source material that depends largely on the focus of the investigation or experiment. In food science, the obvious sources are the many types of food that are analysed by instrumental and sensory methods. Thus, for such work, samples are taken from a defined larger lot or batch, which can be viewed as the original population. In other types of study such as with consumer surveys on food-related issues, another type of population is more recognisable – that of a population of people, that is, the consumers themselves.

In all these examples, values are generated by the particular measurement system, giving rise to many forms of scales and levels of measurement for data. This chapter highlights the importance of concise definitions and specifications for measurement systems, sampling methods and populations so that the conclusions from statistical analysis are valid. These points are discussed below beginning with data and measurement systems.

2.2   The nature of data

Data occur in many forms. In their original 'as-measured' state, data are 'raw' and they may convey readable details, but usually 'processing' is required before they constitute information in the form of results. Commonly, data occur as numbers, but other forms are possible. These can be letters, words, images, symbols, sounds, sensations, etc. In food studies, all of these forms can be found, but the term 'science' is often associated with 'scientific measurements' recorded as physical measures of distance, mass, volume, etc., and powers of these. These measures are expressed as numbers with units of the particular quantity being measured. For

Statistical Methods for Food Science: Introductory Procedures for the Food Practitioner,
Second Edition. John A. Bower.
© 2013 John Wiley & Sons, Ltd. Published 2013 by John Wiley & Sons, Ltd.

Table 2.1   Examples of measurement types in food science.

Measurement/quantity	Example (typical unit)
Weight	Balance reading (gram)
Force	Mechanical firmness (newton)
Absorbance	Spectrophotometer (absorbance)
Colour	CIE units
Extensibility	Extensometer units
Sensory intensity	Intensity scale for bitterness
Rotation	Polarimeter (degree)
Volume	Burette reading (millilitre)
Energy	Calorimeter (temperature, °C)

example, the quantity *weight* is expressed as a number with *gram* units. Beyond this, many scientific instruments in food science analysis have their own units for measurement of parameters such as absorbance, colour units, texture units and retention times (chromatography) (Table 2.1).

Other measures can occur without units and use numbers to express **counts** of the occurrence of objects, for example the number of bacterial colonies on an agar plate, or the number of people preferring one food sample to another.

All data are measured in some way, whether by use of a scale, by simple enumeration or by collection using specific instruments. All of the examples in Table 2.1 involve a scale of some sort.

2.2.1   *Measurement scales*

Scales are characterised by having divisions labelled with anchors, which identify points on the scale. **Anchors** can be numbers or words and they guide selection of each individual measure. For example, when measuring length with a 15-cm metric ruler, the numbered anchors for each centimetre guide the selection of the length measure. Note that for some scales, the investigator may be required to estimate the final reading, as when reading absorbance from an *analogue* meter and the pointer lies between the marked divisions. In other scales, there is no estimation as all parts of the scale are anchored or the scale provides a *digital output*. The format of a scale depends on a number of factors. Instrument scales have their own specific readouts such as a balance (g) or spectrophotometer (absorbance). Scales used in sensory work and in consumer surveys come in a wide variety of forms. They can occur as categories with tick boxes or as graphic lines, etc. More information on scales with examples is provided below.

In all cases, the quantity being measured constitutes a **variable** in experimental design terms. A variable can take any value on the scale of measurement. The scale can be within an instrument or in a questionnaire, or on a sensory assessment sheet. The measures are taken from **objects** in the form of food samples, consumers, different instruments being compared, different process lines, etc. Ultimately, the objects constitute sampling units, which make up experimental units in the

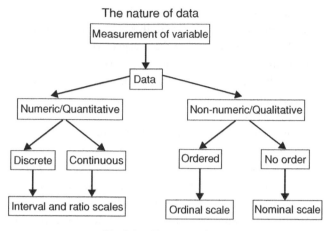

Fig. 2.1   The nature of data.

investigation (Section 2.3.1). There are several aspects to the nature of the data generated by the various measurement systems (Fig. 2.1).

These include whether the data are generated as numbers or non-numbers such as words, the form that these take and finally, the scale on which they are measured.

At this point, a scale can be classified in terms of its *level of measurement*. There are four levels (detailed below) ranging from non-numeric, non-ordered scales (nominal), and those with order (ordinal), to those with a fuller numeric nature (interval and ratio).

2.2.2   *Numeric and non-numeric data*

One way to classify data is to decide whether they are numbers or not and this in turn decides some other features. If numbers are generated, the data are *numeric*.

Numeric data

Numeric data are *quantitative* in that they represent measures of 'amounts'. They can occur in two forms, one as *discrete* numbers, where the variable can take one of a sequence of numbers, typically integers. For example, the number of peas in a pod – there can be 0, 1, 2, 3 ... peas. *Continuous* variables have numbers that can take any numeric value on the scale, depending on the precision of the measurement system. Weighing a sample for analysis with a 'pointer-type' balance could give a continuous number such as 5.385 g, where the last digit was estimated from the position of the pointer on the scale (between 5.380 and 5.390 g). The true weight could be 5.38343672 g, but the measurement system cannot read to this precision. Modern digital balances do not require estimation of the final digit, so although a continuous variable is being measured, the measure becomes discrete in that one of a sequence of numbers must be selected. Thus, for a milligram balance, the weight could be one of 5.380, 5.381, 5.382, 5.383, ..., 5.390 g, but unlike a discrete variable, the variable weight can take any value – it is just limited by the measurement system in this case.[1] The terms continuous and discrete are similar to ***analogue*** and ***digital***, respectively. For instance, a voltage measured on an analogue meter with a scale and a pointer can take any of an infinite number of values that must be estimated visually, but a digital reading consists of a series of integer digits, although admittedly this can be up to the limitation of the device, as seen above.

For continuous variables and in many cases for discrete ones, the data are measured on a scale that is of the 'fuller' numerical type (at least interval in nature, but more likely ratio – see Section 2.2.3). The term ***parametric*** is sometimes used to describe numeric data of interval or higher level of measurement. This actually refers more to the population from which the data originate, in that the population will have certain well-identified parameters (Section 2.4). The term ***metric*** on its own signifies a generalised measure of distance, but it can be used in reference to interval or ratio measures, thus ***metric data***.

Non-numeric data

These data occur as letters, words, etc. (sometimes referred to as 'string' format signifying a string of characters or letters) and they can be described as ***qualitative*** in nature. The words can be *categories* or *classifications* without order in which case a ***nominal (categorical) scale*** will have been used, e.g. 'red', 'green' and 'blue'. If order is present in the categories then an ***ordinal scale*** is evident, e.g. 'poor', 'fair' and 'ok'. A complication is that the numbers are often used with ordinal scales and with ranking tests (ordinal), thus fogging the division between numeric versus non-numeric for ordinal variables. In terms of analysis, coding with numbers is used for both ordinal and nominal data, but this does not change its original nature. Certainly, the source population from which they originate is viewed as ***non-parametric*** in that an assumption of no parameters is made.

While all these examples can be recorded in numeric or non-numeric form, a crucial characteristic of the data is the *amount of information* that they contain. One major aspect of this is given by considering the ***level of measurement*** possessed by the scale or the method used to gather the data.

2.2.3   Levels of measurement

Level of measurement is a specific term that describes the *quantity of information* in data, but it refers to the measurement system used to obtain the data. Thus, the description ***nominal data*** refers to data that were generated by a *nominal scale*. Level of measurement decides the degree to which the data can be manipulated in a mathematical sense and points the way to which forms of analysis are appropriate.

---

[1] Continuous values are therefore estimates which lie between identified limits, but discrete values are accurate in that they are always 'on the mark'.

Table 2.2   Levels of measurement.

Level	Data form	Key features	Example	Analysis	Term for scale use
Nominal	Word, names	No order	Apple variety	Count, mode, crosstabulation	Classifying
Ordinal	Words, numbers (rank)	Order, no magnitude	Preference rank; consumer satisfaction (low, medium, high)	As nominal plus median, quartiles	Ranking, rating
Interval	Numbers	Order, magnitude, equal intervals, arbitrary zero	Sensory intensity (low to high)	As above plus mean, standard deviation, correlation, analysis of variance	Scoring
Ratio	Numbers	Interval, true zero	Fat content	As above plus ratio quantifications, geometric mean, coefficient of variation	Scoring

The levels are shown in Table 2.2. This list displays them from the lowest level (top) to the highest. This division does not reflect the 'goodness' of the data; i.e. it is not 'bad' to use the lowest level of measurement. What is defined is the scale and the choice of the level depends on the requirements for measurement. Thus, for some variables, such as consumers' *country of origin*, the nominal level is appropriate. The variable is non-numeric and it takes values as categories (the names of the countries). It would not be appropriate to use a ratio scale for this – whereas it would be for the quantity 'weight'. In this latter case, the level of measurement is decided by the measurement system, i.e. a weighing system such as a digital balance.

If there is a choice of level of measurement then it is desirable to get as much information as possible by using a higher level. Other factors such as the complexity or sensitivity (confidentiality) of the measure in a sensory test or a survey question may need to be considered in this choice. For example, nominal level questions that are easier to interpret and answer in some respects may be more appropriate for very young or very old consumers; sensitive information such as salary or age are commonly asked for using category (nominal) scales as people may be unwilling to give exact (ratio level) values. Thus, although it is possible to record the salary level as a higher level of measurement (ratio), it is usually set lower as a list of category bands (Fig. 2.2).

Fig. 2.2   Example of a measure in categories from a higher level of measurement.

This scale also qualifies for ordinal level and it could be used as such in a particular analysis, i.e. there is order present in the salary categories.

The terms for describing the manner in which the various scales are used are essentially based on those 'defined' by Land and Shepherd (1988). These terms are by no means 'official' across all disciplines and as the latter authors explain definitions are based on international standards for sensory testing terminology. Although they are used commonly, there is some intermingling of the terms in the literature. Thus, 'rating' and 'scoring' can be used interchangeably irrespective of scale. In this book, they are used for scales with specific levels of measurement, as an aid to comprehension, and to avoid confusion. The term *scoring*, which implies use of numbers, refers to use of a numerical, metric scale (interval or ratio). Note that numbers need not be on the scale itself, but they are used on data analysis. *Ranking* is taken as being used only for a ranking test, and *rating* is for use of a scale with ordinal properties, but which does not specify a ranking procedure. Selection of categories from nominal scales is *classification*. As seen above, an element of confusion is caused by use of the term *scaling*, which is generally applied to all levels of measurement. Usually, the context should make the reference clear.

What decides level of measurement?

A useful guide to level of measurement is given by the format of the measurement or observation during its generation. As seen above (Fig. 2.1), two forms are possible: numeric or non-numeric. This applies to measures taken from instruments or from people in sensory tests or questionnaires. The criterion applies to the data as they are measured and not to any coding with numbers that can occur during data entry into some software packages. This distinction of numeric versus non-numeric gives the first indication of which level of measurement applies.

Nominal data

Data occurring originally as non-numeric strings of letters, e.g. a word or name, are likely to be *nominal*. These data are *categories* or *classes* of objects, e.g. type of potato cultivar (Fig. 2.3).

| Classify the potato tuber sample as: |
| 'Maris Piper'   ☐ |
| 'King Edward'   ☐ |
| 'Golden Wonder'   ☐ |
| (Select ONE variety only) |

Fig. 2.3   Example of a nominal scale.

They are often recorded on a *category* or *attribute* (property or characteristic) choice list in sensory or consumer testing. These lists possess *no order* and *no magnitude* between the categories. Other than counting the incidence of each category, there is very limited mathematical analysis possible by conventional methods. The classifications can be based on one or more attribute descriptions for the object. There can be confusion if there is an implied order given with a nominal scale. For example, 'low', 'medium' and 'high' or 'soft', 'firm' and 'hard' in a texture classification. Such terms are commonly used as anchors in interval scales in sensory descriptive analysis methods. When used as *categories in a nominal scale*, they imply order and the scale is treated as being at least ordinal. Some scientific instruments and tests generate nominal data, for instance chemical or microbiological tests that produce a 'positive' or 'negative' result (hence the term 'qualitative tests'). Nominal data can also be referred to as *attribute* or *category* data.

Ordinal data

These can occur as two forms. One is that generated by a *ranking test*, i.e. a task in sensory and consumer tests where objects are put in *order* of an attribute or characteristic. For example, *'Arrange the following samples in order of preference'* (Fig. 2.4). Here, the data are measured as numbers, but in rank order only.

Ordinal data are also produced by a scale that has some natural order to it. These *ordinal scales* are used to *rate* items on a list of descriptions that have logical order. A common example in consumer surveys and market research is the *Likert item* based on the *Likert scale* (a composite scale containing several individual items), where variables such as level of *agreement, satisfaction, importance, acceptance and goodness* are measured on statements and concepts (Fig. 2.5).

This example is not a ranking test as the respondents to this question would not arrange the scale anchors – they select one only, but they are assumed to use the anchors in an ordinal manner. Some practitioners argue for interval status for such scales, especially when there is more evidence of progression in the category

---

Rank the following items in order of **preference** where 1 = most preferred and

5 = least preferred:

      Sample 385 ☐

      Sample 590 ☐

      Sample 127 ☐

      Sample 078 ☐

      Sample 931 ☐

Fig. 2.4   Example of a ranking test (ordinal scale).

Please state your opinion on the following aspect of the food service:

'The overall quality of the main course'

Poor	Fair	Good	Very good	Excellent
☐	☐	☐	☐	☐

Fig. 2.5   Example of an ordinal scale (individual Likert-item style).

terms as with the '*agree/disagree*' form of the Likert scale (Fig. 2.6; Sorenson and Boque 2005).

Stone and Sidel (1993) give an account of the use of the Likert scale in various formats, including acceptance, for sensory and consumer research.

Both forms (rank and rating scale) of ordinal data have order, but there is no indication of *magnitude* of differences and it cannot be assumed that there are equal divisions between the anchors. Numbers are often used in conjunction with these scales and this may imply equal divisions (i.e. as for an interval scale). This may or may not be the case, but these types of scales of peoples' opinions such as level of agreement and importance are viewed as being used in an ordinal manner (Sirkin 1999), when used as *single items* (the original Likert scale uses a scale containing several items that are then summed and averaged (Clason and Dormody 1994)). Ordinal scales that are anchored with a more obvious continuum (e.g. 'weak', 'moderate' and 'strong') may have more interval properties (Stone and Sidel 1993), but those with more of a subjective feel such as 'good', 'ok' and 'poor' are less convincing (Lawless and Heymann 1998).

The well-known ***hedonic scale*** (Peryam and Pilgrim 1957) used in sensory work (descriptions of '*like extremely*' to '*dislike extremely*') is similar in nature to the Likert scale above, but it is viewed differently. Considerable research has shown that the divisions on the hedonic scale are sufficiently interval in nature to meet one of the assumptions of the more powerful statistical methods (Gacula and Singh 1984; Lawless and Heymann 1998) particularly when the nine-point scale is used.

As stated above, ordinal data are often coded into numbers (or occur as rank order numbers) for analysis, but the values cannot be interpreted in an interval

Please state how much you agree/disagree with the following statement:

'*GM foods are beneficial to health*'

Strongly agree	Agree	Slightly agree	Neutral	Slightly disagree	Disagree	Strongly disagree
☐	☐	☐	☐	☐	☐	☐

Fig. 2.6   Example of an ordinal scale (individual Likert-item style).

manner, for example *'important'* cannot be judged as twice as much as *'slightly important'* on an 'importance' scale, or rank order 2 cannot be viewed as half of rank order 4. Thus, analysis of these data in their basic form is usually limited to a count of the frequency of each descriptor or rank allocations and certain low power significance tests. If several ordinal scales are combined to give a *composite measure* then the data are now more continuous in nature. They can be classed as interval as many more points on the scale are possible than the limited discrete integer nature of each of the original individual scales. For example, the *nutrition knowledge* scale used by Alexander and Tepper (1995) contained ten items, and the *food choice questionnaire* examined by Eertmans *et al.* (2006) contained nine scales with three to six items in each.

Bipolar and semantic differential scales are usually ordinal. In a **bipolar scale**, the anchors at either end have opposite meanings as with the agree-disagree scale and the hedonic scale. Another example is the **just-about-right** scale that has a series of anchors indicating insufficient, ideal and excessive degrees of a particular attribute, e.g. 'not sweet enough', 'ideal sweetness' and 'too sweet'. **Purchase intent** scales are similar with 'definitely not buy' at one extreme and 'definitely buy' at the other.

Semantic differential scales have two reference terms with opposite meaning at either end of an ordinal scale. They include a wide variety of possibilities in consumer surveys. A particular characteristic is rated on the scale referenced by the two opposite terms (Fig. 2.7).

Another complication arises with ordinal scales in that they can be used with anchors linked to category boxes (Fig. 2.5) with various points on the scale. Typically, five points are used, but as low as three and as many as nine are possible. Usually, higher numbers of divisions on a scale (ordinal and above) give better discrimination ability; i.e, it is easier to differentiate items. Another possibility for such scales is the **graphic line** with two or more anchors. For this type, the number of divisions on the scale is theoretically infinite. In Fig. 2.8, scale (a) consists of categories with order. The data produced would be very limited in terms of being able to discriminate between items. Scale (b) has more discriminatory power and although still at least ordinal, it has more of a continuous nature. Lawless and Malone (1986) examined various forms of scale and found a slight advantage in the category box type in terms of discrimination ability, but they indicated that use with five or less categories could reduce the sensitivity.

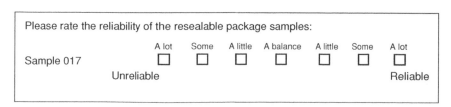

Fig. 2.7　Example of a semantic differential scale (ordinal).

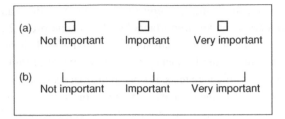

Fig. 2.8   Category box and graphic line scales.

Interval data

Interval scales will always produce numeric data that can be discrete or continuous. There are equal intervals and thus the scale has order and magnitude. Interval level data can be described as originating from a 'limited scale' in that there is an arbitrary zero setting. One example of this is the *degrees Celsius* scale where 0°C happens to be set at the freezing point of water, as this is a useful reference point. This differs from the true absolute zero point of degrees kelvin (–293°C). Analysis that is much more mathematical is possible with interval data although strictly speaking ratios of scale points do not apply. For example, between 5°C and 0°C is the same as the difference between 15°C and 20°C, but 20°C is not four times as hot as 5°C.

Many arbitrary scales in sensory and consumer panel tests can be described as approaching this level of measurement, such as a scale describing a specific graded change in some aspect of food quality and sensory scales of intensity for attributes (Fig. 2.9).

The example uses a graphic line, but this type can be used with category boxes also. A feature of such scales is that they have fixed limits as used – unlike ratio scales, which can go theoretically from zero to infinity.

Ratio data

Any data that involve more concrete physical and scientific measures will usually be ***ratio*** data. This includes most measures of time periods (e.g. freezing time), physical dimensions (e.g. weight, volume and peak area), absolute temperature and many other instrumental measures (e.g. air pressure, flow rate and starch content) that are continuous in nature. Any form of counted objects gives discrete ratio data (e.g. numbers of people in a consumer survey and number of defective packs

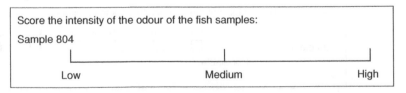

Fig. 2.9   Example of an interval scale.

in a daily batch). Ratio level of measurement is a 'true scale' with equal intervals and a ***true absolute zero*** point. Full mathematical analysis is now possible and multiplication and division result in valid comparisons. For example, a firmness reading of 0.5 N is five times as firm as firmness of 0.1 N.

Interval and ratio scales are more recognised as scales of measurement in the scientific sense and for this reason the resulting data are often treated as 'scale' or 'metric' in some statistical packages. The view adopted in this book is to refer to interval and ratio data as ***metric data***.

NB: In food research, sensory measures on ordinal scales are often treated as interval, which in turn is often treated as ratio for analysis. These procedures depend on a number of factors such as whether the scale has an underlying more continuous nature to it. Other factors such as sample size also have a bearing where large sample sizes can reduce the dependence on assumptions regarding level of measurement and the appropriate analysis. Level of measurement per se does not automatically qualify a measure for parametric analyses as other factors such as distributional form also contribute. More details are provided on this in later chapters.

Changing level of measurement

All levels of measurement above nominal can be converted to a lower level. This may be required for a particular type of analysis, but it does involve loss of information. For instance, in a consumer survey, a question could be *please indicate your frequency of consumption of this product type in the last month*, with a scale given as *0 (none) to 30 (every day)*. The scale in this case is ratio level, but for certain analysis purposes, the data can be formed into divisions based on a specified 'level of consumption group' (e.g. as 'low consumption (0–10)' and 'high consumption (11–30)'). Thus, the original data may have been metric in nature, but as a rule, the grouping process is viewed as putting the data into categories and hence the data are taken as categorical or nominal in this form.

How do data sources differ in level of measurement and other aspects?

Direct measures on food samples or other objects of interest using instrumentation usually employ interval or ratio data scales, e.g. ash content, weight and height of dieters, floor area of retail food displays, number of defects, cost of low fat retail products. These are all numeric (discrete or continuous). With sensory measures, level of measurement depends on the particular test. Difference tests generate choice from two or more samples, e.g. *which sample do you prefer (A or B)?* In this test, the answer would be a non-numeric (A or B), thus level of measurement is nominal. This contrasts with descriptive analysis tests where ordinal and interval profiling scales are more common.

In consumer testing survey questionnaires, where the objects are the respondents or subjects, each question is a data source and a variety of data forms is generated, ranging from nominal to ratio. Focus groups and interviews with consumers will generate recorded words, which are nominal in level. The frequency of occurrence of selected words and phrases can be counted to give a numerical summary, but again this does not change the original level of measurement.

The terms 'instrumental', 'sensory' and 'consumer' along with 'data' are used frequently in this book to indicate the source of data. Usually instrumental data are numeric, continuous and parametric in nature. Sensory data can be similar, but sensory tests include non-numeric ordinal and nominal classification (non-parametric) scales. Consumer data from food testing are more likely to be non-parametric even when using specific sensory methods (due to lack of familiarity with scales, etc.).

Some indication of choice of level of measurement has been given above, but usually it is pre-decided when an instrumental measurement system is chosen. In sensory and consumer test, there can be a choice. Usually, it is better to use higher levels of measurement if appropriate.

For all the above classifications, measures can be ***univariate***, ***bivariate*** or ***multivariate*** in nature. These terms are often applied to the data as well. They signify the number of unique measures which were taken on samples or which are being analysed. Thus, considering a single affective measure such as 'degree of liking' involves a *single univariate measure*, whereas a descriptive analysis test may generate three or more sensory attribute measures (e.g. 'watery', 'grassy' and 'bitter') and is thus *multivariate*. *Bivariate* refers to the circumstance where two measures are made on the objects being examined, such as in the case of a relationship study on measures of 'stale odour intensity' and chemical peroxide content of stored foods. Each food sample would be scored for stale odour level by a sensory panel and receive chemical analysis for peroxide content.

2.3   Collection of data and sampling

During experimentation, the readings or recordings, whether from instrument readout or a consumer survey sheet, provide the data. To obtain these, a sample is gathered from a defined population. Sampling is a fundamental procedure in most investigations, but it needs to be justified and placed in context.

Why do we have to sample?
Sampling is viewed as a process of taking a fraction from a 'parent lot', source or population. Ideally, all of the population would be sampled (100% sample) because in one sense there would be 100% confidence that a 'trustworthy' result had been obtained. Instead of an estimate in the form of a statistic, the actual population parameter itself would be determined. However, with a large population and limited resources this is not possible – it would be too expensive, too time consuming and there are other disadvantages (see below). Thus, the practitioner has to rely on an estimate via a sample and this exposes the determination to sampling error. As will be seen, this is but one error source. The nature of the sample itself involves definition of some terms and fuller explanations.

2.3.1   Sample, sample units and subsamples

A sample is made up of individual items, objects or people referred to as **sampling units**. As indicated above, in food studies, the units can be weighed amounts of food, or individual consumer respondents in a survey, etc. In either case, the total quantity of the food material or the whole consumer population will not be subjected to measurement and a **sample** is taken as a small fraction. Some examples of the form of this procedure are shown in Table 2.3.

The population of interest is any defined group such a large batch of raw material food, or a day's production of food packs, or a consumer group. The sampling units are made up by the individual members of the population. Sometimes this is obvious if the units are discrete in nature, e.g. individual consumers or individual packs of food coming off a production line. In the case of continuous material, the sampling units are usually weighed or measured portions of material – the actual amount depends on the method of analysis and the number of subsamples required. Thus, the nature of the population can be of two forms – one is as *discrete* units such as 450 g cans of a meat product, or individual consumers in a survey, and the other as *continuous material* such as white flour in a large silo. In the former case, the sample and the sample units are as per the population, e.g. individual cans. In the latter, the sample must be defined in terms of a measured amount. The detail of the sampling protocol depends on circumstances and there may be additional stages. For instance, in sampling of foods to obtain food composition data, *primary gross samples* are taken from the population batch. These are then aggregated before *secondary sampling* to obtain *laboratory samples* (Greenfield and Southgate 2003). There is some commodity-related differences in this terminology, e.g. tea sampling involves selection of initial *large lot* samples from the consignments, which are blended before selection of a *primary sample* from each, followed by combining into a *bulk sample* used for selection of *laboratory samples* (BSI 1980).

In some cases, there is further sampling for the **end determinations** or **observations**, thereby generating **subsamples**. Thus, in the examples above, laboratory samples may go through at least one other stage to render *analytical samples*. These latter terms refer to *repeat observations* and the readings for the subsamples are averaged to give a single measure per sample unit, or whatever intermediate sample

Table 2.3   Relationship of population with sample types.

Population	Sample nature	Sampling unit	End determination subsamples
Raw material batch	Weighed/measured amounts	Each amount	One or more parts of the amount
Final products of 1 day's production	Sample of the unit packs	Each pack	One or more measured amounts from each pack
Consumer population of United Kingdom	Sample of consumers	Each consumer	
Food prototype batch	Weighed/measured amounts	Each amount	One or more parts of the amount

Table 2.4   Specific examples of populations and sample types.

Population	Sample size (number of units/amount)	Subsample (number/amount)
(a) Fact finding experiment (% fat content)		
5 000 cans	20/cans of 450 g each	2 of 50 g from each can for fat determination
(b) Comparative experiment (sensory similarity for 2 products with 300 consumers)		
12 000 tubs (100 g) product A	1/75 tubs of 100 g	300 portions of 25 g
12 000 tubs (100 g) product B	1/75 tubs of 100 g	300 portions of 25 g
(c) Survey experiment:		
234 000 consumers	100/consumers	None

Note: Further analysis of subsample data is also performed in some cases in order to ascertain the quality of the data and the measuring system, but it is not valid to analyse the subsamples as true separate sample units (see below). Variations in subsample end determinations give a measure of measurement error assuming homogeneity in the main unit. To assess experimental error and to ascertain differences, etc., experimental unit values (which consist of sampling unit(s)) must be compared. More detail on this concept is given in later chapters (Section 7.3).

type is being dealt with. Table 2.4 gives some specific examples to illustrate this point. Thus, for the analytical example, 20 cans are sampled from the population and two portions of 50 g are taken from each can and are subjected to fat analysis. For the sensory similarity experiment, much greater amount of food material is required and a sample of 75 tubs is required to provide 7500 g of product (the unit) so that 300 consumers can assess a 25 g subsample each. The consumer survey investigation is simpler in that each consumer constitutes a unit and division into subsamples is not required.

Thus, it is important to distinguish between the *source material* from which the units originate and the end determination material, i.e. the subsamples from which measurements or observations are being taken. There is a complication for sensory work when large numbers of assessors (usually consumers) are involved, in that it is not possible for all assessors to be given subsamples from the same individual unit as illustrated in example b in Table 2.4. Bearing these points in mind, the sample taken has two dimensions:

- Sample size or number
- Sample amount or quantity

These dimensions also apply to the subsample size and amount. Calcutt and Boddy (1983) and Meilgaard *et al.* (1991) highlight these distinctions for chemical and sensory data, respectively. Some of the possibilities are given above, but logically the next question concerns how to decide on sample and subsample size and amount.

2.3.2   Sample size

What size of sample is required?
It could be argued that a sample size of 1 is sufficient. If there is one sample such as a single can from a population of 5000, with one reading or measure then statistical

analysis will be simple, but limited. Admittedly, one unit has the great advantage of saving much time and resources, but this is not recommended for several reasons. The main one is that such a procedure would not give a representative value for the population. This is because all objects in a population vary in amount or extent of their various characteristics, such as chemical and physical make-up. If there was no variation then a sample size of more than 1 would superfluous, as each unit would be the same.

Thus, a sample size of 1 is rare. It may be the procedure with some types of routine chemical analysis, but it suffers from another drawback: What if an error was made? Errors could have been made at a number of points. Even if the analyst is confident that these errors were controlled and that the single result can be viewed with confidence, there is still the possibility that the result does not represent the population.

Thus, a single sample does not provide adequate representation and it does not allow for detection of error – a minimum of 2 is required for the latter state. More may be necessary depending on the circumstances of the experiment for a variety of other reasons as explained below.

At the other end of the sampling size question is the upper limit – 'How large should the sample be?' or 'How many unique measures are required?' This will depend on several factors, such as the specific circumstances and the nature of the experiment or investigation involved. The following example illustrates some of the issues:

A food analyst is asked 'What is the protein content of this batch of raw meat?' This is a relatively simple exercise and the objective is to establish new information. What are the possibilities for sample size?

Assume that the meat lot is a deboned joint and it weighs 10 kg. To answer the original question above, one could analyse the whole 10 kg. All the material would have been analysed and there would be a high degree of confidence in the result, barring measurement errors. Unfortunately, at the end of the determinations none of the meat would remain, presuming that a destructive method of chemical analysis was used. The analysis would involve many samples, plus a lot of analysis time and expense. These disadvantages are, of course, overcome by taking a limited number of samples from the original source material, then *inferring back* to the original (much larger) batch. Thus, the investigator is faced with decisions regarding the samples – first the number of samples as units (sample size) and if appropriate, the amount of material in the unit (sample amount).

Factors influencing sample size

The factors having a bearing on sample size are listed in Table 2.5. Some have a consequence for the *error* of any statistic calculated from the data. Ideally, the analyst would wish for no error to be present and that the value obtained from the sample (the *estimate*) would be same as the population value (the *parameter*). For instance, in the meat analysis example above, a sample would provide an estimate of the mean protein content, but the population mean could be obtained

Table 2.5 Some factors affecting sample size and amount decisions.

Factor	Reason
Representation	Influences sampling error; affects accuracy of the measure
Resource considerations – time, finance, personnel	Affects costs of sampling and measurements, etc.
Homogeneity of population (variability of the material)	Masks population value
Chosen method of analysis (instrumental or sensory)	Determines amount required
Other statistical requirement	Specifics of power, confidence interval, significance level, etc.

by analysing all the material. Total absence of error is rarely achieved and the possibilities are that the _estimate_ will either be too low or too high compared with the population _parameter_, i.e. there is an 'over-' or 'underestimate' indicating possible _bias_ (an effect that pulls the estimate away from the 'true' or 'correct' result). This means that the estimates obtained in determinations will be _accurate_ to varying degrees (_accuracy_ is a measure of how 'true' an estimate is). To ascertain how accurate would require further calculations, which are described in Section 3.4.3.

Increasing the sample number and amount can reduce such errors and improve the estimate, but this is balanced against increased costs. Each factor will now be detailed.

Representation

The only 'sure way' to determine a population parameter is to examine all units. As a consequence of taking a sample, error will be present even assuming no errors in determination, etc. This is _sampling error_, which affects the 'trueness' (accuracy) of the result and it is due to the sample and the population being different in number. The error is larger for small sample sizes and will decrease as the size increases, finally reaching zero when the sample equals the population in number (at which point the test result = the true value). To ensure minimisation of this error, certain features must be integral to the sampling procedure:

- Ensure the sample is representative of the population

How can a representative sample and minimisation of sampling error be ensured?
A _representative sample_ is one that closely resembles the original population. This means that the sample should mirror the characteristics of the population. To achieve this for any sample size requires that the sample is selected in a certain way (by a probability method – see Section 2.3.3). Another way is to increase the sample size.

Thus, to ensure a representative sample:

- Use a random sampling method
- Use a larger sample size

As the sample size increases, the estimate approaches the value in the population. **Random sampling** will ensure that variability similar to the population is present in the sample. When the measurements have been taken, the result will be closer to the value of the population parameter and the estimate will possess higher **accuracy**, i.e. closer to the 'true value'. Non-random sampling could result in the sample being **biased** in that not all population units have the same probability of selection.

Unfortunately, adopting the above may be limited by resource implications.

Resources

With unlimited resources in the form of finance, staff, time, facilities and equipment, etc., large numbers of measurements can be taken. Sampling method, sample size, unit amounts and numbers of subsamples can be tailored to get the most efficient experiment. Costs can overcome deficiencies in personnel. For example, a modern instrument enabling much faster analyses with a single operator compared with many operators using long traditional methods. Unfortunately, many research projects have limited resources and some compromise is required, and limitation of sample size is one way of reducing dependence on resource requirements.

Variability of material

Decisions on sample size and amount also depend on the form of the material in the population. For food material, the physical and chemical make-up decides whether the material is **homogeneous** or **heterogeneous**, i.e. uniform in nature or not. Liquid foods tend to be more homogenous than solids. Food meals can contain several ingredients resulting in very heterogeneous material. Raw foods such as meat, fish and crop commodities all vary markedly due to inherent variation and the effects of species, variety, season, geographical region, etc. This leads to problems in experiments where it is not possible to even out variation and the sample unit must be left in reasonable intact form, e.g. as in fish and meat in sensory assessment – which assessor gets which part of the fillet or cut? With consumer populations, many demographic and psychological influences result in very heterogeneous responses in surveys.

Larger sample sizes and amounts counteract high *variability* in the population from which they originate. Thus:

- Sample size and amount should be indirectly proportional to the population homogeneity

That is, more heterogeneous source materials require larger sample numbers. Additionally for food materials then larger amounts can offset heterogeneity. If consumers make up the sample then the practitioner can expect wide variation in opinions, etc., and a larger sample size would be required. This contrasts with a raw material liquid, which may well be very homogenous in chemical composition – consequently, a much smaller number of sampling units, and amount would suffice. In practice, the amount may depend more on the particular method of measurement being employed.

For some measurement systems such as chemical analysis, homogeneity can be improved just prior to sampling by mixing, macerating, etc., but such comminution may not be suitable for sensory testing, as it would destroy the structure and texture of the material. Subsample selection should receive the same treatment and again, increasing the number and amount of subsamples will counteract variability, at the expense of increased costs.

Randomisation and counteracting variability are techniques in experimental design. If the sample is too small for the variability in the population, then *sample bias* can occur.

Measurement system

Resources in the form of the system that performs the end determination or observation can be of a variety of types such as an analytical instrument, a trained sensory panel or a consumer group. This can have a large influence on the number of subsamples and the amount required for analysis, as well as the sample size itself, i.e. the number of units. Many food analytical methods have specified subsample sizes, ranging from less than 1 g to 100 g, but the number of sample units can be limited by the time and cost of analysis. Assessment of food using sensory methods demands larger subsample amounts due to tasting requirements (except in the case of colour and appearance assessment). As seen above, this can cause difficulties as these subsamples cannot be viewed as true sample units. Consumer surveys using questionnaires cause relatively few problems in this sense, as each consumer (unit) receives the same questionnaire. A range of factors affects the 'quality' of the data from these systems and these are detailed in the application chapters.

Statistical requirements

It is possible to calculate the sample size required to achieve desired levels of accuracy and discrimination in experimentation. This demands that the practitioner makes some other decisions in addition to the points above. These procedures ensure that the statistical analysis itself is up to the task set by the experiment, particularly for comparative experiments where significance tests are used.

Calculation of sample size depends on:

- Degree of confidence required
- Effect size of differences

- Significance required
- Power required

The calculations require a measure of variability within the sample. This may be available in an existing databank of analyses. If not, then an estimate can be based on a pilot or trial determination. Essentially, the higher the confidence level, significance level and power and the smaller the effect size, the larger the sample size. These terms and requirements are discussed in later sections and in detail in Chapter 4. A simple method for calculation is explained briefly in Section 3.4.3.

Hough *et al.* (2006) present tables for the number of consumers for acceptability tests based on consideration of the above statistical factors. Also, Gacula and Rutenbeck (2006), in similar work, established the importance of the effect size, i.e. the difference considered important for detection and the effect on consumer sample size for consumer and sensory panels. An indication of typical sample sizes for various applications is given below (Table 2.7), but it is now appropriate to describe methods for selection of the sample.

2.3.3   Sample selection methods

How are samples selected?

The 'quality' of a sample in respect of how it represents the population also depends on how the members of the population were selected. This in turn also governs the extent to which the practitioner can generalise or infer back to the population. A concise definition of the population must be stated and depending on the sampling method, a list of all members can be prepared, referred to as the **sampling frame**.

There are a number of possible methods.

Probability sampling

Sampling techniques based on probability mean that every possible member or part of the population has an equal chance of being selected. Such sampling allows inferences about a population based on the sample as it gives a sample that is representative. Probability sampling minimises sample error and sample bias. Randomisation techniques are used in this form of sampling hence the name 'random sampling'. Randomisation procedures can be set up with random number tables or a random number generator in a calculator or computer.

Simple random sampling

Here, the sampling units are selected randomly from the population. The simplest example is with discrete items. For example, the population is defined as consumers in a city. The sample is randomly selected from the sampling frame in the form of a databank of names such as a census list or telephone directory. The consumers are contacted and asked if they would be willing to participate in a survey. One disadvantage is that such lists are rarely complete (Upton and Cook 2001), but the method is common as evidenced by Hossain *et al.* (2003), who used data from a sample of over one thousand consumers from a population of 97 million by random selection of telephone numbers. Another population with discrete items could be a single production run of frozen fish finger packs. A sample could be taken of the end of the line at random intervals during the run.

With continuous food material taking a random sample means that the material making up the sampling unit and subsamples should be selected in a random way, for example by drawing samples from different (random) locations in the mass of the batch or unit, rather than just from the outer or easily available parts. For large bulk particulate and block foods, there are a variety of probes and devices to enable 'spearing' into the depth of the material to draw a sample. Cork borers can be used to spear samples from retail products such as fruits and blocks of cheese. Smaller batches of particulate powders can be sampled using a 'quartering' technique where the material is spread out thinly, marked into squares and one of these randomly taken (Kirk and Sawyer 1999; Nielsen 2003).

Systematic random sampling

Systematic sampling 'spreads out' the random selection process so that no two random samples are too close together and can avoid any grouping within the random sample. A scheme is drawn up where a random start point is chosen then a fixed gap is inserted between the sampling units. For instance, in consumer sampling, every 100th person in the census list could be selected. Sampling of food during production can be done by this method, but if periodic differences occur, this could produce misleading data (Giese 2004).

Stratified random sampling (cf. quota sampling)

Stratified sampling takes into account the balance of proportion of subpopulations in a population and is best applied to strata that exhibit larger sources of variation. The sample is then drawn randomly, but mirrors this distribution. For instance, in the frozen food production example above, consider several production runs within a day. A simple random sample could result in imbalance according to the number of packs in each run. Identifying the number of packs in each run as the strata can allow make-up of the sample to mirror the percentage distribution within the strata (Table 2.6). In this example, selecting a sample of 20 is achieved in balance with the run strata.

Table 2.6   Stratified sampling method.

Run number (strata)	Packs	% Of total population	Sample (numbers)
1	350	35	7
2	500	50	10
3	150	15	3

Now any source of variation between the runs is spread out with a balanced representation. Other examples include stratification according to commodity variety, geographical source, etc., and to characteristics in consumer populations based on age group, gender and other demographic variables.

Cluster sampling

In cluster sampling, random selection of identified clusters (typically, geographical groups) is performed followed by a further selection within these groups. Similar to stratified sampling, but all the strata are not included and because the clusters are random, inference can be made back to the population of clusters. This method can be employed in consumer surveys where the clusters could be major cities – instead of sampling all cities, a random selection of these is performed initially followed by further random selection within each.

All random sampling methods tend to involve more resource requirements. If these are limited then non-probability methods are the alternative.

Non-probability sampling

With non-probability sampling, there is more likelihood that a larger sampling error is present. The major advantage is that it is much less demanding of resources, but this is at the cost of the results having less generalisation potential to the population.

Convenience

Samples are selected as available or in the case of consumers, those who are willing to participate often at short notice. Another example could be food products, which are conveniently available outwith production or process areas, or new food commodities that are limited in amounts.

Quota

Here, the sample structure is set up to align with the populations in terms of representation of various characteristics that are present. In surveys, these can be demographic variables such as gender, age, socio-demographic group and geographical location. Similar procedures can be used to ensure representation of groups of food products according to quotas such as run, day and machine (see stratified sampling example). A quota sample will ensure that sufficient numbers of each element are included. As the sample is drawn, continuous checks are made so that deficiencies in any element can be made up as sampling proceeds.

Snowball (survey)

This method is applicable in consumer surveys, when access to the population of interest is difficult for reasons of anonymity, confidentiality, etc. The initial contact

Table 2.7   Typical sample and subsample size and amount.

Application	Sample size (number of sample units)	Number of subsamples per unit
Routine instrumental analysis	1–4	2–4
Estimation of bias/precision	4–8 (depends on variance and confidence required)	1–4
Sensory difference test	2–5	10–20
Sensory attribute test	2–8	10–20
Consumer preference test	2–8	30–100
Consumer survey (non-sensory)	50–500	1

is made with a small number of sample units (respondents) who then pass on the measuring instrument (questionnaire) to similar acquaintances, relations, etc. The data sheets are then returned to the researcher.

2.3.4   Application examples

What are the typical sample sizes for the various applications?

Additional details of sampling and replication are given in Part II in the application chapters, but a summary of some possibilities is listed in Table 2.7. In general, samples that are representative and that provide sufficient numbers to enable appropriate statistical analysis in terms of assumptions, etc., ideally should number at least 10 for food sampling circumstances and at least 30 for consumer surveys. In the latter case, many more are readily available, but in the former, especially for complex chemical analysis, resources can limit sample size.

For the protein analysis example, a sample could be two to four randomly selected 100 g cuts (the units) from the meat, which were then homogenised to a fine mince, from which duplicate 2 g subsamples were drawn for the end determinations. Ultimately, it depends on the detail of the design and experiment under study, according to the factors discussed above. Often, ongoing work builds up a databank of sample results so that minimal sample sizes may suffice. The gain in overall trustworthiness of results as dependent on sample size is covered in the next chapter.

2.4   Populations

Appropriate application of certain statistical analysis methods demands that there is some knowledge of the population from which the samples originate. The term 'population' refers generally to *all the objects or subjects that make up a defined group*. While it is sometimes possible to examine all members of a population, this is not common for a number of reasons. Hence, as illustrated in Section 2.3, it is more likely that a sample is taken of much smaller size than the population itself.

Ultimately, it is the actual measurements or readings on the sample units themselves that make up the population under consideration. For example, in examination of the fat content of bacon samples, the fat content (%) values provide the data and a population of these values is assumed. A population of data would be real (rather than assumed) if all the food material within the defined population were to be analysed. Thus, if the source material were 'apples' with a population definition of 'apples grown in the United Kingdom in 2007' then a chemical analysis investigation into malic acid content would generate a sample of acid values. For consumer surveys, the data would be assumed a sample from a population of values from various scales and selections in a questionnaire. For example, the source population could be defined as *'young adult consumers (age 18–25 years) domiciled in London during the time of the research (2006)'*. Assume that a sample of such consumers was surveyed on their opinion regarding use of disposable packaging by means of a Likert scale-type measure. The sample values from the Likert scales would be assumed to come from the population of these values. In addition to these examples, populations can be made up of other sources such as different analytical methods, different process machines or different laboratories depending on the focus of the investigation.

Sample data can be displayed and analysed in various ways to give a view of certain characteristics, which relate to the population. Such knowledge is fundamental to the understanding of sampling and statistical analysis, from organisation of data to estimation of population parameters and significance testing. The first of these aspects relates to the ***distributional form*** of the population.

2.4.1 Population distribution

The term ***distribution*** refers to the form and shape of the degree of variability that the data reveal when summarised and displayed. The displays reveal measures such as the *range* and *frequency* of values within the sample. This can be done in the form of a table, but more conveniently, graphically. The applicability of graph and table type depends on level of measurement, but many scientific instrumental data are at least interval in level of measurement and are continuous in nature. They form a very characteristic distribution shape when plotted in a histogram or a grouped frequency graph (Section 3.2.3); for example, calcium content in a large sample set of a dairy product (Fig. 2.10).

The variation is caused by many small independent ***random errors***[2] resulting in values above and below the mean (Nielsen 2003). This is due to a number of influences including random variations in environment and small unit-to-unit inherent variation in food material (Section 3.4.1).

Increasing the number of values plotted would eventually result in a smoother curve.

---

[2] The normal nature of such distributions refers more specifically to the distribution of these random errors.

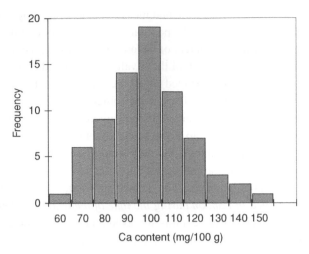

Fig. 2.10   Frequency distribution of Ca content of 74 sample units of dairy product (Excel).

Ultimately, with an infinite number of sample units the distribution forms a symmetrical *bell-shaped curve*, typical of a ***normal distribution*** (Fig. 2.11). Many instrumental measure populations have a similar form of distribution, where the data are continuous. The display shown is the theoretical shape based on an infinite number of samples from the same source (i.e. same material being analysed by the same method and operator). The distribution has a number of important features.

It can be visualised as a map or ***model*** of how the individual data points are spread out in the population. The area of different proportions below the curve shows large variations. It can be seen that about three fourths of the values lie within

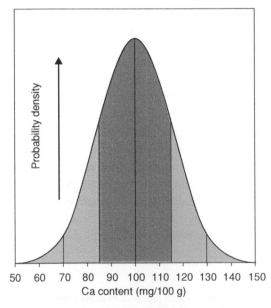

Fig. 2.11   A normal distribution (Excel).

the central region. Much fewer lie within the tails (the ends of the curve at either extreme). When drawing random samples from the population, this proportional spread allows **probabilities** to be assigned to the chances of obtaining certain values. Additionally, certain population **parameters** (Section 3.3) can be located. In the centre, the *mean (μ)*, *median* and *mode* are located. The spread on either side gives a measure of the variability within the population, calculated as the *variance* and *standard deviation*. The area underneath the curve can be demarcated into specific sections (the shaded areas in Fig. 2.11) on the basis of the *population standard deviation (**sigma** (σ))*, which gives the magnitude of the spread as below ($-\sigma$) and above ($+\sigma$) the central average value. Thus, 68% of the values lie within $\pm 1\sigma$, 95% within $\pm 2\sigma$ and 99.7% within $\pm 3\sigma$ (approximate percentages).[3]

Normal distributions can be rendered into a *standard form*, which allows comparison of different distributions. This is achieved by converting the variable (x) to that of the **standard normal variate** (z):

$$z = \frac{x - \mu}{\sigma}$$

Such a distribution has a mean of *zero* and a standard deviation of *one*. Assume that the original distribution above has a mean of 100 mg/100 g of calcium and a standard deviation of 15 mg. Any single calcium value can be standardised as:

$$\text{Original value} = 130 \text{ mg}/100 \text{ g}, \quad z = \frac{130 - 100}{15} = 2.0$$

The **z-value** (or *z-score*) is in 'standard deviation' units, i.e. in the standardised form the value is two standard deviations from the mean (see Section 3.4.3 for determination of *z*-values). In turn, this allows the **standard normal distribution** to be divided up into portions based on *z*-values exactly in the same manner as described above for the original non-standard form.

How is the knowledge of the normal population used?

Knowledge of the shape and character of the normal distribution provides a concise guide to the spread and location of all the data. This allows certain parameters to be located and assumptions to be made concerning how large or small a magnitude the values will attain. Hence, the term **central location** that indicates where to find the mean, median and mode. Variability (variance) can be calculated from the width of the spread on either side of the centre.

When random samples are taken, it can be seen that it is more likely (i.e. there is a higher **probability**) that samples will be from the central region as this region has more area and has a *higher density* of the population units. To qualify as 'normal' in this sense, the distribution must have the general *peaked* shape and

---

[3] Figure 2.11 is also known as a **probability density function** and can be drawn using data produced by Excel's NORM.DIST function (NORMDIST Excel 2003).

Table 2.8   Probability.

Event probability	Chance of happening (%)
0	0 (no chance)
0.05	5
0.1	10
0.75	75
1.0	100 (certain)

must be *symmetrical* as above (or at least approximately so). Deviations from this are assessed by graphing of sample data and by specific tests (Section 3.5).

Finally, a very important consequence follows. It has been found that when *many samples* are taken and averaged, the **distribution of the sample means** is normal, even if the parent population deviates from normality. The proof for this is known as the **central limit theorem** (Chatfield 1992). The **sampling distribution** has the same mean value as the original distribution, but a smaller spread. It enables assumptions of the normal characteristics to be applied to much wider sampling circumstances.

Probability

This concept is used frequently throughout this book and it refers to the chance of events happening. **Probability** can be expressed on a scale of 0–1, or as 0–100%. Hence, if an event has a probability of happening equal to 0.5, then there is a 50% chance that it will happen (Table 2.8).

The events can be choice of one food over another, proportion of consumers liking 'green' foods, an analytical result being different from others, etc. Probability forms the basis of decisions in significance testing and calculation of confidence levels, etc.

Does the normal distribution apply to all data?

The above illustration displays the distribution of a continuous ratio level variable (calcium content) and this form would apply to similar data of interval or ratio level. Thus, many instrumental measurement distributions are normal in nature. However, many data are not continuous and there are other levels of measurement. For discrete data of interval or ratio level, the normal distribution provides an *approximation*. Sensory methods generate a number of different levels of measurement, but descriptive analysis data may possess a normal distribution.

Other distributional forms are possible (Table 2.9). Ordinal data from a ranking test will not be normal as the data have equal occurrences of each rank value. Ordinal rating scale frequencies may give a normal distribution. Hence, for some data, the assumption of normality may not apply and brief detail of some other possible distributions is given below.

Table 2.9 Data and distribution type.

Data	Level of measurement	Type	Distribution
Instrumental	Interval and ratio	Continuous	Normal
		Discrete	Approximately normal
Sensory descriptive	Interval	Discrete	Approximately normal
Sensory difference	Nominal	Discrete	Binomial/chi-square
Ranking test	Ordinal	Discrete	Non-normal
Survey data	Nominal	Discrete	Binomial/chi-square

Derived distributions

The normal population does not always provide the best model for analysis of data. In some cases, it may introduce error when calculating certain statistics. One such instance is given when using the normal population parameters to determine a confidence interval for a mean value (Section 3.4.3). The calculation requires knowledge of the population variability (spread), i.e. the population standard deviation. In practice, this is not available, and it is estimated from the sample data. Using this estimate with the normal population model underestimates the width of the interval. A derived distribution – the *t-distribution* – is used instead. This is similar in character to the normal distribution, but is wider at the base reflecting greater variability. When sample size is *30 or less*, the *t*-distribution should be used for this analysis and for significance testing with small sample comparisons. Above 30, the *t*-distribution gets narrower and approximates more and more to the normal form.

The normal distribution is applicable when the statistic of interest is the mean of continuous data, but due to the central limit theorem it can assumed as an approximate distribution for other types of data when the sample is large enough (Upton and Cook 2001).

Discrete

There are some discrete distributions, the **binomial** and **Poisson**, which exhibit event occurrences. Binomial distributions are important in significant testing of sensory difference tests. With large samples, and when the number of possible events increases, the spread and height of the distribution assumes a shape similar to that of normal also.

Other continuous distributions

If interest is in other parameters, such as population *proportions* as frequencies for categorical (nominal) data, then the normal distribution may not be a suitable model unless the number of categories is large. Another continuous distribution, *chi-square*, provides a suitable model for nominal data when the deviations of observed frequencies from expected frequencies are examined. Chi-square is continuous, thus the variability of frequencies originating from discrete data conforms to an *approximate chi-square distribution*. The minimum number of categories is two and this gives one distributional form. As the number of categories increases in different populations, different chi distributions occur. Ultimately, with large samples, these too will approximate the normal form.

F-distribution

This is another continuous distribution that models the distribution of a ratio (F) of variances. It is used in testing differences in variability between data sets (Section 3.5). It is also employed in significance testing where it compares variability estimates between and within data sets (Sections 5.5.2 and 7.5). The ***F-distribution*** calculation gives maximum values of the F-ratio attainable by chance that is compared with the test value for significance evaluation.

2.4.2   Identification of population distributional form

It is important to identify the nature of the population as it influences the choice of the statistical method. It is possible to check for this using a variety of statistical tests (Section 3.5), but the simplest rule is to *always plot the sample data*. Note that level of measurement itself does not decide whether data are normal in distribution or not, i.e. distributional form is not decided by level of measurement. Usually interval and ratio data will conform to the normal distribution, but there are exceptions. For example, salary level in consumer surveys as pounds or dollars per year. These data are ratio in level as a person's salary can take any value from zero upwards, but unlike a 'natural' characteristic, salary levels are set by other factors and there are many on high salaries and many on low. The distribution of such data is often skewed and bimodal (Section 3.5.1), when gathered from a consumer sample. Salary level is not an inherent property that humans possess (compared with, e.g. height) and it is determined by type of employment and other external factors. 'Age' is another example of a ratio variable that is not normally distributed – there are more young ages than old and the distribution is skewed. In addition, microbial count data can contain values that differ by large magnitudes requiring modelling by non-normal methods (Joglekar 2003).

References

Alexander, J. and Tepper, B. (1995) Use of reduced-calorie and reduced-fat foods in young adults: influence of gender and restraint. *Appetite*, **25**, 217–230.

BSI (1980) *Methods for Sampling Tea*, BSI 5987 (ISO 1839), British Standards Institute, London.

Calcutt, R. and Boddy, R. (1983) *Statistics for Analytical Chemists*. Chapman & Hall, London, pp. 1–4.

Chatfield, C. (1992) *Statistics for Technology*, 3rd edn. Chapman & Hall, London, pp. 114.

Clason, D. L. and Dormody, T. J. (1994) Analyzing data measured by individual Likert-type items. *Journal of Agricultural Education*, **35** (4), 31–35.

Eertmans, A., Victoir, A., Notelaers, G., Vansant, G. and Van Den Berg, O. (2006) The Food Choice Questionnaire: factorial invariant over Western urban populations. *Food Quality and Preference*, **17**, 344–352.

Gacula, M. J. and Rutenbeck, S. (2006) Sample size in consumer test and descriptive analysis. *Journal of Sensory Studies*, **21**, 129–145.

Gacula, M. C. and Singh, J. (1984) *Statistical Methods in Food and Consumer Research*. Academic Press, Orlando, IL, pp. 28–30.

Giese, J. (2004) Advances in food sampling. *Food Technology*, **58** (2), 76–78.

Greenfield, H. and Southgate, D. A. T. (2003) *Food Composition Data*, 2nd edn. Food and Agriculture Organisation of the United Nations, Rome, pp. 63–82.

Hossain, F., Onyango, B., Schilling, B., Hallman, W. and Adelaja, A. (2003) Product attributes, consumer beliefs and public approval of genetically modified foods. *International Journal of Consumer Studies*, **27** (5), 353–365.

Hough, G., Wakeling, I., Mucci, A., Chambers IV, E., Gallardo, I. M. and Alves, L. R. (2006) Number of consumers necessary for sensory acceptability tests. *Food Quality and Preference*, **17**, 522–526.

Joglekar, A. M. (2003) *Statistical Methods for Six Sigma*. John Wiley & Sons, Inc., Hoboken, NJ, pp. 172–173.

Kirk, R. S. and Sawyer, R. (1999) *Pearson's Composition and Chemical Analysis of Foods*, 9th edn. Longman Scientific and Technical, Harlow.

Land, D. G. and Shepherd, R. (1988) Scaling and ranking methods. In *Sensory Analysis of Foods* by J. R. Piggot (ed.), 2nd edn. Elsevier Applied Science, London, pp. 155–186.

Lawless, H. T. and Heymann, H. (1998) *Sensory Evaluation of Food: Principles and Practices*. Chapman & Hall, New York, pp. 208–264.

Lawless, H. and Malone, G. J. (1986) The discrimination efficiency of common scaling methods. *Journal of Sensory Studies*, **1**, 85–98.

Meilgaard, M., Civille, G. V. and Carr, B. T. (1991) *Sensory Evaluation Techniques*, 2nd edn. CRC Press, Boca Raton, FL, pp. 257–267.

Nielsen, S. S. (2003) *Food Analysis*, 3rd edn. Kluwer Academic/Plenum Publishers, New York.

Peryam, D. R. and Pilgrim, F. J. (1957) Hedonic scale method of measuring food preference. *Food Technology*, **11**(9), 9–14 (supplement).

Sirkin, R. M. (1999) *Statistics for the Social Sciences*, 2nd edn. Sage Publications, Thousand Oaks, CA, pp. 33–58.

Sorenson, D. and Boque, J. (2005) A conjoint-based approach to concept optimisation: probiotic beverages. *British Food Journal*, **107**, 870–883.

Stone, H. and Sidel, J. L. (1993) *Sensory Evaluation Practices*, 2nd edn. Academic Press, London, pp. 66–96, 258.

Upton, G. and Cook. I. (2001) *Introductory Statistics*, 2nd edn. Oxford University Press, New York, pp. 80–84.

Chapter 3
Descriptive statistics

3.1 Introduction

Once data are collected, they are in the form of a recorded list or a computer file. Organisation and analysis require that the data are entered or loaded into an analysis package. For the majority of examples in this book, **Excel** is used followed by **Minitab** and reference is made to **MegaStat for Excel**. In all cases, data are entered as a column of numbers or categories (nominal data). A label or heading is assigned to identify the data and if necessary non-numerical data can be coded as numbers for the purposes of analysis (NB: this does not change the level of measurement). For most Excel examples, more detail is provided, as illustrated in Table 3.1. This is the basic worksheet layout so that the reader can enter the data and the analysis functions for future use. Data are blocked in one or more columns, referred to as **array(s)** and the sample size (n) is given.

The name of the statistic is listed on the left and the formula is entered in the next cell to the right. The nature of the function used is displayed in the adjoining cell. Use of the **Analysis ToolPak** and some functions require variation from this, but data will always be identified and some guidance given. In some examples, statistics required in later calculations in the same table are identified by an abbreviated name for use in later formulae. The **ToolPak** is activated in Excel 2010 via the page for managing add-ins, available in the File sub-menu on the Ribbon – thus **File/Options/Add-Ins**, then click the **Go** button and select from the list; for Excel 2003 activation is gained from the top menu bar – click **Tools/Add-Ins** then select. Access to the tools is obtained by **Data/Data Analysis** (Excel 2010) or **Tools/Data Analysis** (Excel 2003).

Excel calculations produce values with many decimal points. In most circumstances, these are excessive considering the preciseness with which the original data were measured. Additionally, readability is reduced. Thus, the majority of values are adjusted to display two decimal places, but any permanent rounding should not be done until the final result is obtained. Data layout for Minitab is

Statistical Methods for Food Science: Introductory Procedures for the Food Practitioner,
Second Edition. John A. Bower.
© 2013 John Wiley & Sons, Ltd. Published 2013 by John Wiley & Sons, Ltd.

Table 3.1   Example of Excel data entry and function display.

	A	B	C
1		Example of Excel worksheet use	
2			
3		Sample data ($n = 5$) moisture content (g/100 g)	
4		65.78	
5		64.38	
6		62.56	
7		63.22	
8		64.79	
9	Result:		Formula:
10	Mean	64.15	=AVERAGE(B4:B8)
11			

similar and this is followed by selection of the particular graph, statistic or analysis method from a menu. Sample sizes are chosen to fit the circumstances wherever possible, but some are limited for reasons of space (typically ten data values). Some analyses receive more of a preamble and introduction and are identified as numbered examples (for instance, Example 3.2-1).

The first stages of analysis are often those that summarise the data. Under this heading comes *descriptive statistics*, which are methods used to summarise the characteristics of a sample, e.g. the average value, but which also includes displays with graphs (charts) and tables. Excel charts are produced by selecting the column(s) of data and then choosing the **Insert** tab on the ribbon to access the **Charts** menu (Excel 2010) or by clicking directly on the **Chart Wizard** button on the standard toolbar (Excel 2003). Note that in some cases, charts have been edited regarding colour, line and fill styles, and labels, etc., to get the particular appearance of the examples in this book. More details of the procedures in Excel are provided for some charts.

3.2   Tabular and graphical displays

Graphical displays in their basic form provide indications of how the sample data are distributed. This is also possible using table forms such as a **frequency table**. Graphs and tables can also be used to display descriptive statistics, which include a number of summary values such as the **measures of central tendency** and variation. Graphs and charts have the advantage of giving a more rapid overview, and they can indicate possible trends, effects and relationships. Graph icons on their own will mean that values will have to be estimated from the axes, unless the software allows a numerical display superimposed on the icons (numerical values with more than one decimal point are better shown in a table).

A large number of possibilities exist, but selection of an appropriate graph or table depends mainly on the level of measurement of the data and whether they are continuous or discrete.

3.2.1 Summarising nominal data (discrete)

Nominal data can be summarised using simple tables or graphs. These will indicate frequency of occurrence (i.e. a count) of the categories in the data. A graph or list of frequencies shows the distribution of the data in the sample. The method chosen for display must ensure that each item of data is treated exclusively and displayed once, i.e. there must be no overlap in the frequency counts.

Tabular displays of nominal data are usually in the form of a ***frequency table***. In Excel, these are produced by using the FREQUENCY() function.

Example 3.2-1 *Summary of nominal data by frequency table*
A food-processing line monitors quality in fruit by recording the incidence of defects in each batch that passes through. The tally or count of such occurrences can be summarised by a frequency table.

Table 3.2 shows an example of the frequency of occurrence of the defects in fruit. There are four categories of defect labelled with codes as 1–4. The coding key is entered on the right side of the data column as shown, and then the data

Table 3.2   Data and frequency table (Excel).

	A	B	C	D	E
1				Frequency table	
2				Categories of defects in fruit	
3	Sample data (n = 10) Occurrence	Defect (label)	Defect code	Frequency table (result)	Formula
4	4	Bruise	1	2	= FREQUENCY(A4:A15,C4:C7)
5	2	Cut	2	3	
6	1	Mould	3	2	
7	2	Fragment	4	5	
8	3				
9	4			Frequency of defect by batch	
10	4			Batch 1	Batch 2
11	1		Bruise	2	6
12	3		Cut	3	1
13	2		Mould	3	2
14	4		Fragment	4	3
15	4				

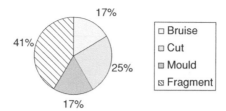

Fig. 3.1   Pie chart of occurrence of defects in fruit sample (Excel).

are entered using the codes. The FREQUENCY() function is entered in the final column by blocking enough cells for the number of nominal categories (in this example = 4, hence cells D4:D7), identifying the data column ('**data_array**'; A4:A15), the code column (the '**bin_array**'; C4:C7), then pressing '**Crtl-Shift-Enter**' to complete (i.e. do not press 'OK' in the function menu). This action inserts the FREQUENCY() function into each cell of the output as an **array formula**. The table includes a final column that displays the formula used in the previous (frequency) column (ignore the lower right table for the moment).

The result is the small table in column D and it shows that defect number 4 ('fragment') is the most numerous.

When there are many categories, such tables can become overlarge and a graphical display is the alternative. *Pie charts* or simple *bar* or *column charts* are suitable for nominal data. The latter two differ in the axis used for the categories (horizontal axis for the column chart) but are essentially interchangeable. Such charts can be used for data with few categories, but many more can be displayed effectively, giving an advantage over a large frequency table. To graph nominal data using Excel, a frequency count is calculated first (as above) and then the values are selected for use in the chart. Each pie slice, bar or column displays the frequency of each category in the nominal data by its size. For the chart of Fig. 3.1, select cells D4:D7.

In Excel 2010, click the **Insert** tab and select the first of the **Pie/2-D Pie** charts, which produces an initial chart and opens the **Chart Tools** menu. This can be used to adjust among many other aspects: colour (**Design** tab), chart and axes titles, data labels (**Layout** tab) and fill styles (**Layout** or **Format** tab). To obtain the name labels for the slices of the pie, click on the chart if not already selected, then **Design/Select Data**. Click the Edit button in the **Horizontal (Category) Axis Labels** part and select the label data (B4:B7). The division of the pie slices by count value in a variety of styles including percentage is turned on via the **Layout/Data Labels** menu. Select **More Data Label Options** to turn off **value** and display **percentage** as shown in the figure. (For charts with axes (not the pie chart), axes scales can be changed in **Layout/Axes** – select the desired axis and then **More Primary. . . . Axis Options**.)

Excel 2003 starts by selecting the data column as above, then the **Chart Wizard** button. Select **Pie** for the chart type followed by the basic sub-type of this chart. Click the **Next** > button and the display shows the chart and the data range used and at this point the **Series** tab can be clicked and the slice labels turned by

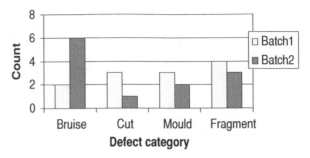

Fig. 3.2   Clustered column chart of batch defects (Excel).

selecting B4:B7 in the **Category Labels** box. Clicking **Next** > again accesses more options including titles for the chart and axes, and choosing **Data labels** can be used to turn on **Percentage** display. Click **Finish** to complete. Fill colour and pattern menus, etc., can be obtained by right-clicking one of the slices and selecting **Format Data Series.** (For charts with axes, right-click an axis then select **Format Axis/Scale** to adjust scale.)

The display on the pie chart agrees with the frequency table (Table 3.2) in that there is predominance of the 'fragment' defect. In this graph, the nominal data have also been converted to percentage *proportions*. Pie charts have another advantage in that the whole sample is shown in the pie with no particular order to the slices, which is reflective of the nature of nominal data. Creation of simple bar and column charts is treated in a similar manner. If required, two or more quantities can be shown within categories on the same chart. This is possible using a form of *compound chart*, referred to as a *clustered bar* or *clustered column* in Excel. For example, examining the relationship of batch number and fruit defects displays both variables (Fig. 3.2), showing that batch 1 has a more even distribution of defects. Frequency counts for both batches are prepared (small table at lower right in Table 3.2), and then these values are blocked with 'the Batch 1/2' headings, and graphed via the chart facilities using a **Clustered Column** chart in Excel.

Proportions and nominal data

Nominal data can be summarised by expressing frequencies as *proportions*. One proportion can be the focus of the investigation, e.g. one particular fruit defect (Table 3.2) could be considered critical – the frequency of this defect as a proportion could be subject to further analysis. The sample proportion provides an estimate of the population proportion (the number of such defects in the whole batch). Proportions can be calculated and displayed in both tables and graphs (see below for summary measures of proportions; Section 3.3.3).

More complex forms of nominal data, such as those produced by a multiple-choice question, require that each possible choice be treated as a category. Selection can be coded as '0 = not selected' and '1 = selected'. The frequencies of the

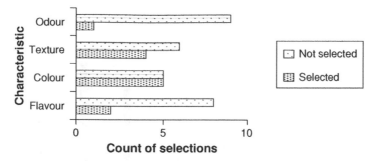

Fig. 3.3   Clustered bar chart of selection of characteristics considered important by consumers for quality in meat (Excel).

selection data can then be plotted as a simple bar, column or pie chart for the occurrence of selected categories or both can be shown in a clustered chart.

Figure 3.3 illustrates use of a ***clustered bar chart*** and shows that consumers selected 'colour' and 'texture' as being important for meat quality, as these characteristics had a higher incidence compared with 'odour' and 'flavour'.

For many categories with low incidence, tables can become too large and graphs contain too many slices or bars, leading to difficulty in interpretation. ***Grouping*** of categories can aid this. For example, data consisting of calendar dates when food items were purchased – instead of having the individual dates, groups of dates per week or month will give a more understandable display.

The graph and table displayed above may clearly identify the predominant category in terms of numerical occurrence, but this can be identified accurately by calculation of the ***mode*** (Section 3.3.1).

3.2.2   Summarising ordinal data (discrete)

With these data, order is present in the scale used as ranks or as points on an ordinal scale. Bar or column charts are more appropriate than pie charts as the order of the original scale is displayed along a horizontal or vertical axis.

Ranking tests

How to summarise a ranking test
A ranking test produces ordinal data. These data can also be summarised by frequency tables and simple bar charts, but the results may be unsatisfactory. Ranking more than three items generates a compound bar or column graph with several bars per item.

Example 3.2-2 *Summary of ordinal data by a compound chart (clustered column)*
As part of a product development study, 25 consumer judges assessed the preference of five fruit-flavoured beverages using a ranking test where 1 = most preferred and 5 = least preferred.

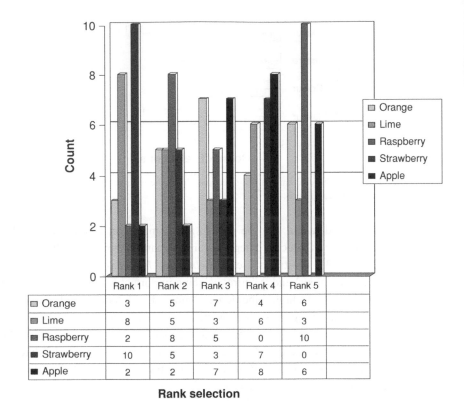

	Rank 1	Rank 2	Rank 3	Rank 4	Rank 5
☐ Orange	3	5	7	4	6
▨ Lime	8	5	3	6	3
▧ Raspberry	2	8	5	0	10
■ Strawberry	10	5	3	7	0
■ Apple	2	2	7	8	6

Rank selection

Fig. 3.4   Clustered column chart of frequency of rank allocations for five fruit-flavoured drinks by 25 consumers (Excel).

Using Excel, frequencies were evaluated as described above and then these values were tabulated with rows as rank 1–5 and columns as the fruit flavours (see data table at foot of graph in Fig. 3.4). These values were blocked and graphed as described above for the fruit defect batch data. Two changes are required to get the display – in Excel 2010, with the chart selected, click **Layout/Data Table/Show Data Table**. For Excel 2003, the data must be plotted as **Rows** in the **Data Range** menu page of the **Chart Wizard** and **Show data table** must be turned on in the **Data Table** of the next page.

In the example (Fig. 3.4), there is a lot to see, but initial observation shows that ten consumers have ranked the 'strawberry' sample as first rank (most preferred) and ten have ranked 'raspberry' as lowest.

Such displays can be difficult to interpret, as would be a frequency table of the rank selections. One solution is to plot separate charts or graphs for each item. Even the simplest ranking task (three items to order) would require three tables or three separate charts to display the data. Such summaries are possible, but as the number of items increases, it can become cumbersome to view several together unless software is used to arrange the items on one page. Accepting these limitations, it may still not be possible to decide with confidence from the results, which item in the ordinal data has received a clear first-choice predominance.

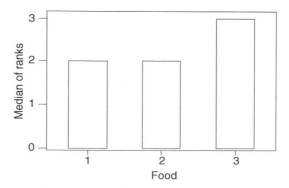

Fig. 3.5   Median bar chart for preference ranking (Minitab).

Ultimately, this can be judged by a significance test, but a summary measure is the *median* (Section 3.3.1). Summaries such as median values for a ranking test can be displayed as a table and as a *median bar chart* (Fig. 3.5). Using Excel would require calculation of the medians separately, but Minitab does this directly from the raw data using the **Chart option**, which can plot summary measures (Fig. 3.5).

In the example, there are three food items, two have equal medians and the third is ranked lower in preference.

Ordinal scales

Scales such as the Likert item are deemed to be used in an ordinal manner (Sirkin 1999). These data can also be summarised by frequency tables and simple bar or pie charts as described earlier but suffer from similar limitations when several items are summarised together (Table 3.3).

Table 3.3   Frequencies for Likert scale type items (5 statements, seven-point scale, 21 consumers) (Excel).

	A	B	C	D	E	F	G
1			*Frequency table*				
2			*Likert-item statement*				
3	Response	Code	s1	s2	s3	s4	s5
4	*Disagree strongly*	1	5	0	1	0	0
5	*Disagree*	2	6	3	4	3	0
6	*Disagree slightly*	3	6	7	6	5	6
7	*Neutral*	4	2	10	5	12	10
8	*Agree slightly*	5	1	1	2	0	2
9	*Agree*	6	0	0	1	0	2
10	*Agree strongly*	7	1	0	2	1	1
11							
		Total	21	21	21	21	21

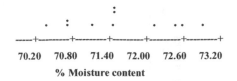

Fig. 3.6   Dot chart of moisture content in raw potato (Minitab).

This table, a form of crosstabulation, was created using the Excel FREQUENCY function for five columns of data (21 values in each). As seen, interpretation of such tables can be difficult, but statement numbers 3 and 5 have most of the 'agree' category. Here, a table or bar chart of median values can sometimes give a more comprehensible overall picture.

3.2.3   Summarising metric (interval and ratio) data (continuous or discrete)

Interval and ratio data can be summarised by some of the methods above, but there is one major limitation. Visualising large data sets cannot be done easily using frequency tables or pie and bar charts, as there are usually too many individual values, particularly if the data are continuous.

With a small number of values, graphs such as the ***dot chart*** (also known as the *cross* or *blob chart*) (Calcutt and Boddy 1983) can be used and can be drawn quickly manually or by using Minitab's **Character graph** (**Graph>Charactergraph>Dotplot**) option and MegaStat's Dotplot. The moisture contents of 11 samples of potato flesh are marked on the chart (Fig. 3.6) where the values are spread out with a slight central peaking.

For larger data sets, some form of *grouped display* of the frequencies is required. Graphically, this is provided by a **histogram** or in table form as a **grouped frequency table**. The grouping process should be such that the group sizes are of the same range or width on the graph (Fig. 3.7).

Excel has a **Histogram** tool in the **ToolPak** menu (**Data/Data Analysis** in Excel 2010; **Tools/Data Analysis** for Excel 2003). Data (% fat content in retail meat cuts) are entered as a column(s) followed by an adjacent column, which lists the

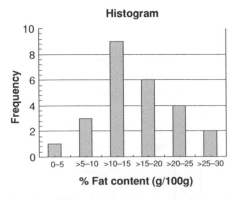

Fig. 3.7   Histogram of % fat in minced beef (Excel).

Table 3.4   Data, grouped frequency table and histogram analysis (Excel).

	A	B	C	D	E	F	G
1	Sample data ($n = 25$) % fat (g/100 g)			Grouping (bins)			
2							
3							Result
4	17.00	11.33	20.81	5		Bin	Frequency
5	4.54	12.55	9.70	10		0–5	1
6	15.56	17.54	7.09	15		>5–10	3
7	10.93	13.31	10.69	20		>10–15	9
8	18.13	12.60	16.44	25		>15–20	6
9	13.38	18.90		30		>20–25	4
10	29.45	12.35				>25–30	2
11	14.00	24.34				More	0
12	27.84	21.63					
13	6.16	21.92					

desired group sizes (referred to as **bins**) plus one empty cell. Each bin represents a group range for the data, namely the bin numbered as '5' will include all values in the range 0 (or less) up to 5, etc. The data (**Input Range**) and the bins (**Bin Range**) are selected in the histogram menu along with **Chart Output**. Label headings can be included in the input but they must be in the cells immediately above the data and the bin columns and the **Labels** box must be ticked (if data are in 2 or more columns then omit labels). The resulting output (Table 3.4) is an example of a grouped frequency table, and it will be accompanied by the histogram (Fig. 3.7).

An alternative way to obtain a histogram is to deselect the chart output option in the **Histogram** menu and use the FREQUENCY() function as above for the bin groups and plot a column chart with the charting facilities. In Table 3.4, the output bin column has been edited to give more detail to the group sizes before plotting as a simple **Column Chart** using the groups as *x*-axis labels (as Fig. 3.7). The majority of fat contents is in the 10–15% region of the data.

Line graphs (Fig. 3.8) are useful for displaying changes over time as in storage or processing changes, etc. The line shows the rate of change of temperature over the period with the product attaining 2°C by 55 min.

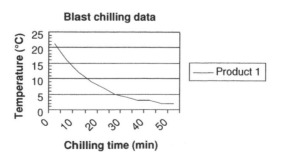

Fig. 3.8   Line graph of temperature change during chilling (Excel).

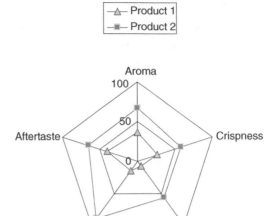

Fig. 3.9   Radar chart of sensory profiling data (Excel).

Radar charts (also known as spider charts or web charts) have a special place in food science – that of displaying multivariate sensory data from descriptive analysis techniques (Fig. 3.9). Product 2 has higher scores for all attributes.

Such graphs appear regularly in the literature as they can display more than one measure for more than one sample product. Chambers *et al.* (2004) used a radar graph to display three samples, 15 sensory attributes along with significant difference indicators.

Boxplots are a graph type that displays several aspects of data including the distribution and summary values (Fig. 3.10).

In the chart (available in Minitab and MegaStat for Excel), which has a vertical layout in this example, the maximum and minimum values (range) of the data are at the end of the 'whiskers' unless extreme values are present; the upper and lower quartiles (Section 3.3.2) are at the top and bottom of the box, respectively. The median can appear as a 'shelf' that divides the box. In this example (a plot of the data of statement 1 of Table 3.3), the values range from 1 to 7 and the median is 2. Boxplots are also able to display outlying or extreme values marked individually with an asterisk. For these data, it can be seen that there is one extreme value.

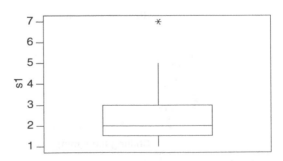

Fig. 3.10   Boxplot of statement 1 (Table 3.3) of Likert scale (Minitab).

Data points of this nature, generally referred to as **outliers**, should be examined for validity (see below). The median is off-centre indicating that the distribution is not perfectly symmetrical and is slanted or *skewed* to one side. Comparison of several boxplots together is one rapid way of comparing data from different samples.

Metric data summary measures can be displayed graphically in a number of ways including summary statistics along with error (variability) indicators.

Example 3.2-3 *Summary of metric data by graph of mean and standard deviation*

As part of a chemical analysis of dehydrated food samples, moisture content was determined for three samples with five replicate determinations. The data were analysed using Excel functions to give the means and standard deviations:

	Sample 1	Sample 2	Sample 3
Mean	1.4	5.6	4.0
Standard deviation	0.49	2.65	0.63

To display the data graphically using Excel requires that the summary measures are calculated first and tabulated as above (see Tables 3.1 and 3.6). The mean values are then blocked and plotted as a simple column chart. To get to the 'error bar' menus in Excel 2010, click on the chart if it is not already selected to access **Chart Tools** then click **Layout**. Click the **Error Bars** button then **More Error Bars Options**. Turn on the **Display** as **Plus** and **Cap** then for the **Error Amount** turn on **Custom** and click **Specify Value**. Select the three standard deviation values as the **Positive Error Value** then click OK followed by **Close**.

For Excel 2003, the 'error bar' option is accessed by a right-click on a bar or point on the graph and selecting **Format Data Series**. Select **Y Error Bars** and one of the error markers. Click the **Custom** button and block the standard deviation values for the data table as the '+' values for the error bar to finish.

Figure 3.11 shows mean values and standard deviations for the moisture content in the samples of dried food. Sample 2 has a higher average moisture content and a much larger standard deviation than the other samples, which agrees with the table values above.

3.2.4   Summarising two variables together

How to summarise bivariate data

These procedures are often carried out in examination of relationships between variables. Two different measures are required on each object or person and these can be of the same level of measurement or different ones.

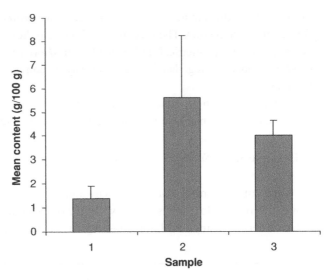

Fig. 3.11 Mean value and standard deviation (Excel).

Bivariate nominal data

Two sets of nominal data can be displayed together by use of a compound column or bar chart (see Figs 3.2 and 3.4). Thus, Fig. 3.2 shows one nominal variable 'defect' with four categories and a second nominal variable, 'batch' with two. Alternatively, a *crosstabulation* table can be used (similar to Table 3.3, which is a '5 × 7' table of 'statement' by 'response'). Here, the data are rearranged and summarised, with one variable in the vertical direction and a second variable in the horizontal direction. A crosstabulation can be described as showing the relationship between all possible values that each variable can take. The simplest crosstabulation would be the '2 × 2' case. The term refers to the number of rows and columns, respectively.

Minitab was used to show an example (Table 3.5) for three lots of raw material from different locations ('source') that were graded into five colour classes ('grade'; **Minitab Stat>Tables>Crosstabulation** option).

The example is a '3 × 5' table. The 'crosstabs' shows that source '2' appear to have more incidence of the colours '3' and '5' and that source '1' has a spread of colours '2–5'. Crosstabulation tables are sometimes easier to interpret if the

Table 3.5 Three-by-five table of location and colour grade (Minitab).

	1	2	3	4	5	All
1	0	11	9	12	8	40
2	0	0	22	0	18	40
3	12	18	9	1	0	40
All	12	29	40	13	26	120

Cell contents – count

Rows, source; columns, grade.

frequencies are shown as percentages, particularly when comparing different tables with differing overall numbers of items.

Bivariate ordinal data

With ordinal data sets, crosstabulation and clustered bar charts can be used, but as more individual points are possible on ordinal scales this can make such tables and charts difficult to interpret. It is possible to use a **scatter graph** with two sets of ordinal data, referred to in Excel as the **XY (Scatter)** chart. The allocation of which variables are X and Y can be arbitrary unless more involved analysis is intended (Chapter 6). One difficulty may arise with superimposed points when the scale is discrete and when there is a large sample size. Some software packages allow display of these cases usually by the size of the point being related to the number of values (point density). This can be alleviated in some software by turning on some form of display for *point density* (see below).

Example 3.2-4 *Summary of bivariate ordinal data by a scatter graph (superimposed points)*

A consumer survey gathered data on peoples' opinion of genetically modified foods by recording their level of agreement with various statements. Two of the statements (Table 3.3) were '*GM foods are beneficial*' (no. 2) and '*I have a reasonable level of scientific knowledge*' (no. 3). Is there evidence of a relationship between these two measures on the basis of a scatter graph?

The data for the two statements were entered in Excel as two columns (these data are summarised in Table 3.3 as frequencies and are available as individual values on a website file). Superimposed points are present in these data and they can be displayed in Excel by first sorting the two columns of data so that points with the same coordinates are listed together.

In Excel 2010, select one column, then click **Data tab** on the **Ribbon** and in the **Sort** menu choose the **Sort A to Z** icon. Ignore the warning message and click **Sort**. Repeat this for the second variable. In Excel 2003, select a column then click **Data** in the top menu bar followed by **Sort**, **Sort** again, then in the menu click OK.

When the data are viewed in this way, there are several instances of superimposed points, three for value 2 and five each for values 3 and 4. The symbols representing these occurrences can be edited to make them larger in proportion to the number of occurrences.

The scatter graph is produced by blocking the two columns of values (to get the display as per Fig. 3.12, copy the data so that statement 3 is in the left-hand column). For Excel 2010 click **Insert**, select **Scatter** from the chart menu and choose the unlined chart. This gives a graph with all points the same size. Identify a superimposed point then click it twice, then double click it to access the **Format Data Point** menu. Point size can be changed using the **Marker Options (Built-in** button). In Excel 2003, use the **Chart Wizard** and selecting the **XY (Scatter)**

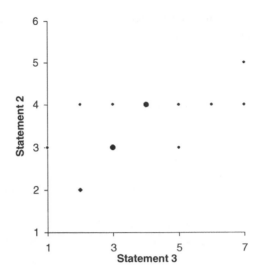

Fig. 3.12 Scatter diagram with point density (Excel).

option (**Chart sub-type** with points only). Superimposed points can be changed by clicking on a point twice then double-clicking as above; the symbol size can then be changed via the sizing box.

For this example, the superimposed values were changed to size 6, 8 and 8, respectively, whereas the other single instance points were set at size 4. In addition to the main set of points, it can be seen (Fig. 3.12) that the superimposed ones display a stronger relationship, which can help interpretation. In the chart, the impression is of a positive relationship – as statement 3 values increase so do those of statement 2. There appears to be a relationship between a person's science knowledge and the perception of GM foods as beneficial: those who are more scientifically knowledgeable are more likely to perceive advantages.

Minitab's character graph option produces a scatter diagram with point density shown as numerical digits for two or more superimposed points.

Bivariate metric (interval and ratio) data

Metric data can have many scale points, especially when continuous and crosstabulation tables are often large and unwieldy to view. A common graphical display is by the *scatter diagram* as in the previous example (Fig. 3.12), but there may be less incidence of *coincident points* (see more examples of scatter graphs in Chapter 6).

Bivariate mixed levels of measurement

This usually requires that one variable is modified to lower its level of measurement to that of the other variable. Thus, nominal plus metric would require that the metric variable be grouped into categories first. Ordinal and metric data can often be displayed effectively by use of a scatter diagram. It is also possible to leave the data as gathered and examine for relationships for mixed levels of measurement using other methods (Field 2006).

The examples above have essentially illustrated the distribution of individual data items or groups data. While useful information can be derived from these types of results, more can be gained by calculating *summary statistics*, which reduce the data to single numerical values. These are the **descriptive statistics** and include such common examples as the mean, range and standard deviation. Most of these latter measures are applicable to metric data, but there are forms available for nominal and ordinal data.

3.3   Descriptive statistic measures

Calculation of these statistics reduces data to key numerical summaries. They are often presented in tables, but graphs can also be used as illustrated in Figs 3.5 and 3.11. Several of these summaries are usually performed in one operation and a list of measures is obtained. A full illustration using Excel is given below, but first some brief definition and detail are required.

3.3.1   Measures of central tendency

These measures are named in this manner because the parameter or statistic obtained is located centrally on the measurement scale between the full span of the data. This can be visualised as being at the centre of the population distribution (Fig. 2.11).

The mean (arithmetic mean)

The **mean** or average value is probably the most familiar statistical measure for a sample of data. It is obtained by summing all of the individual data items then dividing by the number (n) of data: e.g. five determinations of magnesium content to the nearest milligram/100 g:

$$20, 24, 18, 26, 27; \text{mean} = 23$$

The **trimmed mean** gives a statistic based on the middle values, and it can be used as a check on the effect of extreme data.

The median

The **median** is the middle value in a series of values that have been arranged in rank order, e.g. seven ratings on an ordinal scale:

$$1\ 1\ 2\ \boxed{3}\ 5\ 6\ 9; \text{median} = 3$$

When there is an even number of values, the median is the average of the two middle values:

$$1\ 1\ 2\ \boxed{3\ 4}\ 5\ 5\ 8; \text{median} = (3 + 4)/2 = 3.5$$

The mode

The *mode* is the most frequently occurring value in a series of values:

$$2,2,2,3,3,4,4,4,4,4,4,5,6,6,7,7,7; \text{mode} = 4$$

Other measures of location

In addition to the median, which can be located at the centre of the data when they are ordered in magnitude, other measures of location are possible. The ordered values can be divided into groups of various sizes. Thus, the *quartiles* give four groups. The lower and upper quartiles are identified by the last value in the first and third group, respectively, e.g.:

$$\boxed{2,4,5,6,8}, \boxed{12,15,16,20,22}, \boxed{26,28,30,40,46}, \boxed{48,50,52,55,59}$$

lower 25% at 8   upper 25% at 46
(The interquartile range is $46 - 8 = 38$)

Similar procedures can be used to create centiles (10 groups) and percentiles (100 groups). These measures are useful for comparing the spread of data via derived measures such as the *interquartile range* (see below) and for dividing data into similar group sizes for comparison (subgroup comparison). Kolodinsky *et al.* (2003) used a 75th percentile split to divide a consumer group in two levels of concern regarding genetically modified foods, and West *et al.* (2006) compared quartiles of mineral levels in fruit juices to show which sources had more variability.

3.3.2   Measures of dispersion or variation

The measures of central tendency give single value estimates from the sample data. They are limited in this respect and can mislead, as it is possible to have data sets with similar 'central' statistics, but which differ otherwise:

$$44, 44, 45, 46, 47, 48, 48; \text{mean} = 46$$
$$5, 5, 10, 20, 80, 90, 112; \text{mean} = 46$$

Thus, a mean value alone gives insufficient description of a sample. What is required is a measure of 'spread' of data on either side of the mean (or median). There are a number of ways of calculating this dispersion or variation of values in a set.

The range

This is given by subtracting the smallest value in the series from the highest value:

$$34,35,47,56,69,74; \text{range} = 74 - 34 = 40$$

The *range* is limited in that it is calculated from two values, irrespective of the total number in the data, which can be much larger. It can give a false impression, especially if one value is extreme compared with the other data. More powerful measures of variation are given by methods that look at the deviation from the mean for each datum.

Mean deviation

The simplest of these is the ***mean deviation*** (also known as the *average deviation* or *mean absolute deviation*). For this, the absolute deviations (i.e. negative signs are ignored) from the mean are summed and divided by the sample number (n). Similar calculations can be used to obtain a ***median absolute deviation***.

Standard deviation and variance

A more familiar measure of variation is the ***standard deviation*** (sd), which is calculated from its 'parent' measure, the ***variance***. Calculations are similar to the mean deviation, but the deviations are squared before summing and division by n 1. The square root is then taken of the variance to give the standard deviation in the original units of the data. The mean deviation, variance and the standard deviation all give the degree of variation in the sample data. The higher these measures are, the greater the variation judged relative to the mean. A minimum value for any is *zero* (i.e. no variation in the sample values – they are all the same). The standard deviation is more commonly quoted and both this and the variance are more sensitive to wider variation due to the squaring of values in the calculation (O'Mahony 1986).

A distinction is made between the 'population ***standard deviation***' (σ) and the '*sample standard deviation*' (**sd**) in that the divisor $n - 1$ is used for the latter (and is the usual formula in software and calculators). This is because on the basis that the sd is an estimate of σ, the use of 'n' results in a *biased estimate* of the population parameter.

When comparing data sets from different measuring systems with different units for the extent of variance, absolute values can mislead and a *relative measure* is more appropriate. Thus, if two sensory intensity-scaling methods are compared and the first (a scale of 0–30) has an sd of 1 for a mean of 10, this would reflect a relatively large variation. The second method uses a scale of 0–150, and also obtains an sd of 1, but for a mean of 100, indicating a very small spread of values. In these cases, a more valid way of comparing the variation is to calculate a relative form of the deviation: the ***coefficient of variation*** (**CV**; also known as the *relative* *standard deviation*, RSD). It is calculated by dividing the sd by the mean and is conveniently expressed as a percentage by multiplying by 100:

sd	Mean	%CV (= (sd/mean) × 100)
1	10	10%
1	100	1%

The %CV values now reveal the large difference in spread. Consequently, the %CV is appropriate when comparing different analytical methods, sensory cf. instrumental, etc. Strictly speaking, this statistic is only applicable to ratio data, i.e. scales with a true zero, and negative data calculations could mislead (Chatfield 1992). Both deviations can be taken as a measure of how consistent repeated measures are on the samples of the same material – thus they are measures of *precision* (Section 3.4.2).

Standard error

The measures of variation above indicate the spread of the data, but they do not provide measures of confidence about the estimate of the mean value. The *standard error* (se) is calculated from the standard deviation (sd) divided by the square root of the sample number (n). It takes into account not only the spread but also the sample size. The standard error will vary indirectly with the certainty of the mean value, as it gives an indication of how far the estimate is from the true unknown mean (James 1996). It is derived from the *sampling distribution of the mean* (Section 2.4.1) and is used extensively in significance tests and procedures that test mean values. It is a measure of the variation of mean values of repeated samples of size n – similar sds can give rise to different se values due to different sample sizes:

Data		Mean	Sd	se	n
Set 1	2, 4, 5, 2, 5, 6, 4, 7, 1	4.0	2.0	0.7	9
Set 2	1, 3, 6, 3, 5, 6	4.0	2.0	0.8	6
Set 3	2, 4, 6	4.0	2.0	1.2	3

In this example, the smaller standard error for data set 1 signifies more confidence in the estimate. Standard errors are reported in many published experimental results for food studies, usually along with mean values in tables or graphs, e.g. those shown in Martín-Diana *et al.* (2007) for bar graphs of instrumental firmness of vegetables, but the standard deviation is possibly more common.

The interquartile range

Another way to look at spread of data is to consider the range based on the location of points in the data when they are ordered in magnitude. Thus, the ***interquartile range*** (the upper quartile minus the lower quartile) gives a view of the spread of the data away from the extremes of the range itself. This can indicate the presence of extreme values (Fig. 3.10).

3.3.3 *Summary measures for proportions*

Nominal and ordinal (non-numerical) data can be summarised as frequencies, percentages and proportions (Section 3.2.1 and Table 3.7). Descriptive statistics such as the arithmetic mean and standard deviation are not applicable for such data and they have other measures. Proportional balance between two possibilities (e.g. 'yes/no' response in a consumer survey) follows a binomial distribution (Upton and Cook 2001), which, when the sample size is large, approximates to a normal distribution. The measures are illustrated in a simple example (then with reference to the data of Table 3.7).

Proportional mean

A ***proportion*** is calculated from the frequencies of two or more possible outcomes, but only two are considered here. The proportion of interest is identified and this is divided by the total number of samples, e.g. 100 consumers are asked '*Have you modified your food shopping habits due to food scares?*' and they are required to answer 'yes' or 'no'. The result is that 40 say 'yes' and 60 say 'no'. Assuming that the proportion of interest is the number who reply 'yes', then the *point estimate* of this is:

$$\text{Proportion} = 40/100 = 0.4 \text{ or } 40\%$$

NB: as with probability, proportions can be expressed as a decimal fraction 0–1 or as a percentage 0–100%.

The normal approximation allows calculation of an average value of a proportion as the above and as the proportion multiplied by the sample number:

$$\text{Average}_{\text{proportion}} = 0.4 \times 100 = 40$$

Variance of a proportion

The variability of a proportion is given by the product of the two frequencies expressed as proportions:

$$\text{Variance}_{\text{proportion}} = 0.4 \times 0.6 = 0.24$$

The 'spread' of such data is provided by the proportional variance. Hence, with the example above for a 40%:60% result the variance is 0.24. If a 50%:50% split was obtained then the variance would give a maximum value (0.25) and this decreases as the two proportions diverge. Thus, a 90%:10% division would have lower variance (0.09), as more of the sample respondents are in agreement. When all respondents agree (a 100%:0% response), the variance would be zero. A proportional standard deviation and standard error can be calculated from this form of variance and can be used to calculate a confidence interval for a proportion (see below).

Example 3.3-1 *Illustration of descriptive statistics*
Part of the information required in a development trial included various measures on a fruit drink prototype: sugar content by chemical analysis, sensory analysis and a consumer ballot on several aspects of the concept and its selling potential. Three data sets are provided: an instrumental chemical analysis of total sugars in one ingredient (blackberry pulp; 10 sample units from 1 large batch); a sensory analysis on 10 sample units from a prototype formula of the soft drink, evaluated for 'fruitiness' on an intensity scale by a novice panel (Table 3.6); for illustrative purposes these latter data comprise the panel modes rather than the means, arguably more appropriate due to lack of training. The third set are data on one question (Table 3.7) from a consumer survey on '*Would you purchase this product on a regular basis?*' The instrumental and sensory data have received the same analysis, but the consumer analysis includes some additional statistics.

For Excel analysis, data were entered as shown in Tables 3.6 and 3.7. Descriptive summaries were produced by Excel functions. In some cases (range, standard error and confidence interval based on *t*), there is no function and a calculation was required (the **Data Analysis Tools** option **Descriptive Statistics** gives these measures and most of the others directly).

The Excel function formulae are shown for the instrumental data column only, but they have been applied to the sensory data also and can be copied over for readers who wish to prepare the worksheet for their own data, etc. Once calculated, some measures are referred to subsequently by an *abbreviated name* and not by cell reference, e.g. the standard deviation (sd) is calculated by the STDEV.S() function (STDEV() Excel 2003) and in later formulae on the same sheet, it is referenced as 'sd'. Reference to these tables (Tables 3.6 and 3.7) is made below and at other points where appropriate, although additional descriptive analyses are also presented.

Comments on instrumental and sensory data (Table 3.6)

The summary values for the data sets show that in the case of the instrumental analysis, data (continuous) for the mean and the median are the same (when rounded up) indicating symmetry in the distribution of the sample data. There is

Table 3.6 Descriptive statistics of instrumental and sensory data (Excel).

	A	B	C	D
1	Sample instrumental data ($n = 10$)		Sample sensory data ($n = 10$)	
2	Total sugars (g/100g) in blackberry pulp		Intensity (1-9) of 'fruitiness' in fruit drink	
3				
4		6.3		9
5	D	7.2	D	2
6	A	6.2	A	1
7	T	6.5	T	4
8	A	7.3	A	8
9		5.8		5
10		5.9		1
11		6.4		1
12		7.0		9
13		6.8		2
14	Statistic (name):		Formula (column B)	
15	Count (n) (count)	10	= COUNT(B4:B13)	10
16	Mean (mean)	6.5	= AVERAGE(B4:B13)	4.2
17	Median	6.5	= MEDIAN(B4:B13)	3
18	Mode	# N/A	= MODE.SNGL[a](B4:B13)	1
19	Minimum (min)	5.8	= MIN(B4:B13)	1
20	Maximum (max)	7.3	= MAX(B4:B13)	9
21	Range	1.5	= max - min	8
22	Mean deviation	0.43	= AVEDEV(B4:B13)	2.84
23	Standard deviation (sd)	0.52	= STDEV.S[b](B4:B13)	3.36
24	Standard error (se)	0.16	= sd/SQRT(count)	1.06
25	Sample variance	0.27	= VAR.S[c](B4:B13)	11.29
26	Kurtosis	−1.21	= KURT(B4:B13)	−1.54
27	Skewness	0.12	= SKEW(B4:B13)	0.57
28	1st Quartile (1st Q)	6.23	= QUARTILE(B4:B13, 1)	1.25
29	3rd Quartile (3rd Q)	6.95	= QUARTILE(B4:B13, 1)	7.25
30	IQR	0.73	= 3rd Q–1st Q	6
31	Trimmed mean	6.54	= TRIMMEAN(B4:B13)	4
32	Accuracy:			
33	'True value' (true)	6.6	(from reference data)	
34	Error of mean (em)	−0.1	= mean−true	
35	%REM	−0.92	= em/true ∗ 100	
36	Uncertainty:			
37	Confidence Int. $_{(95\%, z)}$	0.323	= CONFIDENCE.NORM[d](0.05, sd, count)	2.082
38	t (0.05, df)	2.262	= T.INV.2T[e](0.05, count-1)	2.262
39	$CI_{95\%, t}$ (CI)[f]	0.373	= t ∗ se	2.404
40	Precision:			
41	Standard deviation	0.52	= sd	3.36
42	%CV(RSD)	7.97	= sd/mean ∗ 100	80.00

[a]MODE() Excel 2003.
[b]STDEV() Excel 2003.
[c]VAR() Excel 2003.
[d]CONFIDENCE() Excel 2003.
[e]TINV() Excel 2003.
[f]Available as CONFIDENCE.T() Excel 2010.

Table 3.7 Descriptive statistics of consumer data (Excel).

	A	B	C	D	E
1	Sample consumer data ($n = 30$)				
2	*Would you purchase this product on a regular basis?*				
3		(1 = no, 2 = yes)			
4	1	1	2		
5	2	2	1		
6	1	1	2		
7	2	2	1		
8	1	1	2		
9	2	2	2		
10	2	2	2		
11	1	2	2		
12	2	2	2		
13	1	2	2		
14		Result:		Formula:	
15		Count (n)	30	= COUNT(A4:C13)	
16		Mode	2	= MODE.SNGL[a](A4:C13)	
17		Minimum	1	= MIN(A4:C13)	
18		Maximum	2	= MAX(A4:C13)	
19		Proportion:			
20		Count (yes)	20	= COUNTIF(A4:C13, 2)	
21		Count (no)	10	= COUNTIF(A4:C13,1)	
22		Count (total)	30	= yes + no	
23		Proportion (p_yes)	0.67	= yes/total	
24		Proportion (p_no)	0.33	= no/total	%
25		Mean prop for yes	0.67	= p-yes	66.7
26		Variance (p_var)	0.22	= p-yes * p-no	22.0
27		Standard error (p_se)	0.09	= SQRT(p-var/total)	9.0
28		CI $_{95\%,z}$	0.17	1.96 * p-se	17.0

[a]MODE() Excel 2003.

no mode, i.e. all the values are unique. For the sensory data (discrete) the mean, median and mode are all different suggestive of a non-normal distributional shape.

Histograms (Figs 3.13 and 3.14) produced by Minitab show that the sensory data are less symmetrical than the instrumental.

These statistics have summarised the data set, no matter how large, to single values, usually located at the centre of the distribution of the data. This depends on how symmetrical the distribution is – if there is non-symmetry or more than one mode is present then these values will not coincide. The measures of *skew* (symmetry about the centre) and ***kurtosis*** ('peakedness'), which are both near zero for a normal distribution, show that both distributions are flat and are positively

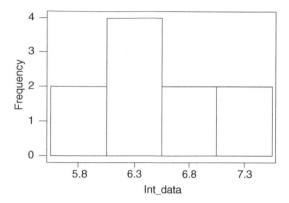

Fig. 3.13   Histogram of instrumental data (Minitab).

skewed, but the sensory data set is more marked in both respects and has a suggestion of bi-modality.

The measures of dispersion for the data sets show quite different results for sensory versus instrumental, but it must be remembered that comparison of absolute values can be spurious with data from different scales. Comparisons of standardised measures such as the %CV are valid: the instrumental data are much less variable.

Comments on consumer data set (Table 3.7)

The summary table provides a number of additional measures to those used for the data of Table 3.6. There are 30 responses of 'yes'/'no' and the mode is 'yes' (coded as 2). The maximum and minimum values provide a simple check on data input error – in this case, there are no accidental entries of double digits or of missing

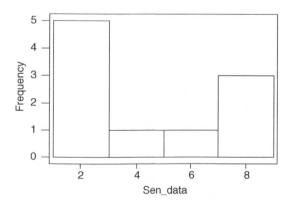

Fig. 3.14   Histogram of sensory data (Minitab).

values. The predominance of 'yes' is confirmed by the count data. Proportional measures were then calculated as shown by simple formulae. It can be seen that the estimate of the population proportion for the 'yes' response (the measure of interest) is 0.67% or 67%. The variance is 'medium' (a 67:33 split) and the value is used to give a 95% confidence interval. This is based on an approximation to the normal distribution as n is relatively high (n multiplied by either proportion should be greater than 5). Expressed as a percentage, this gives $\pm17\%$, approximately. The confidence in the estimate appears very low with a range of 50–84%! This would be unacceptable in most surveys and a larger sample number is required, e.g. increasing the sample size to 300 would reduce the interval to approximately 5%. Further reduction becomes much more costly in terms of sample size – reduction to 3% requires a sample of over 1000 and 2000+ for 2%.

Market research surveys of consumers are often limited by these numbers and tend to compromise on a sample of approximately 1000 with $\pm3\%$ (hence the term *3 percentage points of error*; Upton and Cook 2002; Bowerman *et al.* 2008).

3.3.4   Application of descriptive measures

How to apply and interpret descriptive measures

The applicability of these measures depends on the form of the data (continuous or discrete), the shape of the distribution and the level of measurement of the scale employed. Thus, with nominal data, which do not originate from an ordered scale, there will not be a 'middle' point. An appropriate summary measure is the mode and the data can be examined by a pie chart or frequency table to view the distribution. If nominal data are expressed as proportions then the measures of Section 3.3.3 are suitable. With ordinal data where it cannot be assumed that there were equal intervals on the scale, or where the distribution is skewed, the median and interquartile range for variation are applicable. For metric data (interval and ratio) where symmetry and shape of the distribution approximate to normal, use of the mean, standard deviation and standard error are justified.

Comparison of the median, mean and mode can be made for metric data as a symmetry check, but the mode may not 'exist' (i.e. each value is distinct) in small data sets for continuous data (see Table 3.6) or there may be *bi- or multi-modality* for discrete data, which can be misleading. For some experimental results, other factors such as the nature of the distribution of the data may prompt use of summaries other than the customary methods. Albala Hurtado *et al.* (2001) used medians and quartiles for vitamin K content data due to high variability and non-normal distributions, despite such data being viewed usually as parametric in nature.

It is recommended to examine all data after entry using a variety of graphs, tables and summary statistics to ascertain that no obvious data entry errors have been made and to assess distributional shape (see comments on the data of Tables 3.6 and 3.7).

How to deal with missing and outlying data

Missing values will show up during data entry or in tables and some graphs. They can be handled by most software packages in that the analysis can proceed without the missing data. If not, replacement can be done by one of a number of complex methods, but a simple solution is to replace any missing data using the grand (overall mean) for the whole data set where this is appropriate. Extreme or *outlying observations* may also stand out in tables and especially in boxplots (Fig. 3.10). **Outliers** need to be examined to see if they are due to data input errors, etc. If no accidental reason for the outlying nature can be found then a decision needs to be taken on whether or not the value is 'typical' for the data by examining it in relation to the whole data set. Any values lying outside ±3 sds, or excessively outwith the quartiles may be candidates. The data analysed in Fig. 3.10 have a mean value of 2.6 and a standard deviation of 1.5, thus the extreme value (7) is just on the limit for +3 sds. The scale used for these data had seven integer points and with such scales, particularly with 'noisy' data (Section 8.2) the occasional extreme value is recorded in a valid manner. There are some other tests and identification aids that can be used, such as those available in the **Descriptive Statistics** option of MegaStat for Excel. Analysis may give the same results with and without the outlier(s). Once confirmed that it cannot remain, the outlier can be deleted all together or replaced with the grand mean as above, or replaced with the next largest or smallest value (Chatfield 1992).

Summary statistics give a smaller number of key values to assess the data, and the next stage is to examine the quality of the data in respect of their trustworthiness or integrity and the presence of errors. Some indication of this can be gleaned from the data summaries presented above if some knowledge of 'typical values' is known. To acquire more information, additional calculation is required.

3.4 Measurement uncertainty

How trustworthy are the results?

All types of measurement are subject to *error* sources. Some of these have been touched on in previous sections, and the consequence is that there is a *degree of doubt* or **uncertainty** concerning the results. An example of a laboratory situation demonstrates this (Box 3.1).

As seen, the experimental weight is not the true weight. The process of gathering and measuring data by whatever system is subject to a variety of error sources. Error is often described as 'noise' in that the error makes it difficult to 'hear' any real effect that may be present, hence the expression *noisy data* where there is high error. In the example, the errors have been identified and quantified by simplified means, but proper assessment requires additional experimentation. In more routine situations, the analyst would not carry out such detailed error assessment. Consequently, the experimental result would be viewed as having an error effect above or below the true value. There are statistical measures that can quantify error,

---

> **Box 3.1   Simplified example of error in measurement.**
>
> A food sample is weighed three times on a digital balance and averaged to give experimental weight:
> **EW** = (2.56 + 2.60 + 2.58)/3 = 2.58
> The true weight of the sample is **TW** = 2.3841 **g**
> During the measurements, the lab door was left open and a strong draft made the reading unsteady: **E1**
> A sticker on the balance reads: 'Serviced, calibrated 1982': **E2**
> The balance read to the nearest 0.01 g: **E3**
> Repeated readings were different: **E4**
> Possible results are: **EW** = **TW** – a completely 'correct' or an accurate result
> **EW** < **TW or EW** > **TW** – an inaccurate result
> The reasons for the inaccuracy are due to error(s), thus:
> **EW** = **TW** + **E1** + **E2** + **E3** + **E4** ... [E values can be ±] where,
> Error 1 (the draft) caused the balance pan to lift slightly and under-read the true weight by 0.1 **g**
> Error 2 caused the balance to over-read by 0.3 **g**
> Error 3 rendered the balance unable to weigh to the third and fourth decimal place of the weight
> Error 4 caused repeated readings to vary, thus:
> **EW** = 2.3841 – 0.1 + 0.3 – 0.0041 + (–.02 + .02 + 0) = 2.58 (this is a simplification)

but essentially all experimental results from 'high-tech' instruments to simple questions in a survey, produce a value that is modified by error:

> Measured value = true value + error

Obviously, the practitioner will wish to minimise all error, but first some analysis is required to assess the magnitude of such error. The larger the error, the less the overall trustworthiness of the work and the eventual statistical analysis will be undermined. It is not possible to avoid error completely, but its magnitude should be assessed in some way.

3.4.1   Error types

Errors are sources of variation that arise during all investigations. They are not 'mistakes' for the most part. Types of error are summarised in Table 3.8. Errors begin to accumulate at the sampling stage as discussed in Chapter 2, where *sampling error* arises because the sample is not representative of the population from which it comes – an *estimate* is being made and there is uncertainty within the estimate.

Table 3.8 Types of error.

Error	Nature and effect	Example
Sampling error[a]	Between test result and population parameter; lowers validity/accuracy	Non-representative (biased) sample
Experimental error	Between experimental units; lowers precision	High variability between different units
Gross error	Lowers validity and reliability	Misreading an instrument, scale or survey response; incorrect sample coding or transcription; incorrect/omitted reagent
Measurement error		
Systematic	Determinate, lowers validity/accuracy	Biased method or operator; un-calibrated instrument
Random	Indeterminate, lowers reliability/precision	Variation on repeated readings on same unit or subject

[a]The term is also used in some texts to describe the variation between sampling units.

All the above errors are often described generally as 'experimental error', but this is a distinct effect, which is dealt with in more detail in Chapter 7. Note that *all* measurement systems including questions in surveys are subject to all forms of error variation.

Gross error

Gross errors are made by accident and they can occur at any point. Often they are easily detected as the values will stand out by virtue of their large difference from the other data, i.e. they will appear as outliers. Some marked events such as those in Table 3.8 are typical. When these errors are smaller in magnitude and less detectable, they can have serious consequences and can undermine the whole experiment.

Measurement error

Measurement error can occur during the procedure used to measure the key observations of the investigation – whether it is by use of a spectrophotometer or a sensory attribute scale, the integrity of the data depends on this stage. Contribution to this error can come from the operator and the measuring instrument or method itself, as well as random error.

Systematic error

Systematic errors occur when there is some lack of reference point in the measuring system. They can be difficult to detect unless it is possible to introduce a controlling mechanism. In the example above, a systematic error was introduced because the balance had not been calibrated for over 20 years. This process of **calibration** applies not only to scientific instruments but also to any measurement system. This type of error is also known as *bias* and it will affect all data. Calibration reduces or removes the bias, hence accuracy is increased. It is possible to detect this type of error by introducing a control sample of known nature into the determination. A common example with instrumental measures is 'blank sample', which should give zero or very low value on determination. Others include reference standards with known amounts of analyte, physical properties or sensory stimuli such as liquids with known sugar concentrations. For some instruments and devices, *calibration standards* are provided, e.g. calibration weights for balances, calibration colour plates and temperature tags to calibrate temperature probes and recorders electronically.

Other examples of systematic error include bias in the method of measurement caused by a technique fault by the practitioner, or an inbuilt bias in the method itself, e.g. it is known that certain chemical analysis techniques over- or underestimate. This latter type is a *constant error*, as it cannot be removed unless the method is modified or replaced by another.

Lack of calibration and faulty techniques can be changed and are known as *determinate errors*. With consumer survey procedures bias can be caused by a number of effects including misleading question format, etc. (see below).

Does calibration provide total protection from bias?

Calibrated measurement systems can still give error if they are not sensitive enough to measure the true value being determined. For instance, a milligram balance can measure to the nearest milligram (0.001 g), but if the true weight is beyond the milligram level then an error will be present (see above). Similarly, a sensory panel may be able to measure intensities of sensory attributes to a limited fraction of the scale used, e.g. on a 1–10 intensity scale measurement may be possible to 0.5 of a unit, but not beyond. Findlay *et al.* (2007) used feedback as a sensory calibration procedure for trained panels to enable panellists to see immediately which of their attribute scores were outwith an acceptance region.

Random error

Assuming that the practitioner was able to eliminate the above errors, it could be taken that the measure was now as accurate as possible. However, as shown in Section 3.4, when weighing was repeated on the same unit there was still variation. This is an example of *random error*. It is caused by random variations in environment and electrical supply (Mullins 2003; Nielsen 2003), small unit-to-unit inherent variation in food material and small measurement errors during all the stages of drawing samples for analysis, weighing, pipetting, etc. (James 1996).

Consequently, even when measuring with non-destructive methods, repeated measures are likely to be different, e.g. measuring colour of the surface of a food with a reflectance chromometer – repeated measures at the same point on the food surface under the same settings will give variation. Even with care, these small variations in repeated measures can be still present. As the development of scientific instrumentation progresses, random error will decrease, but it will still be present. For the detection of random error, there must be a least *two* end determinations. End determinations can be averaged to even out this variation, but despite this, they contribute to uncertainty. Random error can also exert an effect in surveys if a repeat measure is performed with consumer respondents. Many psychological and physiological factors can modify a person's response to questions, such as mood, fatigue, etc.

This error type affects the *consistency* of the measure on repeated operation of the method. It is an *indeterminate error* and it cannot be eliminated entirely.

3.4.2 Aspects of data and results uncertainty

There are two specific viewpoints in assessment of the *integrity* or *trustworthiness* of findings:

- Validity
- Reliability

Both terms have several facets to them and general definitions clash with discipline-related forms at points, resulting in confusion and intermingling of the terms. Additionally, the terms are sometimes applied to instruments and sometimes to data, the latter being more appropriate. Essentially, validity has to do with *accuracy* and reliability with *consistency*:

> **Validity** equates with **accuracy**
> **Reliability** equates with **consistency**

Beyond this brief concept, other aspects are best described in relation to the type of measuring system and discipline involved, although the similarity between terms should be evident.

Validity (instrumental and sensory)

Here, it should be envisaged that in one situation the measuring system is in the form of a specific instrumental device (e.g. a spectrophotometer or a mechanical texture-testing machine) or method (e.g. fat content by acid hydrolysis). The other is where a trained sensory panel or a consumer panel assesses or measures food samples for differences in attributes or degree of liking, etc.

Validity in such measurements has two main dimensions: one is *qualitative* in nature in that the measurement system must measure what it is intended to do and the second is *quantitative* validity – 'does the system measure the quantity close to or at the 'true' value (or nearest estimate)?'

For example, if nitrite content is being determined, the method can be valid in two ways: (i) it can determine nitrite and not some other additive and (ii) it will give a nitrite content close to the true value.

The same conditions apply to trained panel and consumer sensory measures – if 'chewiness' is determined by a sensory panel, the measure should determine this attribute and not some other. Quantitative validity can be achieved to a degree

Table 3.9   Measures used to estimate trustworthiness of data.

Measure	Effect
Instrumental/sensory data	
Validity	
Qualitative	Specificity of the measurement system
Quantitative	Accuracy of the measurement system
Population validity	Generalisation of results
Reliability	
Precision	Consistency of results
Survey instrument data	
Validity	
Internal	'Trueness' to underlying construct
Construct	Ability of instrument to demonstrate lack of relationship with theoretically identified non-effect subjects/respondents
Criterion	Closeness (correlation) to established instrument
Reliability	
Internal	Consistency of results within a single survey instrument
Alternate form	Consistency of results between different survey instruments
Test–retest	Consistency of results between repeated surveys

with sensory assessments, although obtaining a 'true' measure of attributes such as 'chewiness' is difficult.

Validity encompasses some wider viewpoints in respect of the quality of experimental findings (Table 3.9).

Qualitative validity can be viewed as the *specificity* of a method, a term used in chemical analyses. Specificity is one criterion in judging the overall quality of data from chemical method analysis, but the concept is applicable to data from any measure. It refers to the extent to which a method measures what it was intended to measure. High specificity implies that a method will be able to 'home in' on the particular analyte, sensory property or consumer respondent opinion under study, despite any interfering factors. For instance, in chemical methods, analysing for specific sugars rather than for total sugars.

For *quantitative validity*, the term refers to how close the estimate is to the true value of the intended measure. In the specific context of instrumental analysis, the term ***accuracy*** is used, but again this view can be extended to cover results from sensory tests and consumer surveys.

Accuracy

A more formal way of defining accuracy, which gives a clue as to how to calculate it, is '*The accuracy of a result is a measure of how close it is to the true value*'.

Similar definitions are given in several texts, particularly the British Standards Institute publications (BSI 1983, 2000). High accuracy is possible for instrumental measures and some sensory measures if ***calibration*** is present. Accuracy can be calculated by comparing the test estimate with a true or defined reference value (see below).

Another aspect of validity is ***population validity***, which refers to the extent to which the results obtained from the investigation can be *generalised*. For instance, can the result be taken as a valid measure of other similar samples under similar conditions? The sampling method (random or otherwise) and how the population and sampling frame have been defined have a major bearing on such issues.

Reliability and precision (instrumental and sensory)

The term 'reliability' as a guide to the quality of data can cause confusion as it is applied in different ways. One is a general measure of the 'dependability' of a measurement system. When used in this way, it considers the ability of the system to 'do the job' for a period of time, etc. (BSI 1983), or as storage stability measure of the products durability and shelf life, e.g. 'how long will a piece of food processing equipment last before it stops working?' The use considered here is more specific where reliability is used to gauge ***consistency*** of a repeated measure and this is described in scientific terms as ***precision*** (BSI 2000).

Precision

The precision of a measurement system is the agreement of repeated measures under specified conditions. This more detailed definition is still a source of confusion, as pointed out by Miller and Miller (1993), due to the exact meaning of the word 'precision' itself. Dictionary definitions of 'precision' do not help as it can be given as 'a measure of accuracy', and the term 'precise' is stated as 'accurate' (as in a 'precision method'). An additional layer of confusion is caused by more recent definitions of accuracy in the BSI 5725 part 1 (BSI 2000) publication, where a general definition of accuracy includes 'trueness and precision'. One way to avoid this confusion would be to use *preciseness* for accuracy and *precision* for reliability. More specific definitions below make the intended meaning clearer.

The basic definition of precision has been expanded to cover different circumstances that arise when measurement takes place in different locations or when done by different operators. This gives rise to questions such as 'Do repeat measures with the same operator (or panel/panellist) on the same material give the same values?', 'Would the same results be produced by the operator at a different time? – Or by different researchers? – Or different laboratories/locations?', etc.

For instrumental and sensory work where repeat measures are performed then precision can take two forms (BSI 2002). The first is ***repeatability***. It is taken to mean precision *within one set of circumstances or conditions*, i.e. within one laboratory or sensory laboratory, or within one set of procedures and with same operator, etc. Taking the definition to cover broader circumstances gives ***reproducibility*** that refers to precision *under different conditions*, i.e. different laboratories, methods, operators and panellists:

> ***Repeatability*** – precision under the *same conditions*
> ***Reproducibility*** – precision under *different conditions*

Precision is essentially measured by calculation of the standard deviation, but calculation of accuracy and precision is detailed below. Note that for instrumental and sensory measures precision and accuracy in whatever form are not automatically directly related. High accuracy does not imply high precision and vice versa (see Section 3.4.3).

Validity and reliability of consumer survey data

The concept of validity and reliability for data obtained by consumer surveys on food issues can be approached in several ways. The measuring instruments in such investigations are the key questions dealing with people's opinions, attitudes, beliefs, motivation, knowledge, satisfaction, etc. In some instances, the researcher may employ **survey instruments** developed by other workers in related fields such as psychology and behavioural sciences, in others new methods may be introduced. The measures can consist of a single question, but more commonly they are multi-component in nature.

These survey instruments attempt to measure a theoretically developed **construct**, which encompasses the 'mind-state' in the consumer subject or respondent; for instance, a set of questions (the survey instrument) that measures the motivation (the construct) of consumers to support 'animal-friendly' food sources.

Ideally, researchers should carry out validity and reliability checks on their own data as well as quoting similar measures from other workers.

The quality of data generated by consumer surveys is also highly dependent on the phrasing of questions and how the process was conducted or performed. Thus, bad grammar, ambiguous or leading questions and use of technical terms without explanation are just some of the effects that could lead to invalid answers. Inadequate control of questionnaire administration can lead to increased error, just as with any lack of experimental control. Another problem is the **Hawthorne effect** (Evenson and Santaniello 2004) caused by respondents reacting to the circumstances of being 'under study'. This can have positive or negative effects, such as respondents supplying answers that they think the researcher wants. Ultimately, the effect can result in invalid data, which is not representative of the true situation. Minimisation of such effects can be achieved to a degree by stressing to respondents the need for honest answers for good research. Accepting that these points are covered, the survey instrument can be judged on other aspects of validity and reliability.

Validity (survey)

Consumer survey measures of people's opinions, motivations, satisfaction and attitudes must, if valid, measure the requested view or characteristic, i.e. survey instruments must possess qualitative validity. The quantitative aspect is more difficult with such less concrete measures and calibration is difficult. These and other psychological factors cannot be measured directly, but must use a constructed scale or a proxy measure, for which it is difficult to have a 'control person' with a known level (zero or other) of the characteristic.

At least three forms of validity are assessable. As with instrumental and sensory data, accuracy or 'trueness' is central to the understanding in each case. *Internal validity* ('content validity', 'face validity') refers to whether or not the survey measure 'covers the ground' in what it is intended to measure, or how well it measures the construct (Malhotra and Peterson 2006); this can be viewed as qualitative validity, i.e. does it include all the necessary dimensions of the measure? For example, in assessing consumer perception of health effects of low-fat diets, a valid construct would need to include all the critical aspects (palatability, energy levels, heart disease, obesity, cost, etc.). Omitting one of these aspects could reduce the internal validity of the data obtained. This type of validity requires assessment by acknowledged experts in the field (the researchers or appointed experts) and is more subjective than a calculated value.

Criterion validity ('concurrent validity') compares the measure with another more established one. This can be achieved by correlation with an established construct run at the same time. This latter form is a type of 'comparison with a true value' and produces a *validity coefficient* (correlation method). It can also be assessed as *predictive validity* (Ruane 2005), e.g. the ability of the measure to predict future actions and behaviours of consumers. This form of validity can be numerically calculated by comparing results (predicted cf. actual; see example in Section 8.4.3).

Construct validity assessment requires examination of how the construct is argued theoretically (Malhotra and Peterson 2006). This can be done as *convergent validity* where the measure is compared (correlated) with other related (but not the same) measures as established by theory and other work. High convergent validity is indicated by outcome of significance correlation coefficients. As *discriminant validity* the measure should be able to discriminate and demonstrate the presence or lack of relationship between group measures (Huck 2004). This could be the ability of the measure to discriminate between identified groups that theoretically should have low and high values (e.g. consumers whose gender suggests that they will have negative or positive attitudes to low-fat diets). In this latter case, gender and attitude should be related if the measure and data are valid.

Population validity applies to the results of surveys also – can they be taken as applicable to other groups of consumers or all the population?

Reliability (survey)

Test–retest reliability is recognised in its similarity to precision. The survey instrument is administered more than once to the same respondents and the results compared by correlation (reliability coefficient). High positive correlation means high reliability or consistency (Section 8.4.3). A simpler form of test–retest can be included in a single run of a questionnaire if one or more questions are presented twice, e.g. at the beginning and at the end so that the respondent is unaware.

Internal consistency reliability refers to multi-item measures and how closely the separate parts measure the same construct. Calculation requires an inter-correlation measure with high values, indicating an acceptable internal reliability.

Alternate form reliability ('parallel form') means 'asking the same question in different ways'. Correlation coefficients can be used to determine to what extent the measures agree. Consistency also applies to the respondents as a group(s) and they can be compared for *inter-rater reliability* when items are rated or scored by calculation of a correlation or *concordance coefficient*.

3.4.3   Determination of measures of uncertainty

How to assess uncertainty of data and results

The term *uncertainty* is primarily applied to chemical analysis data at present, but some of the measures below can be applied to any measurement system. Most rely on the descriptive statistics defined above (Section 3.3), with one or two exceptions, and the formulae and calculations are given in Table 3.6. The methods are described generally and application is covered in more detail in later chapters.

Determination of accuracy and validity

Accuracy (instrumental and sensory data)

The accuracy of a statistical result can be calculated for data of all levels of measurement (such as nominal data proportions), but has been primarily applied with metric data (interval and ratio). The statistic of interest for accuracy calculation with metric data is the mean value. The median and mode can also be used, but this is less common. Based on the definition of accuracy, some measure of the 'correct' result is required, i.e. the true, most probable or any defined value (BSI 2000). The true values for measures on food are likely to be unknown and a 'most probable' substitute is provided by another material of known composition. This can be tailor made or of reference level in character. Hence the terms *calibration standards* and *reference materials* are used, mainly in analytical chemistry. The analyst knows what the contents of these materials are and can use this to calibrate the unknown measures. Similar standards can also be found for physical measures of food properties, e.g. standard materials of known hardness, compressibility or elasticity (food texture) and colour plates (food colour methods), often produced to stringent specifications. Both types of standard (chemical and physical) can also be employed in training and calibration of sensory panels (Chapter 8). For analytical chemistry, these techniques are used extensively in measurement uncertainty trials such as the DTI's VAM (valid analytical measurement) and the Food Standards Agency initiatives and are dealt with in detail in Chapter 9.

A measure of accuracy can be given by the difference between the reference or standard value and the experimental value. The *error of the mean* is obtained by simple subtraction of the true value from the mean (Tables 3.6 and 3.10). It can be standardised to an absolute form for comparison of different methods by

Table 3.10   Measures of accuracy and precision for instrumental and sensory data.

Measure	Calculation
Accuracy	
Error of mean (em)	Difference between the experimental mean (mn) and the true mean (tm): em = mn – tm
%Relative em (%REM)	em divided by tm times 100
Precision	
Mean deviation	'Averaged' deviations of each datum from mean
Standard deviation (sd), variance	'Averaged' squared deviations of each datum from mean
%Coefficient of variation (%CV)	sd divided by mean times 100
Uncertainty	
Confidence interval (CI)	Based on distribution and standard error (level of confidence factor multiplied by the standard error)

division by the mean then multiplication by 100 to give the ***relative error of the mean*** (Ryan and O'Donoghue-Ryan 1989; Table 3.10).

The magnitude of these measures is directly proportional to the level of accuracy. Hence, 'perfect accuracy' would give an em and %REM of *zero*. Positive and negative errors are possible, thus terms as '*overestimate*' and '*underestimate*' can be used. For instance, the sulphur dioxide content of a food sample is determined as 25 mg/kg and the most probable value is 20 mg/kg:

> em = 25 – 20 = 5 mg
> %REM = (5/20) × 100 = 25% (i.e. a 25% overestimate

Confidence interval

Another measure of data quality is given by a ***confidence interval*** (CI). A CI can be viewed as a gauge of precision and to a lesser extent accuracy and is therefore an ideal measure of the ***uncertainty*** in an estimate. The interval expresses a region where the true population value (parameter) is likely to be located. The likelihood is expressed as a *probability*. Narrower intervals reflect a more 'certain' estimate. It should be noted that the population value refers to the population of measurements – if the measurement system and data are *biased* then the 'real true value' may not be at the centre of the interval. It could be off-centre or indeed outwith the interval (see below). CIs are sometimes displayed graphically as:

>   Estimated
> LCL   mean   UCL
> 5.4   7.5   9.6
> |------------o------------|
> $CI_{95\%} = \pm 2.1$

The CI limits are expressed as the lower and upper ***confidence limits*** (LCL, UCL) and the estimated parameter is at the centre. Limits can be quoted as the *whole width* or as ± a *half-width* and the CI expression should include the confidence probability.

CI calculation requires knowledge of the population distribution from which the data originate. In this way, probability can be assigned to the location of the test parameter and the spread or width of the interval calculated. For metric data where the mean value is an appropriate statistic, a **normal population** (Section 2.4.1) is assumed, but there is a complication. The calculation requires knowledge of the variability of the data in the population. If this is not known then it has to be estimated from the sample data. Using this estimate introduces error into the confidence interval itself and underestimates the width of the interval. This is of less importance with large samples ($n > 30$), but for smaller samples the **t-distribution** gives a more valid calculation of the interval. In some applications, sample sizes can be greater than 30, but for instrumental analyses they are usually less than 10; therefore, the *t*-distribution is appropriate.

How to calculate a confidence interval

A confidence interval reflects accuracy only in so far as it spans a region where the true mean is located at the particular confidence level. There are two requirements before the calculation can proceed. Firstly, a measure of the *degree of confidence* must be specified as a probability and usually the 95% limit is the minimum. This indicates that on repeated sampling of the population, 95% of the time, the test mean will be located in the interval. This allows the *level of confidence factor* to be selected from an appropriate distribution.

The second requirement is knowledge of the spread of the data as the standard deviation. This may be known for the population in some cases, but often it is estimated from the sample itself as explained above.

If the population sd, sigma (σ) is known then the calculation (Table 3.6) uses a standardised value (z) representing a 95% content proportion of the region of the distribution spread (Section 2.4) as the confidence factor. This is multiplied by the uncertainty in the form of the **standard error** *(se)*. It can be seen that the larger the se (variation and sample size), the larger the interval:

$$
\begin{aligned}
\text{CI}_{95\%,z}(\text{half width}) &= \text{mean} \pm \text{confidence factor} \times \text{standard error} \\
&= \text{mean} \pm z\ \text{value}_{95\%} \times \text{standard error} \\
&= \text{mean} \pm 1.96(\sigma/\sqrt{n})
\end{aligned}
$$

This form of confidence interval can also be determined by use of the CONFI-DENCE.NORM() function (CONFIDENCE() Excel 2003).

How to determine a standardised z-value

The calculation is possible when σ is known, which allows the use of z-values (Section 2.4.1) from the *standard normal distribution*. It assumes that the sampled values originate from a normal population (a valid assumption for most chemical and instrumental analyses techniques, but less valid for sensory evaluation data). The central limit theorem (Section 2.4.1) allows this assumption to apply to sampling distributions of non-normal sources. The standardised z-value is obtained via the NORM.S.INV() function in Excel (NORMSINV() Excel 2003), thus for a confidence level of 95% (0.95):

$$z = \text{ABS}\left(\text{NORM.S.INV}\left(\frac{1 - \text{confidence level}}{2}\right)\right)$$

$$= \text{ABS}\left(\text{NORM.S.INV}\left(\frac{1 - 0.95}{2}\right)\right) = 1.96$$

Usually σ is unknown and an estimate of it is given by the sample sd. The *CI limits* are much wider and are calculated using parameters from the t-distribution (which has more spread than the normal distribution especially for small values of n). Thus, the confidence factor is represented by a t-value rather than a z-value. Here, t is obtained from the particular t-distribution, which depends on the degrees of freedom (df; equal to sample size minus 1). As n gets small, t becomes large and the interval widens – thus small sample sizes give wide intervals with less certainty in the estimate:

$$CI_{95\%, t} \text{ (half width)} = \text{mean} \pm t \left(\text{sd}/\sqrt{n}\right)$$
$$\text{where } t = \text{two-tailed (5\%) } t\text{-value for } n - 1 \text{ df}$$

Consider four protein determinations with end determinations as 17.0, 19.0, 19.4 and 16.6 (all g/100 g). The sd is calculated solely from the replicate values and the quality (confidence) of it depends on the sample size:

$$\text{Mean} = 18.00, \text{ sd} = 1.4$$
$$CI_{95\%, t} = 18.0 \pm t(1.4/\sqrt{4})$$
$$\text{for } n = 4, \text{ df} = 3, t_5\% = 3.182$$
$$CI = 18.0 \pm 3.182 \times 0.70 = \pm 2.24$$
$$\therefore CI \text{ width} = 15.76 - 20.24\%$$

```
15.76              18.0              20.24
|----------------o----------------|
```

That is, on 95% of occasions when such samples are drawn and a CI calculated, the population mean for %protein lies within the above range. t-Values are obtained from published tables or using the T.INV.2T() function in Excel (TINV() Excel 2003) as shown in Table 3.6. They are based on confidence level (%) and the degrees of freedom. The interval can also be calculated using the function CONFIDENCE.T() (Excel 2010 only).

The maximum value for t at 95% confidence is when there is one degree of freedom, i.e. a duplicate determination. To illustrate this, assume that only two of the %protein determinations were carried out (17.0, 19.0), giving a similar mean and sd:

Mean $= 18.0$ sd $= 1.41$
$t_{(95\%,\ 1)} = \text{T.INV.2T}^{a}\ (0.05,\ 1) = 12.71\ (n = 2,\ df = 1)$
$CI_{95\%\ t} = t*sd/\text{SQRT}(n) = 12.71$
$= 18 \pm 12.71$ CI limits $= 5.29 - 30.71$

| 5.29 | 18.0 | 30.71 |

|- -o- -|

[a]TINV() Excel 2003.

As seen, t is very large as is the interval width, which underlines the low confidence when based on but two determinations. Whether or not the magnitude of the CI is acceptable depends on how confident the analyst is in the validity of sd. In the latter example, it is based on only two determinations and gives a *CI range* of 25.4% – using four replicates cuts this to a fifth.

Larger sample sizes result in a narrower interval, i.e. less error and more confidence. It must be remembered that there is still a chance that the interval does not contain the true value. This is a consequence of using a sample. To become 100% confident would require examination of the whole population. If higher levels of certainty are specified, such as 99%, then the interval will be wider unless sample size is increased. The confidence interval is one of the factors taken into account in sample size calculations. The examples below illustrate these points:

Sample size for confidence of

	95%	99%
Population size		
Large	100	2 500
Small (400)	80	200

As sample sizes decrease, the interval widens and hence the uncertainty increases. The numbers differ due to the second example being a *finite population*, thus requiring an adjustment unless the sample is less than 5% of the population (Czaja and Blair 1996).

CI for proportions
Much of the foregoing has been concerned with numerical metric data, but nominal data proportions can be assessed for uncertainty by calculation of a CI (Table 3.7) as there are measures of proportional variance as explained above (Section 3.3.3).

The assumption is that the data conform approximately to normal when n is high (>30). The binomial variance and proportion estimates are required, then:

$$CI_{prop} = \text{proportion estimate} \pm z \times \left(\frac{\text{binomial variance}}{\sqrt{n}} \right)$$

Thus, assuming a 75:25 variance for 40 consumers where 30 disagree and 10 agree with a statement on *I would shop more from local farms*, and the 'agree' answer is the focus, then:

$$Proportion = 0.25, \quad \text{variance} = 0.75 \times 0.25 = 0.188$$
$$sd = \text{SQRT(variance)} = 0.433, n = 40$$
$$\text{Standard error} = sd/\text{SQRT}(40) = 0.068$$
$$CI_{95\%, \, prop} = 0.25 \pm 1.96 \times se = \pm 0.134$$

The CI width is 0.268 or approximately $\pm 13\%$, which again is high as a relatively small sample was gathered. These calculations are performed using Excel in Table 3.7 for the consumer survey example.

Confidence intervals and sample size calculation

The statistics used in the standard error and CI calculations include use of the sample size used in an experiment or investigation. By rearranging CI formulae, it is possible to produce a method for *sample size calculation*, for both numerical and proportional data. CI calculation requires a measure of variation, the sample size and a statement of confidence level. The standard error is calculated by use of the standard deviation and the square root of the sample size, thus:

$$CI = \text{confidence factor} \times \text{standard error}$$
$$= \text{confidence factor} \times \left(\frac{\text{standard deviation}}{\sqrt{n}} \right)$$

Rearranging the second statement produces:

$$\sqrt{n} = \text{confidence factor} \times \left(\frac{\text{standard deviation}}{CI} \right)$$
Then squaring both sides

$$n = \left(\text{confidence factor} \times \left(\frac{\text{standard deviation}}{CI} \right) \right)^2$$

Sample size (n) can be determined by specification of the required confidence level in the results if the population standard deviation or an estimate is available.

For instance, a food scientist wishes to determine moisture content to $\pm 1\%$, with a confidence of 95%. Previous data exhibited a standard deviation of 1.2%, which is taken as the population parameter. The confidence factor is obtained in the form of a z-value, as described above for the confidence interval. Thus:

$$n = (z_{95\%} \times 1.2/1.0)^2 = (1.96 \times 1.2/1.0)^2 = 6 \text{ (always round up)}$$

Six determinations would be required. Note that the units in this case are percentages. Similar units apply to the error margin (the $\pm$ value) and the standard deviation. There is a slight complication in that because $n < 30$, use of normal distribution z-values would underestimate the sample size, unless there is high confidence in the estimate of the standard deviation. If not, then t-values should be used, but these are less easily incorporated (see above for use of t in a CI). A simple solution is to calculate as above using z, obtain t (df $=$ sample size minus 1) and then repeat the calculation (this process produces $n = 10$).

The same formula can be produced for a proportion. A consumer scientist requires to carry out a survey in which the proportion of consumers who are in favour of switching to non-plastic food packaging is to be estimated to within 5% ($\pm$) with a confidence of 95%. A previous estimate quoted counts of 35% 'in favour' and 65% 'against'. The variance and standard deviation are calculated as explained above, and the sample size can be determined as:

$$\text{Variance} = 0.35 \times 0.65 = 0.23, \sqrt{0.23} = 0.48$$
$$n = (1.96 \times 0.48/0.05)^2 = 350$$

Thus, 350 consumers are required for the survey. This is large compared with the instrumental sample above and is partly due to the level of measurement being nominal.

The above formulae apply to single group samples, i.e. one set of determinations or respondents where a single estimate is intended. If two or more groups are to be compared then modified formulae are necessary. Some software packages provide sample size calculators and there are many available on the Internet. More advanced versions of Minitab include a facility that calculates sample size and power for a range of statistical tests and estimates.

Survey measures of validity

Where a calculated value is possible, these measures use correlation coefficients. These are illustrated in Chapters 6 and 8.

Determination of precision and reliability

Precision (instrumental and sensory)

Any determination of variability between repeated measures will give an indication of analytical precision as defined above (Section 3.4.2). Thus, the standard deviation (sd) and the mean absolute deviation can be used. To enable comparisons between different methods or different circumstances, both these values can be adjusted to give standardised versions (%coefficient of variation). This allows a more valid comparison of variability with different sample sizes and different measuring systems. Other percentage expressions are also possible based on the mean deviation and the confidence interval (in each case divide the measure by the mean and multiply by 100).

Precision of a proportion can be calculated in a similar way by using two or more estimates of the sample proportion.

Reliability (survey data)

Reliability measures require calculation of correlation coefficients. Examples of these are given in Chapters 6 and 8.

Typical values for uncertainty (Tables 3.6 and 3.7)

Some idea of the magnitude of the above measures in different data sets can be seen in the data tables. It is obvious that the instrumental data are more precise and have high accuracy. The sensory data precision is very poor, being more typical of that attained by consumer panels. Both aspects (accuracy and precision) can be improved upon, but it depends on the method and (in the case of chemical analysis) the analyte. Some research workers quote precision cut-off points for acceptance of results such as Ferreira *et al.* (2006), who quoted not accepting duplicate vitamin K (phylloquinone) determinations with a %CV over 15%, contrasting with Glew *et al.* (1999), who achieved CVs of less than 3% in fat determination and Heinonen *et al.* (1997) with acceptable CVs of 5–10% for fat soluble compounds. Quotation of accuracy levels in publications requires that a standard or sample unit of known nature has been tested (as in Table 3.6, where a sugar reference was used). Bower (1989) reported a potassium content study where a reference sample of wheat flour was used to validate the determinations. High accuracy was achieved over a series of duplicate analysis during the course of the study, with values averaging out equal to the reference potassium content. For sensory work, standards may not be available, but accuracy can be judged by the percent correct responses in difference tests (where there is a known correct result) as used by Calviño *et al.* (2005) when monitoring panellist performance. Percentage correct scores of 75–100% were achieved by most participants. More examples of typical values for accuracy and precision are given in Chapters 8 and 9.

Relationship between accuracy and precision

There is no relationship between accuracy and precision, in that the state of one does not imply the other. Both can be high or low or they can differ markedly. Confidence intervals can display the possibilities:

High precision	Low precision
High accuracy	Low accuracy
|---o---|	|---------o---------|
T	T
	(T = 'true' value)

High precision	Low precision
Low accuracy	High accuracy
|---o---|	|---------o---------|
T	T

The width of the interval indicates precision and the confidence in the estimate. Thus, a narrow interval signifies higher precision and a more certain estimate. Uncertainty is also judged by the closeness of the estimate and the 'true value' (T). If T lies within the interval then there is a strong indication of accuracy, but this in turn depends on interval width and the distance between T and the estimate. True values outwith the interval indicate bias of some origin and this can occur even with a narrow interval.

3.5   Determination of population nature and variance homogeneity

Data may also require examination for certain other features prior to further analysis. This will enable application of appropriate method of analysis. Thus, specific methods of analysis can be used to establish such aspects as normality of distribution and variance similarity, etc.

3.5.1   *Adherence to normality*

This can be checked in a number of ways. The simplest is by graphical means using a cumulative frequency plot. The *histogram* or a *dot plot* can achieve this (Figs 3.6 and 3.7). The display should be examined for features that characterise the normal distribution curve (Section 2.4). These include the symmetrical central peak and the tails on either side. Sample plots even from a 'true' normal population will not show perfect adherence, so some deviation is likely. The most common infringement is that of *skew* where the peak is off centre. The peak may be small or there may be more than one peak, as in a bimodal case (Fig. 3.14). There are also calculated values for *skew* and shape in respect of the peak (*kurtosis*). These are available using Excel functions and their magnitude (+ or − to zero) indicates the level of deviation. As pointed out above, Table 3.6 illustrates varying magnitudes of skew and kurtosis with the sensory data values exceeding those of the instrumental data. Thus, there is a lower adherence to normality for the former data.

A very simple check is to compare the mean, median and mode – for a normal distribution they are the same. Additionally, significance tests can be used (Rees 2001), e.g. the *chi-square goodness of fit* test or the *Shapiro–Wilks* test. The goodness of fit test requires large samples (>50), although it can be performed with caution on smaller samples in MegaStat's Descriptive Statistics.

The Shapiro–Wilks test (SPSS) can deal with small sample numbers (10–15), and Minitab provides a test (*Ryan–Joiner*) similar to it in the **Graphs>Normal Plot** menu. This has been done for the data of Table 3.6. The variable (listed in a

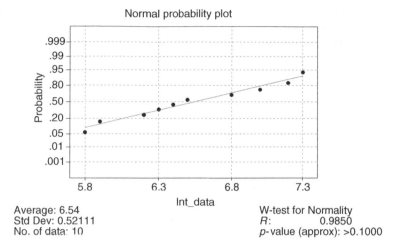

Fig. 3.15 Normal plot and Ryan–Joiner test on instrumental data (Minitab).

column(s)) is entered as the input column in the plot menu and the **Ryan–Joiner test** is selected.

The output includes a ***normal probability plot*** (Figs 3.15 and 3.16). The theory behind this method is that when values from a normal population are plotted on a normal probability paper scale then the points will form a straight line. Assuming all the data are from the same population, deviation from the line (in this application) indicates deviation from normality. A minimum of 16 measures is usually required to give a meaningful impression (Daniel 1959). In the examples, the instrumental data appear adequately normal, but the sensory shows slight divergence and the Ryan–Joiner statistic is smaller, although not significant. This latter test appears weak, in that a Shapiro–Wilks test (by use of SPSS) and an alternative in Minitab (the Anderson-Darling test) on the sensory data were significant – i.e. there is a significant deviation from normality.

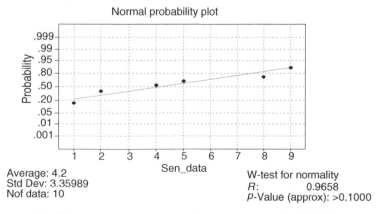

Fig. 3.16 Normal plot and Ryan–Joiner test on sensory data (Minitab).

How large a deviation should there be before data are taken as non-normal?

There will be deviation from 'perfect' normality even when sample are drawn from a truly normal source. Thus, zero values for skew, etc., are not mandatory. Some guidelines are given by Smith (1988) where quotients of skew and the square root of $(n/6)$ and kurtosis $\times$ square root of $(n/24)$ are calculated. If the magnitude of either of these exceeds 1.96 then the population is not normal.

For the analyses in Table 3.6, there is strong evidence (skewed, bimodal, normal plot) that the sensory data do not originate from a normal population. As the sample size is small ($n = 10$) parametric analysis on such data should be viewed with caution, and non-parametric would be more appropriate. Tests of normality are limited by samples size and less than 15 will be less reliable. This comment would appear to apply to the Ryan–Joiner test (application of the test to the Likert data ($n = 21$) of statement 1 in Table 3.3 illustrates a significant 'non-normal' result).

Most instrumental data are continuous and should conform to normal. If discrete categories are on the scale then there may not be adequate room for the spread of the measure, thus skew may result.

What if the assumption of normality is infringed?

As with other measures, the adherence need not be exact. Additionally, large samples sizes obviate the need and some statistical tests are robust to slight deviations. It is possible to transform the data to overcome infringements (Gacula and Singh 1984). If the practitioner is convinced that the data are non-normal and transformation is not performed then tests for non-parametric data are available (see Chapters 4 and 5).

3.5.2   Homogeneity of variance

Some parametric significance tests assume that the variation within sets of data that are being compared is similar and non-compliance is viewed as a more serous infringement than that of normality. Variation can be compared by examining measures such as variances or standard deviations visually (see Fig. 9.1), but a *variance ratio test* (*F*-test) can assess the significance of any dissimilarity. The ratio is calculated by dividing the larger variance with the smaller then examining the *p*-value (or comparing with a critical value). For more than two comparisons, the pair-wise testing can be done, but this introduces the possibility of multiple testing errors (see Chapter 5). The ***F-test*** is available in Excel's **Data Analysis ToolPak** or as the function F.TEST() (FTEST() Excel 2003) for a two-sample variance comparison and some significance tests perform a variance comparison check as part of the process (see Table 4.3 and later examples in applications).

References

Albala Hurtado, S., Vecinia-nogués, M. T., Vidal-carou, M. C. and Mariné Font, A. (2001) Concentrations of selected vitamins in infant milks available in Spain. *Journal of Food Science*, **66**(8), 1191–1194.

Bower, J. A. (1989) Cooking for restricted potassium diets in dietary treatment of renal patients. *Journal of Human Nutrition and Dietetics*, **2**, 31–38.

Bowerman, B. L., O'Connell, R. T., Orris, J. B. and Porter, D. C. (2008) *Essentials of Business Statistics*, 2nd edn. McGraw Hill International Edition, McGraw-Hill/Irwin, McGraw-Hill Companies., New York, p. 295.

BSI (1983) BSI hand book 22. *Quality Assurance*. British Standards Institute, London, pp. 6, 13, 52.

BSI (2000) BS ISO 5725:1. *Accuracy (Trueness and Precision) of Measurement Methods and Results – Part 1: General Principles and Definitions*. British Standards Institute, London.

BSI (2002) BS ISO 5725:2. *Accuracy (Trueness and Precision) of Measurement Methods and Results – Part 2: Basic Method for the Determination of Repeatability and Reproducibility of a Standard Measurement Method*. British Standards Institute, London.

Calcutt, R. and Boddy, R. (1983) *Statistics for Analytical Chemists*. Chapman & Hall, London, p. 19.

Calviño, A., Delia Garrido, D., Drunday, F. and Tamasi, O. (2005) A comparison of methods for monitoring individual performances in taste selection tests. *Journal of Sensory Studies*, **20**, 301–312.

Chambers, D. H., Allison, A. A. and Chambers IV, E. (2004) Training effects on performance of descriptive panellists. *Journal of Sensory Studies*, **19**, 486–499.

Chatfield, C. (1992) *Statistics for Technology*, 3rd edn. Chapman & Hall, London.

Czaja, R. and Blair, J. (1996) *Designing Surveys: A Guide to Decisions and Procedures*. Pine Forge Press, Thousand Oaks, CA.

Daniel, C. (1959) Use of half-normal plots in interpreting factorial two level experiments. *Technometrics*, **1**(4), 311–341.

Evenson, R. E. and Santaniello, V. (2004) *Consumer Acceptability of Genetically Modified Foods*. CABI Publishing, Wallingford, p. 26.

Ferreira, D. W., Haytowitz, D. B., Tassinari, M. A., Peterson, J. W. and Booth, S. L. (2006) Vitamin K contents of grains, cereals, fast Food breakfasts and baked goods. *Journal of Food Science*, **71**(1), S66–S70.

Field, A. (2006) *Discovering Statistics with SPSS*, 2nd edn. Sage Publications, London, pp. 131–134.

Findlay, C. J., Castura, J. C. and Lesschaeve, I. (2007) Feedback calibration: a training method for descriptive panels. *Food Quality and Preference*, **18**, 321–328.

Gacula, M. C. and Singh, J. (1984) *Statistical Methods in Food and Consumer Research*. Academic Press, Orlando, IL, pp. 58–60.

Glew, R. H., Okolo, S. N., Chuang, L., Huang, Y. and VanderJagt, D. J. (1999) Fatty acid composition of Fulani 'butter oil' made from cow's milk. *Journal of Food Composition and Analysis*, **12**, 235–240.

Heinonen, M., Vasta, L., Anttolainen, M., Ovaskainen, M., Hyvönen, L. and Mutanen, M. (1997) Comparisons between analytical and calculated food composition data:

carotenoids, retinoids, tocopherols, tocotrienols, fat, fatty acids and sterols. *Journal of Food Composition and Analysis*, **10**, 3–13.

Huck, S. W. (2004) *Reading Statistics and Research*, 4th edn. International Edition, Pearson Education Inc., New York, pp. 75–98.

James, C. S. (1996) *Analytical Chemistry of Foods*. Blackie Academic and Professional, London, pp. 5–12.

Kolodinsky, J., DeSisto, T. P. and Labrecque, J. (2003) Understanding the factors related to concerns over genetically engineered foods products: are national differences real? *International Journal of Consumer Studies*, **27**(4), 266–276.

Malhotra, N. K. and Peterson, M. (2006) *Basic Marketing Research*, 2nd edn. International Edition, Pearson Education, Upper Saddle River, NJ, pp. 274–280.

Martín-Diana, A. B., Rico, D., Fías, J. M., Barat, J. M., Henehan, G. T. M. and Barry-Ryan, C. (2007) Calcium for extending the shelf life of fresh whole and minimally processed fruits and vegetables: a review. *Trends in Food Science and Technology*, **18**, 210–218.

Miller, J. C. and Miller, J. N. (1993) *Statistics for Analytical Chemistry*, 3rd edn. Ellis Horwood, Chichester, pp. 101–141.

Mullins, E. (2003) *Statistics for the Quality Control Chemistry Laboratory*. Royal Society for Chemistry, Cambridge, p. 8.

Nielsen, S. S. (2003) *Food Analysis*, 3rd edn. Kluwer Academic/Plenum Publishers, New York.

O'Mahony, M. (1986) *Sensory Evaluation of Food – Statistical Methods and Procedures*. Marcel Dekker Inc., New York, p. 70.

Rees, D. G. (2001) *Essential Statistics*, 4th edn. Chapman & Hall, London, pp. 258–260.

Ruane, J. M. (2005) *Essentials of Research Methods*. Blackwell Publishing Ltd., Oxford, p. 64.

Ryan, P. and O'Donoghue-Ryan, F. (1989) Weighing up accuracy and precision. *Laboratory Practice*, **38**(6), 29–33.

Sirkin, R. M. (1999) *Statistics for the Social Sciences*, 2nd edn. Sage Publications, Thousand Oaks, CA, pp. 33–58.

Smith, G. L. (1988) Statistical analysis of sensory data. In *Sensory Analysis of Foods* by J. R. Piggot (ed.), 2nd edn. Elsevier Applied Science, London, pp. 335–379.

Upton, G. and Cook, I. (2001) *Introductory Statistics*, 2nd edn. Oxford University Press, New York, pp. 301–302.

Upton, G. and Cook, I. (2002) *A Dictionary of Statistics*. Oxford University Press, New York, pp. 76–78.

West, B. J., Tolson, C. B., Vest, R. J., Jensen, S. and Lundell, T. G. (2006) Mineral variability among 177 commercial noni juices. *International Journal of Food Sciences and Nutrition*, **57**(7/8), 556–558.

Chapter 4
Analysis of differences – significance testing

4.1 Introduction

The summary methods dealt with in the previous chapter are relatively simple forms of analyses. Graphs, tables and summary statistics such as the mean or a proportion can suggest differences and trends, but to establish *statistical significance* more involved methods are required. Commonly, the intention is to compare an experimental result with a standard value or set to see if they differ, or identify *significant differences* due to the effect of some factor highlighted in the experiment. For example, the effect of the factor 'process temperature' on thiamin retention in processed food. When differences are found, it is usually the case that some association between variables is present. Thus, as 'process temperature' rises, thiamine retention falls. Association and correlation effects can be analysed specifically for significance of effect in the form of a relationship rather than a difference. In some experiments, *cause and effect* structures can be established by use of an appropriate design. A more general approach is taken in this chapter, which gives a commentary on the nature of significance testing. Relationship analysis is covered in Chapter 6 and experimental design in Chapter 7.

The *effect* is the focus of interest and it can be viewed as a measure of the extent of any difference (or relationship) observed in an experiment or a survey. Assessment of the effect will involve at least one of the following:

- Display and calculation of the magnitude of the effect
- Establishing the significance of the effect

Display and magnitude of effects can be done using graph and tables. These can be applied to summary measures, comparison of means or whichever summary measure is being compared. Calculation of a specific *effect size* can be done simply by gauging the magnitude of the difference between summary measures or by calculating a standardised measure to allow general decisions on specific

Statistical Methods for Food Science: Introductory Procedures for the Food Practitioner,
Second Edition. John A. Bower.
© 2013 John Wiley & Sons, Ltd. Published 2013 by John Wiley & Sons, Ltd.

statistics (Cohen 1992; Field 2006). One standard effect size measure is the linear correlation coefficient (Chapter 6). As it is standardised, comparisons between co-efficients from different experiments are valid. Some software packages included the summary measure and effect size statistics as part of the significance test output.

The purpose of the analysis is according to the circumstances of the experiment that generated the data. In cases where a specific design and specific research question were posed and hypotheses have been formulated, a well-established procedure is followed as detailed below.

Assume that the purpose is to ascertain whether or not significant differences are present. At the basic level, the practitioner may pose research questions such as:

> *'Do these 2 products differ significantly in fat content?'*
> Or,
> *'Is product A preferred over product B?'*

These examples could be simple investigations or they could be based on a deductive approach (Section 1.4.1) and to be able to answer the research questions hypothesis testing would be required.

A fundamental aspect of these comparisons is the importance of the effect size. Ideally, the magnitude of an *important effect size* should be specified at the design stage, allowing details of sample size and other statistical requirements to be set up (Huck 2004). These points are dealt with in later sections, but for the moment, the significance testing process requires some more description.

4.2 Significance (hypothesis) testing

This procedure is also known as **hypothesis testing** as hypotheses are stated in the formal stages of the tests. Significance tests are used in experiments where the practitioner wishes to establish that differences or a relationship are of (or are not of) sufficient magnitude to qualify for statistical significance. This will enable confident decisions to be made about products and processes. The analyst should be aware that for the significance testing procedure itself to be valid, several factors need to be considered:

- Inference nature
- Not 100% sure
- Statistical significance not practical importance

The first of these is that significance testing is an inference procedure. It assumes that the analyst is collecting data from a sample then inferring back to a population (hence another term 'inferential statistics'). On this basis, the result can never be

100% sure or 'true' – this would require sampling the whole population (Section 2.3.3). Thus, *all* significance-testing results are less than 100% sure. The big questions are 'how much less?' and 'what is acceptable in these terms?' The concept of probability (Section 2.4) is fundamental to understanding significance testing. While 100% confidence (probability is certain) is not attainable when using samples, lower levels near to this are—95, 99 and 99.9% being conventional (loosely, probabilities of 95–99.9% 'sure', respectively).

An important distinction relates to the difference between *statistical significance* and the *practical importance* of the outcome of the analysis. While it is possible to establish the former, the practical significance may be nil. For instance, in an experiment comparing hedonic ratings for two prototype food formulations, a significant difference of 0.1 (5.9 cf. 6.0, on a 1–9 scale) was found between the average values. This difference is very small and is unlikely to be of consequence in the retail market situation. Contrast this with the level of certain toxins or additives in food, where small differences may be of critical safety or legal importance, respectively.

4.2.1   The method of significance testing

How does significance testing work?

The foregoing chapters have established that all data are likely to have a spread of values around some central measure. For many measurements based on interval and ratio scales, the spread is around the mean value, and it usually originates from a normal population or an approximation of it, i.e. the bell-shaped curve (Section 2.4). For nominal data, there are binomial and chi-square distributions, which can also approximate the normal distribution when sample sizes are large. A number of consequences arise from the model provided by a characteristic distribution:

- A sample from a **single population** will show a range of values
- Two or more samples from the **same population** may have different summary values
- Distribution of values is 'mapped' as probabilities
- Random samples will reflect the probability
- Separate populations can overlap

Thus, when drawing *random samples* from two or more populations for comparison, the degree of difference between them has a bearing on the outcome, as one of a number of circumstances will apply:

1. The populations are almost identical
2. The populations are each distinctly separate
3. The populations overlap

If the populations are close to identical in terms of the measure being determined (circumstance 1), random samples from the different populations are in essence a sample from one population. There is a possibility that differences could be found, but unless samples are very large, significance would be absent. Circumstance 2 will allow detection of significant differences with small samples. Circumstance 3 can allow detection of significant differences, but larger samples would improve the possibility.

The difference between any two population summary values gives a guide to the **effect size**. Circumstance 2 has a relatively *large effect size*. Significance testing calculates the probability of getting the result achieved by the analysis, *by chance alone*. The higher the chance probability, the less likelihood there is of a *real difference*. Differences are compared on the basis of typical spread for the population as estimated by the sample, i.e. it is known that there is a spread of values – how wide is it likely to be? This information is provided by the distributional model, which gives probabilities for variance (normal), deviation from fixed points (chi-square), etc., and is estimated via the sample.

If the differences are large and the sample variability is low then calculated statistics will be large and are unlikely to be due to chance. The test statistic can then be assessed for significance. Hypothesis testing will establish whether or not the difference is statistically significant under the conditions of the test. If the probability of the result is below a preset level then the result is unlikely to be due to chance. Depending on the scale of the investigation, application of significance testing ranges from simple fact-finding experiments to research projects based on previous work, with hypotheses being put forward.

4.2.2   The procedure of significance testing

How to carry out a significance test

The process of carrying out a significance test involves several clearly defined stages. Most statistics textbooks (e.g. Rees 2001) include these and Clarke and Cooke (1998) give a fuller account. The main parts deal with statements relating to the hypotheses followed by details regarding the significance test and conditions of testing:

1. State the objective of the test
2. State the statistical null and alternative hypotheses
3. State the significance level and assumptions
4. Select an appropriate test and analyse the data to obtain a statistic plus descriptives
5. Assess the magnitude/significance of the statistic
6. Decide to retain or to reject the null hypothesis
7. State conclusion considering the significance and effect size

Each stage is discussed below in detail with reference to an example investigation. The full significance test is presented for this example later in the chapter (Section 4.4).

The objective of the test

The origin of a significance test lies with a research or investigational requirement in the form of a question or problem. When differences are under study, this can be between two or more populations or between a single population and some standard value. The investigation (experiment, survey, etc.) will generate data, which on analysis may provide an answer or a solution to the problem. Thus, the starting point can be in the form of a ***research question*** or a statement of the intent.

To illustrate this, consider a relatively simple fact-finding type of experiment where a food manufacturer receives a new supply of vegetable material (beetroot). It is required to ascertain the sodium (Na) content of the new source in comparison with existing supply. This particular piece of information is required for calculation of sodium content in proposed products. If there is a difference then this could have consequences for labelling of products that contain one or both of the vegetable sources (more details are given on this analysis in Example 4.4-1). Thus, the research question could be:

Research question

Do these two sources of vegetable differ in sodium content?

Alternatively, it could be in the form of a general statement or ***research hypothesis*** of the expected or postulated result based on deductive reasoning from previous work or theory. A published example of this was given by Øygard (2000), who based a research study on an existing theory ('Bourdieu's theory'), the basic tenet of which was that social position affects food taste.

In the instance of the sodium example, the reasoning could be that the two sources of vegetable were grown in different soils, containing different sources of sodium and therefore are likely to differ in sodium content. Note that a reasoned, logical and *testable* research hypothesis should be put forward. Usually, this hypothesis is or becomes the ***test hypothesis*** (the alternate hypothesis, see below), and it is usually in the form of a statement of the possible outcome, i.e. an *expectation*. Hypotheses can also include reference to other aspects such as *cause and effect* circumstances.

Examples occur in many research publications, e.g. Jæger *et al.* (1998) postulated that mealiness in stored apples would have a negative effect on consumer liking. They compared fresh and stored samples, tested these for sensory properties related to mealiness and measured liking with a sample of 127 consumers using a hedonic scale. The stored samples had higher sensory scores for attributes linked to mealiness and their degree of liking was lower, thus providing support for the research hypothesis.

In another study, Moerbeek and Casimir (2005) put forward the hypothesis that a higher knowledge of genetic modification would result in a lower acceptance of the technology, based on previous supporting evidence. They analysed secondary data where measures of knowledge and acceptance had been obtained from a very large sample of consumers (12 000+). It was found that people with higher knowledge scores tended to have higher acceptance of genetically modified foods, thus the research hypothesis was disconfirmed.

The null and alternative hypotheses

The *statistical hypotheses* are now stated as the *null* and *alternative hypotheses* with specific detail in terms of population parameters. These hypotheses are the focus of the test calculation and form the core of the whole process. An initial condition is assumed and it is stated as the **null hypothesis**, symbolised by H_0. The null statement can be viewed as the 'no effect' hypothesis, although this is not always the case. It is prefaced in the style – 'There is no difference/no effect/no relationship . . . ', etc. It is stated in terms of the population mean (μ) or other parameter being tested.

Thus, for the current example, the null hypothesis is *The products do not differ in sodium content*:

H_0: $\mu_a = \mu_b$ (the mean sodium content of vegetable source A does not differ from that of source B)

Secondly, the **alternative hypothesis** (H_1) is stated. The alternative states that there is an effect – 'There is a difference/an effect/a relationship . . . ', etc. More specifically the detail of what differs is provided in the actual statement and here a complication arises.

Choice of one-tailed/two-tailed

The alternative hypothesis must be declared as *directional* or *non-directional*. This is referenced as **one-tailed** or **two-tailed**, respectively. These terms specify in which direction the difference (if present) will lie. A **directional difference** will lie above or below the null hypothesis:

$$H_1 : \mu_a > \mu_b \text{ or } H_1 : \mu_a < \mu_b$$

A **non-directional alternative** simply specifies a difference either way:

$$H_1 : \mu_a \neq \mu_b$$

Thus, there are three possible alternatives, but only one can be put forward for testing. The choice initially is between *one-tailed* and *two-tailed*. The chance probability differs markedly by these two possibilities and it is 'easier' to get a significant result with a one-tailed test (probabilities are halved). However, the one-tailed choice must be supported by logical argument and by previous work, such that a *prediction* can be made of the outcome in one direction or the other if the null hypothesis cannot be retained. If there are doubts then two-tailed should be used.

O'Mahony (1986) discusses this and points out that statisticians themselves disagree on the issue, some advocating two-tailed only. The decision must be made by the food practitioner. Apart from the reasons above, outcomes of experiments can surprise. Certainly, for some investigations, results are difficult to predict because of unrecognised factors arising and a conservative two-tailed approach is the recommended alternative. Predictions can be made in certain types of sensory difference tests where a 'correct result' can be identified, and one-tailed tests are appropriate.

In the present example, assume for now that there is no support for a prediction of which vegetable source will have the higher sodium content, thus:

$H_1 : \mu_a \neq \mu_b$ (the mean sodium content of vegetable source A differs from that of source B)

Only one of the statistical hypotheses can remain in force after the test – if the null is not viable then the alternative is accepted.

Testing assumes that the *null hypothesis is true* unless otherwise demonstrated. On this basis, results close to the null would be found. The *observed* experimental data are analysed in comparison to see if any difference (effect) shows a large enough deviation from the null statement. Thus, the objective is to provide evidence that the null hypothesis cannot 'hold water'. This is done indirectly by analysing the alternative hypothesis via the data then examining the probability that the results could have arisen by chance. If this probability is above a pre-chosen level then the null is deemed false and it is rejected. If not, then assuming control of measurement errors, any effects are purely due to sampling variation and not due to real forces.

The conclusions reached in significance testing are therefore based on probability. Because of examining a sample rather than the whole population, it is not possible to be 100% certain about rejecting a null hypothesis. There is a chance that it could be wrong and an incorrect conclusion is made, i.e. the probability of *decision error* exists.

Significance testing errors

There are two types of error that may occur during treatment of the null hypothesis:

Type I error:
Rejecting a null hypothesis that is correct ('rejecting a true null')
Type II error:
Failing to reject a null hypothesis that is incorrect ('retaining a false null')

These errors are controlled during the analysis by selections of probabilities for the test.

The Type I error: The analyst wants initially to have a high probability of being correct when rejecting a null hypothesis. This is conventionally a probability of 0.95 (95%) or 0.99 (99%). Thus, the Type I error is controlled in that it has a low probability of occurring. The practitioner now has a 95% (0.95) chance of being correct and hence the probability of being wrong, and committing the Type I error, is 5% (0.05). This probability is known as the **significance level** of the test (**alpha**, symbolised as α).

If this margin of error still appears to be insufficient and too conservative then α can be decreased to 1% (0.01) or less, to reduce the chance of committing the error. There would be a number of consequences, one being that the effect sizes detectable as significant under $\alpha = 0.01$ would be *larger* than those at $\alpha = 0.05$. That is, reducing the chance of committing the Type I error reduces the *sensitivity* (the reaction to a given effect size) of the test. For example, assume that with $\alpha = 0.05$, a significance test can detect differences of 1.5 as significant on the measurement scale used – on changing to $\alpha = 0.01$, it can now detect differences of a minimum of 2.0 – differences of 1.5 will 'pass through' undetected. What is happening is that the practitioner is ensuring that the null hypothesis is not rejected unless there is a real (i.e. large) effect present. It becomes a matter of balancing up these two opposing factors.

There are other ways to reduce the Type I error; these include increasing the sample size, and other factors related to calculation of sample size, such as the required confidence level. These procedures should of course be taken into account when planning the experiment. A common mistake is an inadequate sample number to detect the envisaged effect sizes.

The Type II error: This error refers to the risk of retaining a null hypothesis, which is false, i.e. the risk of missing a real effect. The probability of this error is **beta** (symbolised as β). It is not specified at the testing stage, but it can be taken into account during sample size calculation. It can be reduced by increasing sample size. Probabilities for β are usually higher than α, e.g. 10% or 20%. This is because researchers usually want to be sure that they do not commit Type I error and do not report results that are false as in one sense there is less to lose by committing the Type II error. There are circumstances when the converse is true and β must be lowered (see Chapter 8). An important term derived from β is **power** (power $= 1 - \beta$). This refers to the ability of a significance test to detect real effects.

Power curves can be drawn up for all significance tests for different sample sizes, different settings of α, and differing magnitudes of the effect. All such curves show an approximate sigmoid shape. They are a type of **operating characteristic curve** (Montgomery 2001).

It can be seen (Fig. 4.1) that for a fixed α and a small effect difference, a sample size of ten gives higher power than a sample size of one. A larger effect difference means that power is higher for both sample sizes, but again the larger sample size is higher in power. To an extent, the Type I and II error risks depend on one another. The value for α decides how large a difference is needed for

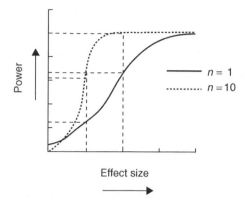

Fig. 4.1   Power curve example (Excel).

significance, and this effect size determines the value of β – as α falls, β rises and power falls.

In practice, α is set and β depends on this, sample size and other factors. They are both lowered when large effect sizes are present in the samples.

Higher significance levels can be used for exploratory experiments, e.g. $\alpha = 10\%$, and it can be lowered when sample sizes increase so that only large differences are detected. Tepper and Trail (1998) used a significance level of 0.01 as a very large sample size was gathered, which would tend to produce many significances.

In any significance test, the above hypotheses are always implied, but they may not be stated. In some investigations, a question is posed and a test done, without reference to the procedure above, but statement of the hypotheses can clarify the conclusion made.

Significance level and assumptions

Bearing in mind the points in the previous section, the setting for α must be decided at this stage. α also relates to the level of confidence in the results of the analysis and is similar in concept to confidence level (Section 3.4.3). Conventionally, significance levels of 5%, 1% and 0.1% are used, corresponding to being confident that when drawing similar random samples, 95%, 99% or 99.9%, respectively, of the time similar results will be found. For the sodium content example, α is set as 0.05 (5%).

A population distribution must be assumed at this point regarding the parameter specified in the null hypothesis. The significance level (α) determines a region in the assumed distribution outwith which the result would not be likely (low probability) to be due to chance.

The sodium content variable is ratio level and the data can be assumed to be from normal populations. Checks on this assumption and that of homogeneity of variance (Sections 3.5.1 and 3.5.2) can be performed at this point (see details later). It is also assumed that a random selection of samples is drawn from the population (in this case each lot of vegetables; see Section 2.2.3).

Selection of test and calculation of statistic and descriptives

The appropriate significance test will depend on a number of factors. One (assumed) distribution will have been considered above and other features such as the number of groups can now be specified in the selection. More detail on test selection factors is dealt with below (Section 4.5). Data are then analysed to calculate a statistic from the sample data. Depending on the software used, the analysis output can include tables and graphs of the summary measures, effect size calculations, etc. No matter which test is used, all outputs are similar in that a *test statistic* is produced. This will often be accompanied by a *probability value* – this is the chance probability for the results and is known as the *p-value* or 'sig', 'prob'.[1] Alternatively, some software will also display a *critical value* for the test.

For the current example, it has been decided that the variable is ratio level, and on the basis that assumptions are confirmed, a *parametric test* is appropriate. There are two groups being compared and as they are distinctly separate, a *two-group, independent test* is required, namely, the *independent t-test*. The data are summarised to produce descriptives (Table 4.3) and analysed by the *t*-test (Table 4.4) resulting in a *p*-value of 0.013.

Significance assessment

At this point, visual examination of the summary measures can reveal the magnitude of differences, effect size and hints of possible significance. On determining this, the calculated statistic is then assessed via the critical value or *p*-value.

The decision regarding significance depends on these values with the *p*-value providing the simplest decision rule:

> *If the p-value is less than the significance level then significance is present.*

For example, if the *p*-value is 0.03 then the result is significant when $\alpha = 0.05$. If the software does not give a *p*-value then the practitioner must examine the *critical value* for the statistic. This is the maximum value that would be attained by chance for the circumstances of the test (*degrees of freedom*), then:

> *If the test statistic is greater than the critical value then significance is present*[2]

Degrees of freedom
Degrees of freedom (**df**) refer to the number of terms that are independent under the conditions of the test. This feature is linked to the number of samples within

---

[1] Terminology varies according to the software used.

[2] There is some variation on this rule with some authors stating that the *p*-value requires to be 'equal or less' than α, or that the statistic requires to be 'equal or greater' than the critical value (e.g. Daniels 2005).

specific groups in the analysis and the experiment. Typically, the magnitude of the df is equal to $n - 1$, where n is the sample number. In planning experiments, sufficient degrees of freedom must be available to enable appropriate analysis. The more involved the analysis, the more degrees of freedom that are required. For example, on analysis, a main effect can be assessed with minimal sample sizes and relatively low degrees of freedom – if interaction effects are to be assessed as well then more degrees of freedom are required and this can be attained by increasing replication, etc.

Decision on the null hypothesis

Two outcomes are possible:

> '*If the p-value is* **less than** *the significance level then* **reject the null hypothesis**'.
> OR,
> '*If the p-value is* **equal to or greater** *than the significance level* **then retain the null hypothesis**'.[3]

And, for critical values:

> '*If the test statistic is* **equal to or less** *than the critical value then* **retain the null hypothesis**'.
> OR,
> '*If the test statistic is* **greater than** *the critical value then* **reject the null hypothesis**'.

Thus, all values at and below the critical values mark a region within which the test statistic is not of sufficient size to justify rejecting the null. The critical value(s) is the *maximum* that the statistic can attain due to chance at the chosen significance level.

These 'rules' need not be followed exactly depending on the circumstances of the experiment. Thus, there can be 'marginal' significance, e.g. a p-value of 0.06 (Hassan 2001). One useful guide is to take any p-value between 0.01 and 0.1 as the 'inconclusive region' (Kvanli *et al.* 1989), although this may 'go against the grain' with some researchers eager to report a significant difference. It should be noted that if the null hypothesis is retained, this does not mean that is 'correct' or that it has been 'proved'. Rather, there is insufficient evidence to reject it. It is not possible to prove the null hypothesis – it can only be disproved.

---

[3] Very low p-values reported as 0.000 are not at zero, but will have one or more significant figures if expressed with sufficient decimal points, e.g. a p-value of 0.000 could actually be 0.000006.

For the illustration with sodium content, the decision is that the *p*-value (0.013) is less than α (0.05) and therefore significance has been found: the null hypothesis is *rejected*. By convention, significant results are reported as one of three possibilities: 'significant' ($p < 0.05$), 'very significant' ($p < 0.01$) and 'highly significant' ($p < 0.001$), often accompanied by $*$, $**$ and $***$, respectively. Absence of significance is referred to as 'not significant', abbreviated as 'ns'.

State conclusion considering the significance and effect size

The conclusion from the test can be stated as a fuller account of the particular hypothesis that is still present. If significance is obtained and assuming $\alpha = 5\%$ then there is confidence that when a similar samples drawn such differences will be present in 95% of samples. The conclusion for the example analysis is that there is a difference in sodium content and vegetable source B has a higher content than A. The effect size is small and this has a bearing on the practical significance (see below).

4.3   Assumptions of significance tests

Significance tests differ in a number of respects regarding the data to which they can be applied. This has been referred to at some of the stages in the testing procedure above. One of the main assumptions relates to the source of the data, in terms of population distribution, for certain parametric significance tests (Table 4.1).

For these more powerful tests a metric level of measurement and normality are assumed plus a continuous nature to the variable being analysed. This may not be apparent in some arbitrary scales and some discrete scales. Some of these latter types can be treated as continuous if the scale can be divided and fractionated even though they are not presented in this way. For example, a sensory scale that uses 10 integer points, which could be divided up into 100 points (both types are found in sensory work). The 100 points could be divided further in 200 half points and so on. In addition, counts of objects theoretically could be taken as half or quarter objects, etc. Sometimes this is less easy to visualise with some rating scales. The well-known Likert and hedonic scales are used as 5 point, 7 point and 9 point, where the description of the anchors on the scale is made increasingly precise. On this basis, they could be treated as continuous, but they are not necessary used in an

Table 4.1   Assumptions of significance tests.

Test type	Assumption
General	Random sampling
	Independence of samples
Parametric	Equal variance
	Normality
	Continuous

interval manner, except for the hedonic scale (Section 2.2.1). With larger samples ($n > 30$), the assumptions are less strict and use of tests reliant on approximations to normal is possible.

Establishing significant differences also assumes that the method of measurement is the same throughout. Thus, all the details of instrumental, sensory or consumer measures must be the same unless the method or the different groups themselves are under study. This procedure itself ensures that another assumption regarding the spread or variance of each group is similar.

The more sophisticated aspects of parametric tests make additional assumptions regarding measurement error and other aspects, which are dealt with in Chapter 7 on experimental design. Further details are available in most statistical texts (e.g. Steel and Torrie 1981).

All significance tests also assume that the sampling was done randomly and that each sample was independent of the others. Hence, each sample can only be used once, or each reading is one per sample. Some instruments take multiple reading on each sample – these cannot be treated as separate independent results – they must be averaged first. A similar situation arises in some sensory tests – many assume that each reading is generated from a single assessor. Repeat measures by each assessor cannot be treated as independent – they must be averaged or alternative forms of analysis employed.

4.4   Stages in a significance test

A full account of the significance testing procedure is presented below. It contains much detail, but it will not be possible to provide this for all the tests in this book and abbreviated steps are used hereafter.

Example 4.4-1 *Worked example of a significance test (t-test)*
A manufacturer requires data on sodium content in raw material vegetable supplies for product labelling purposes. Batches of each source of vegetables (beetroot) are delivered and a random selection of sample units is drawn from each lot of vegetables (Section 2.3.3). As this is a new raw material, 10 samples of 200–300 g containing approximately 10–15 individual units are selected from each consignment, prepared in a similar manner to the form that they would be used in products, then blended to a form a homogeneous mass. Duplicate subsamples are taken and analysed for sodium content. Average values for each sample are reported.

The background to the test is based on a fact-finding type of investigation. The first stages of the test are present in Box 4.1, with the sequence of analysis in following tables, figures and Box 4.2. Checks on assumptions for the data were carried out using Minitab (histograms and N plots) and Excel (descriptive statistics and *F*-test). This latter test employed the **ToolPak**, which was also used for the significance test itself, the independent *t*-test.

Box 4.1   Initial stages in a significance test.

1. State the objective and rationale of the research hypothesis.
Research question: '*Do the two sources of vegetable differ in sodium content?*'
Rationale: The two sources of vegetable were grown in different soils, containing different sources of sodium and therefore are likely to differ in sodium content.
Based on this reasoning, the research hypothesis is:
'*There will be a difference in the sodium content*'
2. State the statistical null and alternative hypotheses.
The **null hypothesis** (H_0) is that there is *no difference* in the Na content.
H_0: $\mu_a = \mu_b$
The **alternative hypothesis** (H_1) is that *there is a difference* in the Na content.
H_1: $\mu_a \neq \mu_b$
Thus:
H_0: the mean sodium content of vegetable source A does not differ from that of B.
H_1: the mean sodium content of vegetable source A differs from that of B.
This is a **two-tailed test**, as the direction of the difference is not specified – there is no specification of whether a higher amount will be found in A or in B – either result could occur.
3. State the significance level and distributional assumption.
$\alpha = 0.05\%$ or 5%
Checks are performed on the sample data (Table 4.2): Histogram, N plot, Ryan–Joiner, skew, kurtosis and an *F*-test for variance homogeneity. The results are presented in Figs 4.2, 4.3, 4.4 and 4.5 and Table 4.3. Based on these results, the assumption of normality and equal variances is accepted (the normality plot has some points off the line but the Ryan–Joiner statistic is large – indicative of conformance).

Table 4.2   Sodium data.

Sodium (mg/100 g) to nearest whole mg

Unit	Source A	B
1	50	53
2	47	57
3	48	52
4	53	60
5	52	55
6	58	57
7	54	58
8	49	61
9	57	53
10	54	58

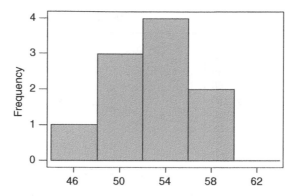

Fig. 4.2   Histogram of sodium content – source A (Minitab).

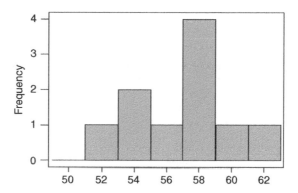

Fig. 4.3   Histogram of sodium content – source B (Minitab).

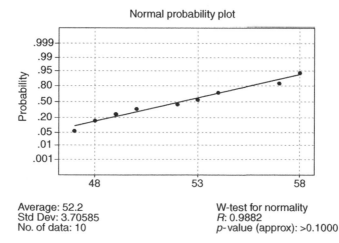

Fig. 4.4   Normal plot of sodium content – Source A (Minitab).

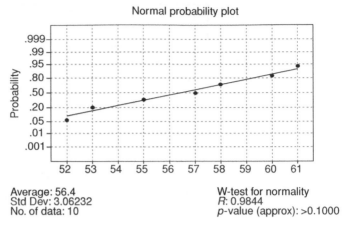

Normal probability plot

Average: 56.4
Std Dev: 3.06232
No. of data: 10

W-test for normality
R: 0.9844
p-value (approx): >0.1000

Fig. 4.5 Normal plot of sodium content – source B (Minitab).

Table 4.3 Summary statistics and variance test for sodium data (Excel).

14		Source A	Source B
15	Statistic		
16	Count	10	10
17	Mean	52.2	56.4
18	Median	52.5	57.0
19	Mode	54.0	53.0
20	Standard deviation	3.71	3.06
21	%CV	7.10	5.43
22	Kurtosis	−1.04	−1.18
23	Skew	0.16	−0.07
24	Min	47	52
25	Max	58	61
26	1st Quartile (1st Q)	49.3	53.5
27	3rd Quartile (3rd Q)	54	58
28			
29	*F*-test 2-sample for variances		
30		A	B
31	Mean	52.2	56.4
32	Variance	13.73	9.38
33	Observations	10	10
34	*df*	9	9
35	*F*	1.47	
36	$P\ (F \leq f)$ one-tail	0.29	
37	*F* critical one-tail	3.18	

Box 4.2   Final stages of the significance test.

4. Select an appropriate test and analyse the data to obtain a statistic plus descriptives.

The gathered data and tests are presented in Tables 4.2 and 4.4. These are metric data of a continuous nature, thus a parametric test is required, based on the assumption of a normal population. Samples sizes are small ($n < 30$), thus the t-distribution is used as an approximation. Two independent groups are being compared, so an independent test for two groups, the t-test is selected, which produces a summary of the data and the t-statistic. An F-test was performed first and found to be non-significant so an 'equal variance assumed' t-test is used. The data were analysed using Excel (Table 4.4). The summary statistics show that the means differ by a small margin.

5. Assess the magnitude/significance of the statistic.

It can be seen from Table 4.4 that the two-tailed significance for the t-statistic is less than 0.05 (it is 0.013), i.e. the p-value is less than 0.05 ($p < 0.05$), The Na levels do differ and the difference is *large enough to be significant* under the conditions of the test.

6. Decide whether or not to reject the null hypothesis.

Thus, based on the sample taken and the significance level of the test, there is evidence to justify **rejection** of the null hypothesis, and the *alternative hypothesis is accepted.*

7. State conclusion (statistical and practical importance).

It is concluded that there is *a significant difference* in the sodium content of the vegetable sources. The research hypothesis statement has been found to apply and the answer to the research question is '*Yes – the sodium, contents differ significantly and vegetable source B has a higher content*'. The difference is small and in practical terms, the difference will not result in important changes to total sodium content in products, but given recent health issues sodium in food must be monitored and food manufacturers are under pressure to reduce levels.

The Excel analyses were conducted by arranging the data as in Table 4.2, with the variable with the larger variance in the left-hand column ('variable 1'). This can be confirmed via the summary measures (standard deviation; upper Table 4.3), which were calculated using the methods of Table 3.6. The F-test was obtained using the **Data Analysis** tool, the F-**Test Two Sample for Variances**, and entering the two column arrays (A and B) as the data ranges. The output (lower Table 4.3) gives a one-tailed p-value – double this for two-tailed. The Excel function F.TEST(variable 1 array, variable 2 array) (FTEST Excel 2003) gives the two-tailed value directly. Based on this analysis, i.e. the variances are not significantly different, the t-**Test: Two Sample Assuming Equal Variances** was used (Table 4.4) to compare means with data entry in a similar manner, concluding via the two-tailed p-value (set **Hypothesized Mean Difference** to 0). This test can also be performed using the

Table 4.4 The independent samples *t*-test (Excel).

t-Test: two sample assuming equal variances		
	a	*b*
Mean	52.2	56.4
Variance	13.73	9.38
Observations	10	10
Pooled variance	11.56	
Hypothesised mean difference	0	
df	18	
t-Stat	−2.76	
$P\,(T \leq t)$ one-tailed	0.006	
t-Critical one-tailed	1.734	
$P\,(T \leq t)$ two-tailed	0.013	
t-Critical two-tailed	2.101	

function T.TEST(variable 1 array, variable 2 array,2,2) (TTEST() Excel 2003), which gives the two-tailed *p*-value directly.

4.5 Selection of significance tests

Significance tests analyse data to determine if there is enough evidence to be able to reject a null hypothesis. There are many significance tests, but they all operate in a similar manner. The researcher's problem is to determine *which test to use*.

Which factors determine the significance test to use?
The choice of significance test depends on several factors related to population assumption and the measure used. One is the nature of the data:

- Level of measurement
- Data form – continuous or discrete
- 'Normal'
- Variance between groups

And secondly:
The circumstances of the experiment:

- Number of groups
- Independence of groups

4.5.1 Nature of the data

Several of these characteristics classify the data as parametric or non-parametric. Checking the data for conformance can be done using the methods described in

Section 3.5 and the examples above for the sodium investigation. Parametric data are assumed to be at least interval in nature and continuous (at least theoretically) even if discrete as discussed above. Variances are assumed to be similar for the groups compared.

4.5.2 Circumstances of the experiment

Number of groups

This aspect requires further clarification.

What are 'two-sample', 'three- or more sample', paired and matched tests?

The number of groups examined for difference in significance tests can be defined in two ways. The first is simply the number of groups examined or compared, which comprise one, two and 'three or more'. The second is where any relationship between the groups is specified. Some groups are separate and *independent* and in such cases, the number of groups may be specified in the name of the test. Thus, 'two-sample tests' indicates that two separate groups or sets of data are compared within the significance test. There are also 'one-sample' tests where one set of data is compared with a population or standard value. For three or more sample group tests, they may be referred to as such along with the term *independent*. Thus, tests that identify numbers of sample groups usually imply that the groups are separate and independent from another.

This contrasts with the terminology for groups that are *dependent* on each other or are *related* in some way, where 'paired' refers to two groups and 'three or more related groups' or 'matched groups' refers to larger numbers of groups.

The number of groups will usually be identified and sometimes the related or independent status is obvious, but in other circumstances, it can cause confusion. Choice of test depends on this so more consideration is required.

Related cf. independent groups

How to tell if sample groups are related or independent

This can be difficult and it is easy to get wrong, resulting in choice of the wrong test. What is required is an examination of the data, a check on how they were gathered and identification of what is being compared. From this, the status of the groups can be judged.

Independent groups

Independent groups can be identified clearly in certain applications. Consumer sampling is the easiest in this respect as grouping can be done by a number of criteria – e.g. demographic measures. Thus, *gender* gives two independent groups, i.e. they are distinctly separate. A grouping based on socioeconomic class would give six or more distinct independent groups depending on the classification system used.

Table 4.5   Independent data sets.

| Group example | | | | | | | | |
| Consumer | | Sensory | | Instrumental 1 | | Instrumental 2 | | |
Male	Female	Food x	Food y	Method A	Method B	Food k	Food l	Food m
y	n	p1	p4	rep 1	rep 1	rep 1	rep 1	rep1
y	n	p2	p5	rep 2	rep 2	rep 2	rep 2	rep 2
y	n	p3	p6	. . .	. . .	. . .	. . .	. . .
. . .	. . .	. . .	. . . .	. . .	. . .	. . .	. . .	. . .

p1–p6 are individual panellists.

Provided the consumers themselves are being compared, then *independent tests* are required. For example, male and female consumers state whether or not they consider themselves concerned on health issues and food with a 'yes/no' answer (Table 4.5 consumer example). If the gender effect is examined then an independent test should be used. In sensory evaluation, independent groups could be based on comparison of results from different panels. For instance, testing a product in one city using one consumer panel, and the same product with a second panel in another city (Stone and Sidel 1993). The two groups of people making up the panels are quite separate geographically and are independent.

The difficulties arise when attempting to apply these criteria in other applications. Thus, in the circumstance of chemical analysis comparing two methods of analysis on one food, or applying one method to three different foods – are these independent or related groups being compared? Here, looking at the source of the data and what is being compared can clarify. Laying the data out in tabular form can help illustrate this (Table 4.5). Thus, in the former case the two methods of analysis are being compared ('Instrumental 1') – two sets of data are generated which are distinct due to two different independent methods – the results of method B do not depend on method A, thus a two-sample independent test is required.

Another way to clarify the perception is to compare the *within-group* and *between-group independence*, and to base the decision on the *between-group state*. In the 'Instrumental 1' case, the within independence is low – they are all repeat samples of the same material. The between independence is high – two separate methods – there is no evidence of a relationship; therefore, an independent test is appropriate.

In the second case ('Instrumental 2') there is no variation in method, but three separate independent food materials provide the data, so a comparison of the results would require an independent 'three or more' sample test. Admittedly, the same method is applied to all three foods, the within-group independence is very low – two pieces of information contribute to this. One is that within each group there are repeat samples of the same food (i.e. they are related and independence is low); secondly, the same method is applied to each of these – another relationship and again independence is low. On the other hand, there is only one feature of the between-group independence – there are three different foods; therefore, it is high. When these factors are compared, the decision is weighted in favour of the groups being viewed as independent.

With independent groups, there is no cross relationship with the data sets, i.e. value 1 in the first set does not have any relationship with value 1 in the next set – they are quite independent. For independent group comparison, group sizes do not require to be the same, which often happens during instrumental analysis (a gross error in one replicate) or when consumers or panellists are not able to be recruited or to attend in equal numbers. Independent groups can be markedly different. If they are used as assessment or 'determination' tools then 'between-group' variation can be high. It may be the objective of the experiment to determine the extent of this variation, otherwise using the alternative design feature of related groups may be more desirable.

Related sample groups

Groups which depend on each other are sometimes more difficult to identify, except when a single group is being tested in a 'before' and 'after' situation. With more than one group, a clue can be the presence of some feature that indicates that the groups are '***matched***' in some way, i.e. they have something in common. Again, in sensory and consumer work it is more apparent. Whenever *all* the assessors or consumers rate or score two or more samples, this is an example of related groups of data. The latter phrase is used because the food samples are being compared in these circumstances (Table 4.6).

The data are sourced by the product samples. For assessor 1, it can be seen that rating for samples all come from the same assessor – thus, they are related. If there are but two groups, they are referred to as ***paired groups***, otherwise they are called ***matched groups***.

In the sensory example above, instead of comparing results based on separate panels only one panel is involved. This means that variation caused by independent groups is not present. The focus on the product differences is heightened and smaller differences can be detected for similar sample sizes cf. independent circumstances. The within independence is high (same product, but different assessors) cf. the between independence being low (three different products, but from the *same* assessor). Thus, the decision is that for these particular circumstances,

Table 4.6   Related groups.

	Product samples		
	A	**B**	**C**
Assessor 1	Rating A	Rating B	Rating C
Assessor 2	Rating A	Rating B	Rating C
Assessor 3	Rating A	Rating B	Rating C
Assessor 4	Rating A	Rating B	Rating C
Assessor 5	Rating A	Rating B	Rating C
. . .	. . .	. . .	. . .

for the groups being compared, the related state is greater. An instrumental example of the 'related samples' case would be with two or more different methods analysing several different food materials (a single analysis on each material by each method).

Usually, it is desirable to use a related design. Groups are paired or matched and are similar in respect of one or more features. Thus, the treatment effects are more easily identified and detected and the experiment has greater precision. More examples of the above differences are given in the next chapter.

Other differences between paired, two-group and 'three or more group' tests

Two group and paired tests are distinct from tests that allow comparison of three or more groups. With two-sample and paired tests, the decisions and conclusions are relatively straightforward: if significance is found then the two groups differ. With the three or more group case, the tests identify whether significance is present or not – but they *do not locate specific differences*. These must be found by examining the summary measures for the data and by following up two group tests. It is not recommended to do a series of two group or paired comparisons directly for experiments with more than two groups (i.e. instead of a 'three or more' group analysis). The reasons relate to the increased probability of committing a Type I error during such **multiple testing**. This is well recognised and discussed in most statistical texts. A significance test at the 5% level means that there is a 1-in-20 chance (5% in 100%) that the test is incorrect. This is stipulated for one test – performing others on the same data means that more chances are being 'used up'. Ultimately, when large numbers of paired comparisons are carried out, the chance of an error becomes high. For example, for seven groups or treatments, 21 comparisons are required so one error is certain. To avoid this, probability levels can be adjusted and tests have been designed which take this into account (Section 5.5.2).

4.6   Parametric or non-parametric tests

The above features, which depend on the nature of the data, essentially reduce significance test selection to two possibilities: *parametric* or *non-parametric* tests (Table 4.7).

As shown, the choice of test depends on the factors discussed above. The choice between them is often done incorrectly. One reason is that results obtained can differ markedly between the methods, and practitioners favour the more powerful

Table 4.7   The nature of parametric and non-parametric tests.

Type of test	Level of measurement	Form	Normality/distribution	Variance
Parametric	Interval, ratio	Discrete, continuous	Normal	Equal
Non-parametric	Ordinal, nominal	Discrete	Any	Any

methods. These are more efficient at detecting differences and are often the desired choice. The conditions above must be met or assumed, at least approximately, otherwise the practitioner should consider use of a non-parametric test. As sample size increases this condition can be relaxed. Non-parametric tests can be used for any level of measurement when there are doubts about how the scale was used, when there are outliers in the data, uneven group sizes or when samples sizes are small (<10).

References

Clarke, G. M. and Cooke, D. (1998) *A Basic Course in Statistics*, 4th edn. Arnold, Hodder Headline group, London, pp. 248–283.

Cohen, J. (1992) A power primer. *Psychological Bulletin*, **112**(1), 155–159.

Daniel, W. W. (2005) *Biostatistics: A Foundation for Analysis in the Health Sciences*, 8th edn. John Wiley & Sons, Inc., Hoboken, NJ, p. 222.

Field, A. (2006) *Discovering Statistics with SPSS*, 3rd edn. Sage Publications, London.

Hassan, M. (2001) *Role of p-Values in Decision Making*, MSc Technical Report. Department of Statistics, University of Karachi, Karachi.

Huck, S. W. (2004) *Reading Statistics and Research*, 4th edn. International Edition, Pearson Education, New York, pp. 200–205.

Jæger, S. R., Andani, Z., Wakeling, I. N. and MacFie, J. H. (1998) Consumer preferences for fresh and aged apples: a cross-cultural comparison. *Food Quality and Preference*, **9**(5), 355–366.

Kvanli, A. H., Guynes, C. S. and Pavur, R. J. (1989) *Introduction the Business Statistics – A Computer Integrated Approach*, 2nd edn. West Publishing Company, St. Paul, MN, pp. 273–276.

Moerbeek, H. and Casimir, G. (2005) Gender differences in consumers' acceptability of generically modified foods. *International Journal of Consumer Studies*, **29**(4), 308–318.

Montgomery, D. C. (2001) *Introduction to Statistical Quality Control*, 4th edn. John Wiley & Sons, Inc., New York, p. 91.

O'Mahony, M. (1986) *Sensory Evaluation of Food Statistical Methods and Procedures*. Marcel Dekker, New York, p. 70.

Øygard, L. (2000) Studying food tastes among young adults using Bourdieu's theory. *Journal of Consumer Studies and Home economics*, **24**(3), 160–169.

Rees, D. G. (2001) *Essential Statistics*, 4th edn. Chapman & Hall, London, pp. 139–160.

Steel, R. G. D. and Torrie, J. H. (1981) *Principles and Procedures of Statistics – A Biometrical Approach*, 2nd edn. McGraw-Hill International Book Company, Singapore, pp. 167–171.

Stone, H. and Sidel, J. L. (1993) *Sensory Evaluation Practices*, 2nd edn. Academic Press, London, p. 123.

Tepper, B. and Trail, A. (1998) Taste or health: a study on consumer acceptance of corn chips. *Food Quality and Preference*, **9**, 267–272.

Chapter 5
Types of significance test

5.1 Introduction

Chapter 4 has given an account of the basis and procedure of significance testing. The tests themselves can now be described in detail. They are dealt with below according to level of measurement and population distribution, as these are initial factors deciding their selection. Indication is also given of other selection criteria, but readers should consult Chapter 4 (Section 4.5) for a full explanation if required. There are many different types of test, but only a limited number are included of the ones likely to be encountered by the food practitioner. Software applications are given for most tests and illustrations that are more specific are given in later chapters of the book. Within all tests, the significance testing procure is assumed but will not be stated in full and the significance level is set at 5%. Checks on assumptions are also assumed in this chapter but are provided in applications.

5.2 General points

It should be borne in mind that whilst the tests are done on *sample data*, any significance applies to the *population* itself. Tests applicable for a particular level of measurement or distributional form can be applied in a 'bottom to top' manner to any higher level of measurement where circumstances dictate, such as doubts about how the scale was used, etc. (Section 4.6). In these cases, parametric tests can be replaced with the non-parametric *distribution-free* alternatives (Huck 2004). Thus, tests for nominal and ordinal data can be applied to interval and ratio data. Data may require rearrangement or grouping for these cases. Note that 'distribution-free' does not mean 'assumption-free' and some non-parametric tests have additional notes referring to certain conditions that must apply (Rees 2001).

Another consideration is sample size. As sample size gets larger, the assumptions for the more powerful parametric tests matter less. Parametric methods have

Statistical Methods for Food Science: Introductory Procedures for the Food Practitioner,
Second Edition. John A. Bower.
© 2013 John Wiley & Sons, Ltd. Published 2013 by John Wiley & Sons, Ltd.

the advantage of locating differences more effectively, but if there are doubts regarding assumptions the reader should consult Section 4.5.

5.3   Significance tests for nominal data (non-parametric)

Nominal data are non-parametric in nature and make fewer assumptions regarding the parent population from which the data originate. Data are in the form of category groups or classifications (two or more). The tests have a variety of applications in food studies, mainly in consumer and sensory testing. These data are often generated in consumer survey tests where demographic variables are gathered. Some sensory difference tests also generate nominal data (Table 5.1). They are usually tested using *chi-square* (sometimes written as χ^2) methods. Another commonly used test for dichotomous two-group nominal data is the *binomial test*.

5.3.1   Chi-square tests

Essentially *chi-square tests* compare the proportions of observed occurrences with those of the theoretical expected ones, assuming that the null hypothesis is true (typically, if the null is true then the proportions will be equal). The calculated statistic is assessed via the critical value or *p*-value. This test is very commonly used in analysis of questionnaire or difference data as a *test of proportions* or *distribution* in that it tests the 'goodness-of-fit' of experimental data to expected data. Another frequent use is in a *test of independence* with data in a contingency table (Section 6.2). Expected values are based on existing knowledge or theory and are defined as 'all categories are equal' or any pre-defined proportion (Table 5.2).

Chi-square tests are recommended to meet the following conditions to avoid a weakening effect:

- Frequencies not percentages
- >5 in each category
- Sample size > 40
- Formula can include continuity correction for two categories

Generally, chi-square requires a minimum count of 5 in each group or 'cell' (i.e. not 5%, but an absolute count of 5) otherwise the test is weakened. This is because

Table 5.1   Significance tests for nominal data.

Test	Number of groups	Group type	Application
Chi-square	Two or more	Independent	Proportions, just-about-right, difference tests
McNemar	Two	Related	Before and after comparison
Cochran's Q	Two or more	Related	Multiple-choice comparison
Binomial	Two	Independent	Paired difference tests

Table 5.2   Chi-square tests of proportion for nominal data categories.

Test	Number of groups/type	Application
Chi-square	Two or more, independent	Proportions: compares proportions of categories in nominal choices in surveys and difference tests, or for pooled categories from Likert scales and just-about-right scales
		Distribution: compares proportions of sample group occurrences with a particular distribution of groups
McNemar	Two, related	Nominal choice in a before and after comparison
Cochran's Q	Two or more, related	Multiple nominal choice comparison

the sample statistic X^2 (discrete) is an approximation to the chi-square population distribution, which is continuous (Owen and Jones 1994). This approximation is valid for large sample sizes but is less valid as n falls, but this is acceptable provided cell frequencies are at least 5. For research studies, total numbers of observations should reflect this requirement, especially with survey experiments, where the sample may be formed into subgroups for chi-square analysis.

If the frequencies are low then categories can be pooled before comparison, e.g. with 'just-about-right' scales and for Likert data (Czaja and Blair 1996), provided such scales are treated as *nominal categories* for the purpose of analysis (see Chapter 8 for more on analysis of these scales). Other scales have been treated similarly. Vickers and Roberts (1993) pooled both categories of frequency of consumption and hedonic responses for popcorn at three levels of salt to ensure large enough cell sizes before using chi-square. Further adjustments can be made to the chi-square calculation for two-group comparisons in contingency table analysis (Section 6.2).

The chi-square test of proportion (goodness of fit)

How to do a chi-square test of proportion
The chi-square test can be done in Excel by summing the squared differences between the observed and expected frequencies (Table 5.3). The expected frequencies are usually taken to be equal and are obtained by dividing the total count by the number of categories (non-integer values are possible in some cases). The differences are then summed to give the chi-square statistic, and the p-value is calculated using the CHISQ.DIST.RT() function (CHIDIST() Excel 2003). If the original data are large in number then frequencies can be calculated using the FREQUENCY function described in Section 3.2.1. Degrees of freedom are equal to the *number of categories* minus 1.

Example 5.3-1 *Significance test on nominal data (chi-square on proportions)*
Twenty-four consumers were asked to select which of two samples (X and Y) they preferred: 10 chose X and 14 chose Y. Does this indicate a significant difference in preference?

Table 5.3   Chi-square test of proportions (Excel).

	A	B	C	D	E
1			**Chi-square test of proportion**		
2	**Sample data ($n = 24$) paired preference choice**		**Result**		
3		**Observed**	**Expected**	**Chi-square term**	**Formula**
4	Number choosing product X	10	12	0.33	$= ((B4-C4)^2)/C4$
5	Number choosing product Y	14	12	0.33	$= ((B5-C5)^2)/C5$
6	Total	24	Chi	0.67	$= SUM(D4:D5)$
7			df	1	1
8			*p*-value (two-tailed)	0.41	$= CHISQ.DIST.RT^a$ (D6,D7)

* CHIDIST() Excel 2003.

Data were entered as shown in Table 5.3 and the analysis stages were followed as described above. The expected frequencies were taken as half the total count of consumers ($=12$). The null hypothesis was that both choice frequencies were equal, the alternative was that they are not equal (two-tailed). The analysis reveals that they do differ, but not to a large enough extent and the *p*-value is >0.05. H_0 is retained: there is no difference in preference.

Readers who consult the Excel Help facility on this function may be confused as it is described as returning a one-tailed probability, although the test is declared as two-tailed in nature. A simple explanation is that according to the criteria of Section 4.2.2, the alternative hypothesis is non-directional, i.e. there is a difference between observed (O) and expected (E) frequencies but there is no specification of which one dominates the difference. However, the chi-square statistic can only increase in one direction as the difference between O and E gets larger, hence one-tailed. It is possible to use the test for a one-tailed alternative in which case the CHISQ.DIST.RT() (CHIDIST.()) probability is halved (Section 8.5.5).

These calculations must be done on the original count data, but the results can be expressed as percentages if desired. The number of categories can be two or more for the chi-square test and Table 5.3 can be expanded to allow more.

When more than two groups are tested, a significant chi-square test does not mean that all groups are significantly different. This can be ascertained by examining a table or graph of the frequency of occurrence of each group. A more conclusive way is to apply the two-group chi-square significance test to each two-group comparison, but there is a danger of multiple testing errors (see Section 5.5.2). For instance, when testing three categories of frequencies 10, 15 and 24, a chi-square test is significant – at least two of the categories differ in frequency. A series of two-group tests show that only 10 cf. 24 show *p*-values below 0.05.

Chi-square tests of the above type are used extensively in consumer food survey literature. Hashim *et al.* (2001) used chi-square to compare proportions of consumers selecting between beef production labelled as traditional and irradiated and found a higher incidence of choice for non-irradiated. Reed *et al.* (2003) measured a number of demographic characteristics in a sample of consumers (702) including gender, age, urban, rural, etc. Chi-square was used to compare acceptance for, and levels of consumption of, chilled ready meals between groups according to these factors. Higher levels of consumption were found for urban male and younger consumers.

The McNemar test

This test analyses for differences in a single group, but in a 'before' and 'after' condition. An example could be comparing a dichotomous response such as 'y/n' before and after consumers received information about foods.

How to do the McNemar test

The McNemar test can be calculated for chi-square by subtracting 1 from the absolute difference between each case where there has been a change in the 'after' condition. This value is then squared and divided by the total number of changes (Upton and Cook 2002). Cases that do not change are ignored (Table 5.4). The test was applied to an experiment where 50 consumers were asked to accept or reject a product on tasting without a brand label: 20 accepted and 30 rejected. Later, a second test with the same source of samples and consumers, but with brand labels, resulted in 15 who did not change after being given information, 14 now accepting the product and 21 now rejecting it. Hence, the two parameters required were:

> **Change1** *(changed from reject to accept)* = 14
> *and*
> **Change2** *(changed from accept to reject)* = 21

This level of change was not significant ($p > 0.05$) and the brand information has not had a significant effect on acceptance.

Table 5.4   McNemar test (Excel).

	A	B	C	D
1		McNemar test		
2	**Sample data (n = 50) consumer choice before and after brand label**		**Result**	
3	**Change 1**	**Change 2**	**Chi-square**	**Formula**
4	14	21	1.03	= ((ABS(A4-B4)-1)^2)/(A4+B4)
5				
6		df	1	1
7		*p*-value (two-tailed)	0.31	= CHISQ.DIST.RT[a] (C4,C6)

[a]CHIDIST() Excel 2003.

Cochran's Q test

Cochran's Q test analyses the **multiple-choice** test situation and compares the incidence of choice for multiple items. The null hypothesis states that all items in the list are chosen with equal frequency. Cochran's test will identify significance, if at least two items differ in selection incidence. An example could be where consumers select one or more items from a list of four possible forms of processed vegetable (canned, frozen, dried and chilled). As this involves more than two objects, the data in the form of a table or a graph must be examined to see which choice(s) dominates (e.g. as per Fig. 3.3). Bi (2006) illustrates use of Cochran's test for multiple matched proportions, where the analysis is performed on the incidence of correct and incorrect responses of a series of tests.

This statistic can be calculated using the formula given by O'Mahony (1986), who also gives worked examples. The test was used for the consumer choice of four forms of processed food where accept = 1 and reject = 0 (Table 5.5). This sheet is larger than usual and formulae are entered on left and right and in some cases underneath the calculation cell. A limited number of consumers (10) participated in the test. The column totals in the 'Column Calculations' box reveal the maximum number of choice for each form of processing. The highest incidence was for 'chilled' followed by 'frozen', but the differences are not large enough for this small sample and the test is not significant ($p > 0.05$): there is no difference in the incidence of the choices.

5.3.2   *The binomial test*

The **binomial test** calculates a statistic using the general binomial theorem. It calculates the probability of getting the result by chance in a choice between two objects (product sample/methods of production, etc.). The total number of occurrences of the choice is viewed as the *trials*. For example, in choice of one out of two food product samples (A and B) by 20 consumers or trained sensory assessors, there are 20 trials. The choice results in two groups – those selecting A and B, respectively. In some applications, there is a correct answer in the choice and these are designated as the number of *'successes'*. In other applications, one of the choices (the larger) is designated as the number of *agreements.* In all circumstances, the chance probability with two objects is a 50:50 split. If the test probability is well away from this (as gauged by the significance level) then a significant choice of one sample over the other has occurred.

The test statistic can be calculated from the binomial terms as done is several texts (e.g. Bower 1996), but software and tables are faster. Cumulative probabilities must be calculated, i.e. all the events including and above the ones which qualify as a 'success' or as an 'agreement'.

In food science, the main application is for sensory difference and consumer preference tests. More than two objects are involved in some tests, but the chance probabilities are all calculated on the two groups of choice, e.g. a triangle test has

Table 5.5 Cochran's Q test (Excel).

	A	B	C	D	E	F	G
1							
2	Number of items or categories =	4		Cochran's Q test	1 = selected, 0 = not selected		
3		Canned	Frozen	Dried	Chilled		
4	Consumer 1	0	1	1	0		
5	Consumer 2	0	1	0	1	D	
6	Consumer 3	1	0	1	0		
7	Consumer 4	1	1	0	1	A	
8	Consumer 5	0	1	0	1		
9	Consumer 6	1	0	1	1	T	
10	Consumer 7	0	1	0	1		
11	Consumer 8	0	0	0	0	A	
12	Consumer 9	1	0	1	1		
13	Consumer 10	0	1	0	1		
14		Row calculations					
15		Sum row	Row sum squared				
16	= SUM(B4:E4)	2	4	= B16^2			
17	= SUM(B5:E5)	2	4	= B17^2			
18	= SUM(B6:E6)	2	4	= B18^2			
19	= SUM(B7:E7)	3	9	= B19^2			
20	= SUM(B8:E8)	2	4	= B20^2			
21	= SUM(B9:E9)	3	9	= B21^2			
22	= SUM(B10:E10)	2	4	= B22^2			
23	= SUM(B11:E11)	0	0	= B23^2			
24	= SUM(B12:E12)	3	9	= B24^2			
25	= SUM(B13:E13)	2	4	= B25^2			

#		Sum row total	Sum row squared			Column total sum	
26		Sum row total	Sum row squared				
27		21	51				
28		(= SUM(B16:B25))	(= SUM(C16:C25))				
29							
30		**Column calculations**				**Column total sum**	
31	Column total	4	6	4	7	21	= SUM(B31:E31)
32		(= SUM(B4:B13))	(= SUM(C4:C13))	(= SUM(D4:D13))	(= SUM(E4:E13))	441	= F31^2
33							
34	Column total squared	16	36	16	49	117	
35		(= B31^2)	(= C31^2)	(= D31^2)	(= E31^2)		(= SUM(B34:E34))
36		**Result**					
37	df		3	= (C2-1)			
38	Q (chi-square)		2.45	= (C2-1) * (C2 * (F34) - F32)/(C2 * F31-C27)			
39	p-value two-tailed		0.48	= CHISQ.DIST.RT[a] (C38,C37)			

[a] CHIDIST() Excel 2003.

Table 5.6   Significance tests for ordinal data.

Test	Number of groups	Group type	Application
Friedman's	Three or more	Related	Ranking test, ordinal rating
Wilcoxon test	Two	Related	Paired ratings or scores
Sign test	Two	Related	Paired ratings or scores
Mann–Whitney *U*	Two	Independent	Two group ratings or scores
Kruskall–Wallis	Three or more	Independent	Three or more groups ratings or scores
Rank sums	Two or more	Related	

three objects, but the results are expressed as 'correct' and 'incorrect'. Specific examples of the binomial test are given in Sections 8.4.2 and 8.5.5.

5.4   Significance tests for ordinal data (non-parametric)

Ordinal data are generated by *ranking tests* and when groups of consumers or sensory assessors rate items on a scale that has some logical order (Section 2.2.3). A number of tests are available (Table 5.6).

These tests compare the median of groups and some take into account the symmetry and shape of the parent distribution. Details that are more specific are given by Bower (1997a,b).

5.4.1   *Related pairs and groups*

Wilcoxon and sign

When pairs of related items are rated by the same assessors using an ordinal or higher level of measurement scale, there is choice between the **Wilcoxon-signed rank test** (matched pairs) and the **sign test**. Here, the data are unlikely to be in the form of ranks, as only two items are being assessed, and the values will be generated by an ordinal scale.

The **sign test** is simpler in that it compares the incidence of negative and positive differences between pairs. By doing this, the magnitude of the difference on the scale is lost and power is reduced. The **Wilcoxon** alternative does maintain some magnitude of differences and is thus more powerful. This test assumes more of an interval scale than the sign test and is thus an ideal test for data that are at metric level of measurement, but which are non-normal in distribution.

Example 5.4-1 *Significance test on paired ordinal data (sign test and Wilcoxon-signed rank test)*
A study on new varieties of vegetables asked consumers to rate two factors ('home-grown' and 'price', f1, f2) for 'level of importance in making a choice' on a five-point scale (1 = not very important, 5 = very important). Is there evidence that consumers view the factors differently in term of the importance for choice?

Table 5.7   Wilcoxon and sign tests using Minitab.

Sample data (n = 10)			Result					
f1	f2	Sign			Sign test			
5	3	2	SIGN TEST OF MEDIAN = 0.00000 VERSUS N.E. 0.00000					
3	4	−1						
3	4	−1		*N*	BELOW	EQUAL	ABOVE	*P*-VALUE	MEDIAN
5	4	1	Sign	10	4	0	6	0.7539	1.000
2	3	−1						
3	1	2						
4	3	1			Wilcoxon signed ranks test			
2	3	−1						
5	1	4	TEST OF MEDIAN = 0.00000 VERSUS MEDIAN N.E. 0.00000					
5	4	1						
			N FOR WILCOXON ESTIMATED					
				N	TEST	STATISTIC	*P-VALUE*	MEDIAN
			Sign	10	10	39.0	0.262	0.5000

Excel cannot perform these tests directly, but both are available in Minitab and in MegaStat for Excel (although the sign test can be analysed using binomial probabilities in Excel). For Minitab, data are entered in columns (Table 5.7) and the **Calc** option is used to calculate the difference and stored in a third column. The difference column is then analysed by selecting **Stat**>**Nonparametrics** then the **one-Sample Sign** or **One-Sample Wilcoxon test**. The 'test median' (= 0) is selected, i.e. the null hypothesis is that the signs sum to zero. There are options for one- and two-tailed alternative hypotheses – select 'not equal to' for a two-tailed alternative (Table 5.7).

Both tests are not significant with $p > 0.05$, thus the ratings do not differ and there is no difference in level of importance. It is also possible to compare a reference value with single sample data for both these tests in the Minitab versions. MegaStat can do both tests and reaches the same conclusion.

Friedman's test

How to analyse a Ranking Test

Ranking tests involve at least three items, which are dealt with by all assessors in related groups. The tests specify a criterion, which is used to order the items. A common objective is to identify the item that is ranked first, e.g. 'where 1 = most preferred'.

The *Friedman test* is used for analysis of ranking tests and for any three or more related group comparisons where items are rated on an ordinal or higher level scale. Its fuller name is *Friedman's two-way analysis of variance* (ANOVA) with two factors. These are referred to as *treatments* and *blocks*, respectively, and Friedman's test uses a *randomised block design* (this terminology is detailed in Chapter 7). Usually, only the treatment (sample) effect is of direct interest and

Table 5.8   Friedman's test on ranking data (Minitab).

Friedman's test
Sample data ($n = 5$) preference ranking of five products

Consumer	s1	s2	s3	s4	s5
1	5	3	1	2	4
2	3	4	1	2	5
3	3	4	1	2	5
4	5	4	2	1	3
5	2	3	4	5	1
6	3	1	5	2	4
7	4	3	2	1	3
8	2	3	4	1	5
9	5	1	2	3	4
10	5	4	2	1	3

Result

Friedman test of ranking by sample blocked by consumer

$S = 10.26$, df $= 4$, $p = 0.037$

$S = 10.31$, df $= 4$, $p = 0.036$ (adjusted for ties)

Sample	N	Estimated median	Sum of ranks
1	10	4.100	38.0
2	10	3.300	30.5
3	10	2.100	24.0
4	10	1.900	20.0
5	10	3.600	37.5

Grand median $= 3.000$

the contribution of the block (assessor) effect is removed giving more precision and power.

Friedman's test is available in Minitab, where the data are entered as a single column, and two other columns identify the samples (treatments) and the assessors (blocks). Table 5.8 gives an example of ten consumers ranking five food samples in order of preference, where 1 = most preferred and 5 = least preferred. The data are given in separate columns ('**unstacked**') to save space – these require to be rearranged into one column using Minitab's **Manip**>**Stack** function. Select the sample columns (s1–s5) and enter a new column (name it 'rank') to store the stacked data. Tick the '**store subscripts**' box and provide a second new column (name it 'sample') to store the identifiers for each part of the stack – in this case, it is the sample (1–5) identifier. Two new columns are produced, but a third (name it 'consumer') is required to identify the block – this is provided by entering the consumer identifier number for the first sample then copying below for the other samples. These three columns are required for the **Stat**>**Nonparametrics**>**Friedman** option, where '**Response**' = rank, '**Treatment**' = sample and '**Blocks**' = consumer.

In this case, the p-value is <0.05 – at least two of the samples differ in preference. Looking at the estimated median, sample 4 has received the 'lowest' rank; i.e. it has been ranked as 'more preferred', whereas sample 1 has received more high rank values (low preference). These samples will differ significantly, but location of all possible sample differences requires *multiple pair-wise testing* (see below).

The significance of the Friedman statistic depends on its magnitude and the number of blocks and treatments. If these latter numbers are small (3 samples plus 2–9 assessors or 4 samples with 2–3 assessors) then *Friedman's special tables* (a version is available in BSI 1989) are required (Minitab does not appear to take this into account). This test is the one recommended in the BSI publication (BSI 1989) for a sensory ranking test. MegaStat produces the same result as above and provides a multiple comparison value for pair-wise testing.

5.4.2   Ordinal scales

How to analyse ordinal scales of Likert format

Questionnaires commonly use ordinal scales to rate statements on food issues with the **Likert** 'agree–disagree' scale'. In addition, consumer tests on food samples often employ the **hedonic rating scale**, which can be regarded as ordinal if the analyst wishes to be conservative regarding the assumption of interval level for this scale (Section 2.2.1).

Such data can be analysed by non-parametric methods, with the test depending on the number of groups being compared and the number of items ('full' Likert scales are often used in sets of individual item scales numbering 5–10).

Likert format scales (survey)

Likert scales were originally intended to be used as a set of statements all measuring the same underlying construct. Examples could be a 10-item scale on nutritional knowledge or a 25-item scale on attitudes to environmentally friendly foods. Individual items are summed and averaged to give a *composite measure* that would qualify as interval level of measurement (see the next section). If the composite measure data are still viewed as ordinal, they can now be analysed for summary measures, etc., and for significance testing, typically, by comparing the composite median values between independent groups. This is done when looking for demographic variation effects in consumer samples – gender effects, age group or geographical location, etc. The choice of methods is between *Mann–Whitney U* and *Kruskall–Wallis* tests (see below).

Likert-type items

Individual item scales of various Likert formats can also be analysed. Although such scales are ordinal, there are few readily available ways of analysis that are recommended and that retain the level of measurement in the data. The objective may be to determine whether or not consumers 'agreed' or 'did not agree' with a statement, or did they consider the item 'important' or 'not important'. A more conservative analysis can be achieved by treating the data as nominal, provided that the scale is in the form of categories boxes. The frequency of each category on the scale can be compared using a chi-square test (Section 5.3). This is recommended for survey data (Czaja and Blair 1996) and can be taken further by pooling of the descriptors on the scale to give fewer categories such 'agreeing' cf. 'not agreeing'.

Hedonic rating (consumer)

The hedonic scale is used to compare degree of liking between different food samples. Usually, the analysis will provide answers to questions such as 'do the samples differ in degree of liking?' and 'which food was liked most?' The non-parametric analysis is achieved using Friedman's Test for three or more food samples and the Wilcoxon for two (see above and Chapter 8 for an example of the Friedman's test on hedonic data). Again, if hedonic data between different groups are being compared then independent non-parametric tests are the choice (as below).

5.4.3   Independent groups

Mann–Whitney U and Kruskall–Wallis tests

When groups are independent the non-parametric, **Mann–Whitney U** and **Kruskall–Wallis** tests are appropriate for two and three or more groups, respectively. Minitab and MegaStat provide both tests.

Mann–Whitney U test (Minitab)

For this test, the data are entered as two columns – one for each sample. These are selected in the **Stat>Nonparametrics>Mann–Whitney** menu. Choose 95% for the confidence interval and a two-tailed alternative hypothesis ('not equal to'). To illustrate the test, a single Likert-item scale (seven-point, 'disagree strongly' (1) to 'agree strongly' (7)) was rated by male and female consumers. The statement was '*British beef is safe*'. The research question was '*do male and female consumers differ in their view on this statement?*' The ratings for both groups were entered into Minitab and analysed by the test (Table 5.9). For this analysis, it can be seen that the medians are both on the 'agree' side of the scale, but they are very similar. The test is not significant ($p > 0.05$) and the confidence interval contains zero confirming this. MegaStat analysis gives a p-value directly rather than a confidence interval, and it agrees with the non-significance. Thus, male and female consumers have similar views on the statement on British beef, both being on the 'agree' side of the scale. If an analysis of the ratings of the whole sample was required in order to establish 'did they agree or disagree with the statement?' then a different analysis would be used such as comparing the frequency of the categories (with pooling if necessary) by using chi-square as explained above.

Table 5.9 Mann–Whitney U test on two groups (Minitab).

Sample data ($n = 10$), comparison of level of agreement with statement (Likert-type item)		Result
Male	Female	Mann–Whitney Confidence Interval and Test
7	5	
6	4	Male $N = 10$ Median $= 5.500$
4	5	Female $N = 10$ Median $= 5.000$
6	5	Point estimate for ETA1–ETA2 is -0.000
5	3	95.5% CI for ETA1–ETA2 is $(-1.000, 2.000)$
2	6	$W = 109.5$
6	4	Test of ETA1 $=$ ETA2 versus ETA1 $\cong$ ETA2 is significant at 0.7624
4	5	The test is significant at 0.7570 (adjusted for ties)
7	5	
3	7	Cannot reject at *alpha* $= 0.05$

Kruskall–Wallis test

This test is used for comparison of rating between three or more groups. Data are entered as one column and then a 'Factor' (sample identification) column. For this example, three groups of consumers of different educational level ('edugrp' where 1 = 'low', 2 = 'medium' and 3 = 'high') are asked to indicate their level of importance (1 = 'not important' to 10 = 'very important') that they attach to price of food products during normal shopping (Table 5.10).

For these results, the medians are different, in that those of 'medium' educational level appear to attach more importance than the other groups. However, this large

Table 5.10 Kruskall–Wallis test on three-group comparison of importance scale (Minitab).

Sample data ($n = 5$), importance rating for price according to education		Result				
edugrp	Rating	Kruskall–Wallis 'rating' 'edugrp'				
1	7					
1	3	LEVEL	NOBS	MEDIAN	AVE. RANK	z, VALUE
1	5	1	5	4.000	6.1	-1.16
1	2	2	5	8.000	10.7	1.65
1	4	3	5	5.000	7.2	-0.49
2	3	OVERALL	15	8.0		
2	8					
2	9	$H = 2.88$, df $= 2$, $p = 0.237$				
2	5	$H = 2.93$, df $= 2$, $p = 0.232$ (adjusted for ties)				
2	8					
3	1					
3	6					
3	4					
3	7					
3	5					

difference is not sufficient (for this small sample) to exhibit significance ($p >$ 0.05). Once again, if a significant ANOVA was found then additional testing to locate the significant difference would follow (see below).

NB: Friedman's and *Kruskall–Wallis* tests are the non-parametric equivalents of the parametric *ANOVA randomised complete block (RCB)* and *completely randomised design (CRD)* methods. If ANOVA is significant then the data must be examined to locate significant differences. The significance can apply to each possible pair or only to one. A significance test can be performed to confirm evidence from a summary table or graphs of the rank sum or medians. Suitable tests for Friedman's test are the sign test and Wilcoxon, and for Kruskall–Wallis, the Mann–Whitney *U*.

Ranked and ordinal data analyses appear regularly in published work. Purdy *et al.* (2002) provide an example of Kruskall–Wallis test use with a follow-up Mann–Whitney *U* test when significance was found in a study of socioeconomic status and salt intake. They used quota sampling to get representative numbers (based on census data) in each of six groups ($n = 360$) and the final sample balance compared with the census report.

Albalá-Hurtado *et al.* (2001) analysed data that they deemed as non-parametric using Kruskall–Wallis and Mann–Whitney *U* tests. They compared a range of infant milks for vitamin content in powdered and liquid form ($n = 41$ and 18) finding significant differences in riboflavin and thiamin.

Friedman's test has an alternative for pair-wise testing described in BS 5929 (BSI 1989), which is also detailed in Danzart (1986) and Sprent (1993). A non-parametric equivalent of *Tukey's honestly significant difference* (HSD) test is also possible (Meilgaard *et al.* 1991). A comparison of these methods is given by Bower (1998a). Multiple comparisons can incur risk of committing a Type I error, which is discussed below in the context of parametric tests.

Many of these tests (except the sign test) perform ranking on the data in some manner and a problem arises when ranks are tied. Usually a relatively high incidence ($>25\%$) of ties has a weakening effect on the tests. Ties can be avoided by specific instructions to respondents to avoid them (in ranking tests) or by using as many points on the scale as feasible. Scales that have three or five categories are more likely to result in ties (Minitab and MegaStat take this into account and adjust the *p*-values for ties).

5.4.4   Other non-parametric tests

Page and Jonckheere tests

These tests compare the experimental rank order with a reference set of known rank, such as chemical content. For instance, when a sensory panel ranks solutions of known sugar or salt content, etc. The **Page test** is for related samples and is detailed in BS 5929 (BSI 1989). The **Jonckheere test** is an independent samples form of the test. Neither of the tests is available in Excel or Minitab, but a full explanation and worked examples are given for the Page method in the BSI reference. These tests appear relatively rarely in publications, but Tepper *et al.* (1994) used the Page test to analyse rankings of perceived fat content against actual fat content – generally consumers were able to perceive fat levels in common foods effectively (the Page statistic was significant).

Rank sums

An alternative to full calculation of ranked data as above is to calculate *rank sums* then consult published tables. This has the advantage of a simpler calculation, and for small data sets and routine small sensory panels, it can be done quickly. Different sets of tables are available in various texts, but not all are valid and comprehensive. Bower (1998a) gives some explanation of the various forms, but the tables produced by Basker (1988a,b) have been cited in a later publication (Lawless and Heymann 1998), although those of Newell and MacFarlane (1987) are also useful. Christensen *et al.* (2006) examined both sets of tables and viewed them as being suitable for comparing the highest and lowest sums, but deemed them as too conservative for multiple comparison of all. They provide expanded tables for this latter purpose.

Rank sums have the advantage of speed and give direct results without having to carry out ANOVA followed by multiple comparisons. The sums must be calculated first manually, or by computer or calculator then the tables are consulted.

Proportions

Proportions can be compared for significant differences within one sample by chi-square and binomial methods (Section 5.3) or by approximation to normal using the proportional estimates detailed in Section 3.4.3. Consider a situation where a local grocery magazine states that '*only 25% of food shoppers buy organic*'. Assume that a group of researchers disagree with this as their analysis of sales figures suggests a higher proportion is likely. To test this, they ballot a quota sample of shoppers and ask '*do you buy organic (y/n)*'. They find that 33 out of 100 say 'yes'. Is this estimate (33%) evidence of more than 25%? Using chi-square as detailed in Table 5.1, the observed frequency of 33 and 67 are compared with the expected of 25 and 75. The *p*-value is 0.065 so the higher proportion is not significant, as it is not large enough to warrant discounting the magazine statement. More advanced versions of Minitab and MegaStat include a proportional test that can be applied to the previous example (see **MegaStat>Hypothesis Tests>Proportion vs. Hypothesized Value**). There is also another test for comparison of *two independent proportions*, e.g. the proportion of faults in product A cf. with those in product B (**Hypothesis Tests>Compare Two Independent Proportions**).

5.5   Significance tests for interval and ratio data (parametric)

In cases where the data are at a higher measurement level than ordinal and distributions conform to normality, the more powerful parametric methods can be

Table 5.11   Significance tests for metric data.

Test	Number of groups	Group type
One-sample t	One	—
Paired t	Two	Related
Two-sample t	Two	Independent
Analysis of variance	Three or more	
Completely randomised design (CRD)		Independent
Repeated measures		
Randomised complete block (RCB)		Related

used. In these applications, an assumption is made regarding distributions and thus *checks on assumptions* are required (Sections 3.5 and 4.3). If the checks indicate a wide divergence from the assumptions then a non-parametric test can be used, or the practitioner can proceed with a parametric one with caution (e.g. increase the significance level). The tests available are more elaborate with several forms of parametric *analysis of variance* and the *t-test* (Table 5.11).

These tests compare the mean values of the measure and the appropriate test depends on the number of groups being compared and whether or not they are independent. As with the non-parametric equivalents above, the tests can be used to compare the items that are scored on the scale or to compare group differences. These tests are universally used in analysis of instrumental data. For these applications, the groups will consist of food items or methods of analysis. For example, two varieties of orange are compared for ascorbic acid content using a *t*-test (the independent version). The significance of any difference can be obtained by such analysis. Comparison of this type for three or more food items would demand use of a CRD form of ANOVA.

These parametric tests are also commonly used on sensory and consumer test data in some research applications. In sensory evaluation, attribute scaling data generated on intensity scales would qualify. Here, the panel assessors are viewed as an instrument and they themselves are not necessarily under scrutiny (except in panellist performance testing – see Section 8.5.5). Thus, different food samples can be compared for intensities of an attribute such as 'saltiness' using a related parametric ANOVA. This form of analysis is referred to in a number of ways (Bower 1998b), all essentially being *repeated measures designs*.

Demographic variations can be examined by comparing two and three or more groups. In addition, the test value can be compared with a value obtained from the literature or some other source, ideally a 'population value' or a normative value. The test used is classed as a *1-sample test*.

5.5.1   *t-tests*

t-tests can be used in a number of forms. *t* is measured in units of standard error (se) – the distance between group means or a mean and a standard value is divided by the se. Larger differences (effect sizes) with low variation (se) will give rise to large *t*-values that are more likely to be significant. *t*-Tests are used for small samples (<30), but with larger samples, the *t*-distribution is similar to normal and the *z*-test (see below) can be used under certain conditions. All *t*-tests are available in Excel except the one-sample *t*, but this is simple to calculate.

One-sample t-test for parametric data

How to do a one-sample t-test

Example 5.5-1 *Significance test for 'one-sample' metric data (one-sample t-test)*

A new fruit variety is claimed to have more ascorbic acid content than the current main variety. A series of sample units were analysed and compared with the reference value (35 mg/100 g; based on an established average value for the main variety). Do the data provide support for the claim?

The data were entered as a column (Table 5.12) and the required values were obtained via Excel functions and formulae.

Table 5.12   One-sample *t*-test (Excel).

	A	B	C
1		One sample *t*-test	
2			
3		Sample data (*n* = 7) ascorbic acid in fruit (mg/100 g)	
4		35	
5		42	
6		37	
7		43	
8		41	
9		45	
10		37	
11	Result:		Formula:
12	Count (*n*)	7	= COUNT(B4:B10)
13	Mean	40.00	= AVERAGE(B4:B10)
14	sd	3.70	= STDEV.S[a](B4:B10)
15	se	1.40	= B14/(SQRT(*n*))
16	Reference value	35.0	
17			
18	*t*	3.58	= ABS((B13-B16)/B15)
19	df	6	= *n* − 1
20	p-value (one-tailed)	0.006	= T.DIST.RT[b](B18,B19)

[a]STDEV() Excel 2003.
[b]TDIST(B18,B19,1) Excel 2003.

Summary statistics are calculated by methods used in Table 3.6. Based on the previous evidence and reasoning, the alternative hypothesis in this test is *one-tailed* – if the null hypothesis is rejected then it can be predicted that the new variety will have a higher content of ascorbic acid. The value of the *t*-statistic is obtained by dividing the absolute difference between the test mean and the reference value by the standard error. The *p*-value is calculated using the T.DIST.RT() function with parameters as the *t*-value and the degrees of freedom (TDIST() for Excel 2003 with an extra argument: '1' for one-tailed). The mean value for the new variety is indeed greater than the reference value and it is significant ($p < 0.05$). There is evident to support the claim.

Two-sample t-test for parametric data (independent groups)

For *t*, the two-sample and the paired test can be obtained from Excel's **ToolPak Data Analysis** menu or by using the functions. The data are entered in two columns and the type of test selected.

How to compare two groups with parametric data

For a two-group comparison, the Excel T.TEST() function (TTEST() Excel 2003) can be used or the independent *t*-test in the **ToolPak (Data/Data Analysis** in Excel 2010 or **Tools/Data Analysis** for Excel 2003). Two forms of the test are available: one for an assumption of equal variances and one without (a full example of this test is provided in Chapter 4 that includes a variance check prior to the analysis). One advantage of the independent test is that group sizes do not need to be the same. An example of this was presented by Aguilar and Kohlmann (2006), who used *t*-tests to compare two independent groups, one of producers ($n = 19$) and the other of consumers ($n = 101$) for several aspects of 'willingness to consume and produce transgenic bananas'. They found highly significant differences between farm manager producers, who could see benefits, and consumers who did not.

z-test (Excel)

When *both* sample sizes are larger (>30) and the population standard deviation is known, a **two-sample z-test** can be used in a very similar manner to the above with the Z.TEST() function (ZTEST() Excel 2003) or the **ToolPak**, testing that the difference between the means is zero. All the original data must be used in the basic Excel *z*-test, but MegaStat (**Hypothesis Tests>Compare Two Independent Groups**) allows a summary input, which can be more rapid if mean values, and the population standard deviations are known.

Paired t-test (Excel)

For paired data, the **paired t-test** is used. Take the instance of a trained panel scoring the intensity of 'stewed' flavour in two baked apple preparations, using a line scale (0–100). One lot of each was prepared, assessed on sub-units

Table 5.13   Paired *t*-test (Excel).

	A	B	C
1		**Paired *t*-test**	
2			
3	**Sample data (*n* = 8)**		
4	**Preparation 1**	**Preparation 2**	
5	74	57	(Assessor 1)
6	35	47	(Assessor 2)
7	75	70	(Assessor 3)
8	83	33	(Assessor 4)
9	58	45	(Assessor 5)
10	47	73	(Assessor 6)
11	89	42	(Assessor 7)
12	57	39	(Assessor 8)
13		**Result**	
14	t-test: paired two sample for means		
15			
16		Variable 1	Variable 2
17	Mean	64.8	50.8
18	Variance	345.36	211.64
19	Observations	8	8
20	Pearson correlation	−0.23	
21	Hypothesized mean difference	0	
22	Df	7	
23	*t* Stat	1.52	
24	$p\,(t \leq t)$ one-tailed	0.09	
25	*t*-Critical one-tailed	1.89	
26	$p\,(t \leq t)$ two-tailed	0.17	
27	*t*-Critical two-tailed	2.36	

and the panel scores analysed by a **paired *t*-test** using the **Data Analysis** tool (Table 5.13).

This *t*-test examines the data on the basis that the differences sum to zero; i.e. the null hypothesis is true. As seen from the summary values, there is quite a large difference in the scores, but there is a high variance within the sets showing variability due to the panellist and possibly between the sub-units within a preparation. This brings a two-tailed *p*-value of 0.17, which is not significant. Preparation one has more of a stewed flavour by a large margin of the scale, but until variability is reduced, it would be difficult to demonstrate this consistently and significantly.

5.5.2   Analysis of variance (ANOVA)

How to compare three or more groups with parametric data

This test is a form of analysis of variance (ANOVA) and it is used to test hypotheses concerning comparisons of three or more groups (thus being an independent sample test, i.e. there are three or more separate groups). In the simplest form, it is an extension of the *t*-test above. A dependent variable (usually continuous and parametric) and an independent variable (usually categorical) need to be identified. If an overall significance is established then the location of specific differences can be indicated by examination of the mean values, but a *multiple comparison test* is required for location of significant differences. Forms of ANOVA are discussed in more detail in Chapter 7. For the moment, an example of *one-way ANOVA* (i.e. there is one independent variable) is illustrated.

Example 5.5-2 *Significance test on metric data with many groups (one-way ANOVA)*

A cereals laboratory is examining baking effects on products and requires to answer a question that arose from an investigation into oven conditions: '*Does temperature of processing cause changes in brittleness of baked biscuit products?*' Four lots of each product dough were baked in random order at a range of temperature settings (t1–t4). Instrumental texture readings of 'brittleness' (newtons) were taken on four random biscuits from each lot, and the replicate values averaged. Do the viable oven temperatures used have a significant effect on product brittleness?

This type of experiment is viewed as a completely randomised design (CRD; Section 7.4) where more than two independent groups are compared. It can be analysed using Excel's one-way ANOVA option in the **ToolPak**. For this analysis, the data are entered as shown in Table 5.14. The data values and the column headings (t1–t4) are blocked and the analysis proceeds by selecting **Data/Data Analysis** (Excel 2010) or **Tools/Data Analysis** (Excel 2003) then selecting **Anova: Single Factor**. The blocked data and label constitute the **Input Range** (tick **Labels in First Row** box). Select an **Output Range** location (a single cell of the top left-hand corner of the output box) to complete.

The summary values show that the average values do differ – but are these due to real effects or simply sampling and other error effects? In this example, the *p*-value of the *F*-ratio (foot of Table 5.14) is above 0.05 and the null hypothesis is retained: there is no difference in the brittleness at the temperatures used.

Assume that the temperature range is changed to three new settings (t5–t7) and that the experiment is run again (Table 5.15).

Now the mean values show more difference and the *p*-value is significant ($p <$ 0.05), although in the 'inconclusive' region which has ramifications for location of the significant effect(s). The null hypothesis is rejected and the alternative accepted: at least two of the brittleness values differ. The ANOVA does not reveal this as such – all the means differ, but is each pair-wise comparison significantly

Table 5.14 Analysis of variance (Excel).

	A	B	C	D	E	F	G
1	**Parametric ANOVA**						
2							
3	Sample data ($n = 4$) ('brittleness' (N))						
4	t1	t2	t3	t4			
5							
6	0.12	0.27	0.16	0.33			
7	0.16	0.19	0.21	0.23			
8	0.20	0.30	0.34	0.17			
9	0.25	0.25	0.24	0.11			
10							
11	**Result (data analysis tool)**						
12	ANOVA: single factor						
13							
14	Summary						
15	Groups	Count	Sum	Average	Variance		
16	t1	4	0.73	0.18	0.0031		
17	t2	4	1.01	0.25	0.0022		
18	t3	4	0.95	0.24	0.0058		
19	t4	4	0.84	0.21	0.0088		
20							
21							
22	ANOVA						
23	Source of variation	SS	df	MS	F	p-value	F-critical
24	Between groups	0.012	3	0.0038	0.77	0.53	3.49
25	Within groups	0.059	12	0.0049			
26							
27	Total	0.071	15				

different? Multiple comparisons are required, in this case three, as t5 cf. t6, and t6 cf. t7 and t7 cf. t5. As indicated above with respect to the non-parametric ANOVAs and as discussed below there is a choice of which test to use for comparison.

Multiple comparison tests after ANOVA (three or more groups)

This procedure is performed only if a significant F-ratio for ANOVA is obtained.

It is subject to multiple testing risks as explained in Chapter 4. For parametric tests, there are many possible *'post hoc' tests* and there is much debate on the issues raised in their selection in statistical literature. Readers should consult texts such as Gacula and Singh (1984) for a fuller account. The tests calculate a *range value* against which the differences between treatment means are compared. Table 5.16 shows some common tests.

Table 5.15 Analysis of variance (b) (Excel).

	A	B	C	D	E	F	G
1				Parametric ANOVA			
2							
3	Sample data ($n = 4$) ('brittleness' (N))						
4	t5	t6	t7				
5							
6	0.23	0.36	0.42				
7	0.27	0.40	0.29				
8	0.32	0.38	0.37				
9	0.29	0.29	0.39				
10							
11				Result (data analysis tool)			
12							
13	ANOVA: single factor						
14							
15	Summary						
16	Groups	Count	Sum	Average	Variance		
17	t5	4	1.11	0.28	0.0014		
18	t6	4	1.43	0.36	0.0023		
19	t7	4	1.47	0.37	0.0031		
20							
21							
22	ANOVA						
23	Source of variation	SS	df	MS	F	p-value	F-critical
24	Between groups	0.019	2	0.0097	4.29	0.049	4.26
25	Within groups	0.020	9	0.0023			
26							
27	Total	0.040	11				

The tests are performed at the same significance level as ANOVA. Choice of test depends on assumptions regarding the data as for application of a parametric test. The more powerful ones risk committing the Type I error and conversely the weaker ones the Type II. In some, the number of comparisons causes the significance value to rise directly with the number (the *LSD test*). *Tukey's test*

Table 5.16 Some common multiple comparison tests.

Test	Relative power (approximately)
Least significant difference (LSD)	High (liberal)
Duncan's multiple range	High (all possible pairs)
Newman-Keuls' range	Medium (all possible pairs)
Bonferroni LSD	Medium (all possible pairs)
Tukey's HSD (honestly significantly different)	Low (all possible pairs)
Scheffés	Low (conservative; all possible pairs)

avoids this but has less power for detection of true differences as numbers increase. *Duncan's test* calculates a range value, which varies according to the number of comparisons. The ***Bonferroni LSD*** adjusts the significance level according to the number of tests. This technique, the ***Bonferroni adjustment*** (Huck 2004), can be applied simply by adjusting *alpha*. If there are three comparisons to be made then for an original *alpha* of 0.05 that for the comparison is 0.05/3 = 0.017, for six comparisons 0.05/6 = 0.008, etc. This method appears ideal, but when ANOVA significance is 'borderline' (*p*-value between 0.04 and 0.05), it can be difficult to locate any pair-wise differences with the Bonferroni adjusted *p*-value.

Choosing a test depends essentially on the confidence in the data. Instrumental data will usually qualify for the more powerful tests. For sensory and consumer data, it may be more advisable to use a more conservative alternative. Consideration of the error types, e.g. in cases where missing a true difference is of high consequence, a more powerful test is appropriate. A compromise is a 'medium' test such as Tukey's HSD. There are many examples in food research literature of use of this entire test set for a wide variety of experimental circumstances.

How to perform a multiple comparison test

Excel in its basic form does not provide multiple comparison tests, although Carlberg (2011) describes an involved implementation of some procedures including the Scheffé test. With caution and with adjustments such as the Bonferroni procedure above, independent *t*-tests can be used to give an indication of where the significant difference(s) lie. Doing all possible *t*-tests can be viewed as equivalent to the LSD test (Table 5.17). Ideally, analysis software that includes multiple comparison procedures as part of the ANOVA should be used. MegaStat provides multiple comparison *t*-tests and a form of Tukey's test, as does Minitab and SPSS ANOVA menus include a wide choice.

With the ANOVA example above and basic Excel the only option for multiple comparison of means is to perform *t*-tests. Equal variance can be assumed or

Table 5.17   Use of *t*-test as follow-up comparison after a significant ANOVA (Excel).

t-Test: two-sample assuming equal variances		
	t5	t6
Mean	0.28	0.36
Variance	0.0014	0.0023
Observations	4	4
Pooled variance	0.0019	
Hypothesised mean difference	0	
df	6	
t-Statistic	−2.62	
$P(T \le t)$ one-tailed	0.02	
t Critical one-tailed	1.94	
$P(T \le t)$ two-tailed	0.039	
t Critical two-tailed	2.45	

checked using F-tests (Section 4.4). This was assumed for the data of Table 5.15 and t-tests were performed using the appropriate tool in the **ToolPak** to compare the means (Table 5.17). If α is taken as 0.05, then mean t5 with mean t6 and t5 with t7 were significantly different ($p < 0.05$). Comparison of the means for t6 and t7 was not significant ($p > 0.05$). Thus, ANOVA has identified the presence of at least one difference, and the pair-wise comparison test revealed that two pairs are different. However, if the Bonferroni adjustment is made to these tests (p-value = 0.017) then none of the tests are significant. If Tukey's test is performed on these comparisons, mean t5 cf. mean t7 are the only pair that approach a significant difference – this is a consequence of using the weaker test (see below). MegaStat's 'One factor ANOVA' agrees with the main analysis and the t-test comparisons above and Tukey's test finds no significant differences.

Minitab provides one-way ANOVA and comparisons from its **Stat** menu (Table 5.18). The ANOVA output is similar to that of Excel except for the confidence intervals of the mean values. These show in graphical form that mean t5 is more distinct than those for t6 and t7. **Fisher's LSD** and Tukey's test were selected for the comparisons. The output requires more study to interpret – significant differences are located by a confidence interval (*difference interval*) that does not contain zero, i.e. if the difference between a pair of means is at or near zero the

Table 5.18   Use of LSD and Tukey's test follow-up comparison after ANOVA (Minitab).

MTB > Oneway 't all' 'factor'; SUBC > Tukey 5; SUBC > Fisher 5.
Analysis of variance on 't all'

Source	DF	SS		MS	F	p
Temp	2	0.01947		0.00973	4.29	0.049
Error	9	0.02042		0.00227		
Total	11	0.03989				

Individual 95% CIs for mean based on pooled STDEV

Level	N	Mean	STDEV	—+——+——+——+—
1 (t5)	4	0.27750	0.03775	(———*———)
2 (t6)	4	0.35750	0.04787	(———*———)
3 (t7)	4	0.36750	0.05560	(———*———)

Tukey's pair-wise comparisons   Family error rate = 0.0500
Critical value = 3.95   Individual error rate = 0.0209
Intervals for (column level mean) – (row level mean)

	1 (t5)	2 (t6)
2 (t6)	−0.17409	
	0.01409	
3 (t7)	−0.18409	−0.10409
	0.00409	0.08409

Fisher's pair-wise comparisons   Family error rate = 0.113
Critical value = 2.262   Individual error rate = 0.0500
Intervals for (column level mean) – (row level mean)

	1 (t5)	2 (t6)
2 (t6)	−0.15620	
	−0.00380	
3 (t7)	−0.16620	−0.08620
	−0.01380	0.06620

MTB >

means are less likely to differ. On this basis, Tukey's test does not reveal any differences. For instance, comparison of mean t5 with means t6 shows an interval of –0.17409 to +0.01409, which lies on either side of zero, although t5 cf. t7 is 'close' to significance.

The above results and observations are a consequence of the use of tests with differing levels of power, plus the fact that the ANOVA *p*-value is very close to the decision point (0.049 cf. 0.05). Thus, for ANOVA results that are 'just' significant, the less powerful multiple comparison tests may not locate differences.

The more powerful LSD (Fisher's) test locates two significant differences (1 cf. 2, 1 cf. 3) as only one comparison (2 cf. 3) has an interval (–0.08620 to +0.06620) containing zero. Note the importance of using the summary table of mean values and confidence interval to aid interpretation of comparisons after ANOVA.

An idea of the possibilities with ANOVA can be seen by the research done by Péneau *et al.* (2006) on attributes affecting perception of apple freshness. The researchers employed three-way ANOVA where the factors were age, gender and level of consumption and they measured importance on a simple scale (five categories). Bonferroni tests were used to compare pairs when a significant ANOVA was found. Freshness rated high in importance and well above other influences such as 'organic'. Of the factors, gender was most significant in that females gave higher ratings on most attributes.

References

Aguilar, F. X. and Kohlmann, B. (2006) Willingness to consume and produce transgenic bananas in Costa Rica. *International Journal of Consumer Studies*, **30**(6), 544–551.

Albalá-Hurtado, S., Vecinia-nogués, M. T., Vidal-carou, M. C. and Mariné Font, A. (2001) Concentrations of selected vitamins in infant milks available in Spain. *Journal of Food Science*, **66**(8), 1191–1194.

Basker, D. (1988a) Critical values of differences among rank sums for multiple comparisons. *Food Technology*, **42**(2), 79–84.

Basker, D. (1988b) Critical values of differences among rank sums for multiple comparisons by small panels. *Food Technology*, **42**(7), 88–89.

Bi, J. (2006) *Sensory Discrimination Tests and Measurements*. Blackwell Publishing Ltd., Oxford, pp. 80–81.

Bower, J. A. (1996) Statistics for food science III: sensory evaluation data. Part B – discrimination tests. *Nutrition and Food Science*, **96**(2), 16–22.

Bower, J. A. (1997a) Statistics for food science IV: two sample tests. *Nutrition and Food Science*, **97**(1), 39–43.

Bower, J. A. (1997b) Statistics for food science V part A: comparison of many groups. *Nutrition and Food Science*, **97**(2), 78–84.

Bower, J. A. (1998a) Statistics for food science V. Part C: non-parametric ANOVA. *Nutrition and Food Science*, **97**(2), 102–108.

Bower, J. A. (1998b) Statistics for food science V. Part B: ANOVA and multiple comparisons. *Nutrition and Food Science*, **98**(1), 41–48.

BSI (1989) BS 5929 ISO 8587: *Methods for Sensory Analysis of Foods Part 6. Ranking*. British Standards Institution, London.

Carlberg, C. (2011) *Statistical Analysis: Microsoft® Excel 2010*, Pearson Education, QUE Publishing, Indianapolis, IN, pp. 277–286

Christensen, Z. T., Ogden, L. V., Dunn, M. L. and Eggett, D. L. (2006) Multiple comparison procedures for analysis of ranked data. *Journal of Food Science*, **71**(2), S132–S143.

Czaja, R. and Blair, J. (1996) *Designing Surveys: A Guide to Decisions and Procedures*. Pine Forge Press, Thousand Oaks, CA.

Danzart, M. (1986) Univariate procedures. In *Statistical Procedures in Food Research* by J. R. Piggot (ed.). Elsevier Applied Science, London, pp. 19–60.

Gacula, M. C. and Singh, J. (1984) *Statistical Methods in Food and Consumer Research*. Academic Press, Orlando, FL, pp. 84–93.

Hashim, I. B., McWatters, K. H., Rimal, A. P. and Fletcher, S. M. (2001) Consumer purchase behaviour of irradiated beef products: a simulated supermarket setting. *International Journal of Consumer Studies*, **25**(1), 53–61.

Huck, S. W. (2004) *Reading Statistics and Research*, 4th edn. International Edition, Pearson Education, New York, pp. 486–512.

Lawless, H. T. and Heymann, H. (1998) *Sensory Evaluation of Food: Principles and Practices*. Chapman & Hall, New York, pp. 690–691.

Meilgaard, M., Civille, G. V. and Carr, B. T. (1991) *Sensory Evaluation Techniques*, 2nd edn. CRC Press, Boca Raton, FL, pp. 261–268.

Newell, G. J. and MacFarlane, J. D. (1987) Expanded tables for multiple comparison procedures in the analysis of ranked data. *Journal of Food Science*, **52**(6), 1721–1725.

O'Mahony, M. (1986) *Sensory Evaluation of Food – Statistical Methods and Procedures*. Marcel Dekker, New York, pp. 105–106.

Owen, F. and Jones, R. (1994) *Statistics*, 4th edn. Pearson Education, Harlow, pp. 428–250.

Péneau, S., Hoehn, E., Roth, H.-R., Esher, F. and Nuessli, J. (2006) Importance and consumer perception of freshness of apples. *Food Quality and Preference*, **17**, 9–19.

Purdy, J., Armstrong, G. and McIlveen, H. (2002) The influence of socio-economic status on salt consumption in Northern Ireland. *International Journal of Consumer Studies*, **26**(1), 71–80.

Reed, Z., McIlveen-Farley, H. and Strugnell, C. (2003) Factors affecting consumer acceptability of chilled ready meals on the island of Ireland. *International Journal of Consumer Studies*, **27**(1), 2–10.

Rees, D. G. (2001) *Essential Statistics*, 4th edn. Chapman & Hall, London, pp. 161–178.

Sprent, P. (1993) *Applied Non-Parametric Statistical Methods*, 2nd edn. Chapman & Hall, London, pp. 143–151.

Tepper, B. J., Shaffer, S. E. and Shearer, C. M. (1994) Sensory perception of fat in common foods using two scaling methods. *Food Quality and Preference*, **5**, 245–251.

Upton, G. and Cook, I. (2002) *A Dictionary of Statistics*. Oxford University Press, New York, p. 225.

Vickers, Z. and Roberts, A. (1993) Liking of popcorn containing different levels of salt. *Journal of Sensory Studies*, **8**, 83–99.

Chapter 6
Association, correlation and regression

6.1 Introduction

Previous chapters have concentrated on several examples of detection of differences between treatments. Differences are not the only way in which effects can manifest themselves. Effects that cause differences often exhibit an *association* or *relationship* between two variables (Section 4.2) provided a sufficient range of readings are taken. The current chapter interest is focused on the *strength of the relationship*, which gives a measure of the effect size. Variables that exhibit this feature are considered to depend on one another in some way. This cannot be taken that one causes an effect in the other. This would require a more detailed examination and attention to experimental design and other features but for cause and effect, association must be present.[1] Both terms are interchangeable, but commonly, association is used with respect to nominal or categorical variable data (non-numeric) and correlation with ordinal (numeric) and metric data (Rees 2001).

For both circumstances, two measures are required on each sampling unit, be it a food product, processing machine or a consumer subject. Circumstances affecting the quality of the analysis include the range of values within each measure and their independence; i.e. each reading must not be affected by any other. A sufficient range of values in the measure (correlation) and independence are required. Association can be displayed by tabular methods, correlation graphically and both can be assessed for significance by a significance test. The strength of the relationship is assessed by calculation of a *coefficient*. Regression is a technique

---

[1] It is possible to have significant association between two variables and significant differences in one according to variation in the other, but not always. This can depend on the magnitude and linearity of the differences, e.g. temperatures of 15, 20, 25°C give yields of 5, 25, 35% (case 1 – both significant); case 2 yields of 5, 6, 7% – significance correlation, non-significant difference; case 3 yields 2, 29, 5% – significance difference, but non-significant correlation – non-linear.

Statistical Methods for Food Science: Introductory Procedures for the Food Practitioner,
Second Edition. John A. Bower.
© 2013 John Wiley & Sons, Ltd. Published 2013 by John Wiley & Sons, Ltd.

that combines correlation with some of the ANOVA methods of Chapter 5 and is of great importance in prediction models.

Studies of relationships are commonly applied in food science investigations with sensory and instrumental data and in consumer surveys where demographic variables are examined for association with factors such as preference and purchase intention for food products. Ultimately, *regression* may be employed in order to allow the strength of predictive power in these relationships to be established.

6.2   Association

Analysis of association is done on two nominal variables. These can be nominal originally or can be transformed by dividing a higher level of measurement into two or more categories. Thus, each variable can have two or more categories, although the 'two-by-two' case is easier to interpret. Sample size must be sufficient to allow a minimum count of 5 in each category cell of the analysis, if statistical analysis of significance is to be performed (chi-square test – Section 5.3).

Association is examined initially by displaying the two nominal variables in a special table known as a *contingency table*, also known as a *crosstabulation* (Section 3.2.1). The counts of the occurrence of each variable are split between that of the other. The two-by-two case is the simplest, giving four cells in the table. A *Pearson's chi-square test* is often used with two sets of nominal data to test the significance of association between the variables. This is a *test of independence* between the variables. The chi-square test establishes whether or not the counts differ between the divisions of each category. Sufficient deviation from the expected frequencies may point towards dependence. A test result with significant dependence *does not imply cause and effect* and the research question should reflect this.

Calculation of a coefficient can be done to measure the strength of the association. There are a number of possible measures for this, but *Cramér's V* (available in MegaStat and SPSS) appears to be most useful as it covers more than two categories in either variable and it allows comparison across different experiments. This ranges in magnitude from zero (no association) to a positive unity (maximum association). It is readily calculated from the chi-square statistic using a formula given in various texts (e.g. Upton and Cook (2002)). In Excel, a simplified formula gives $V = \textbf{SQRT}(X^2/(n * (k - 1)))$ where n = grand total and k = lower of the variable category counts for the rows and columns (e.g. for a '3 × 4' table, $k = 3$ and for a '2 × 2', $k = 2$).

Association analysis by 'crosstabs' can be done using Excel pivot tables and by a more detailed use of Excel's functions and formulae (Middleton 2004; Carlberg 2011), which in the latter case involves use of the CHISQ.TEST() (CHITEST() Excel 2003) functions. Minitab and MegaStat provide a full crosstabulation analysis with much simpler data input. Both packages can analyse the original data and Minitab also allows direct entry of frequencies.

How to carry out a crosstabulation analysis with a significance test

Example 6.2-1 *Crosstabulation analysis and test of independence*
A consumer survey examined data for a possible association between the sex of the respondent and their propensity for low-fat foods. A sample of consumers were asked '*If given a choice would you choose low-fat foods over conventional? (y/n)*'. Gender and answers were recorded and collated as frequencies. Does gender appear to have an association with a propensity for low-fat foods?

The null hypothesis is that gender has no association with a tendency to choose low-fat foods and the alternative (two-tailed) is that association is present. Using Minitab, the data require entry as columns, one for the gender of the respondent and the other for 'low fat' response (y/n). If frequencies are used then only the categories for each variable are required and the counts go into a third column ('freq'). **Stat>Tables>Cross Tabulation** is used to obtain the analysis (Table 6.1).

The expected frequencies are calculated on the basis of the null hypothesis being true and the chi terms are summed to get the statistic. Minitab does not provide the *p*-value directly and it must be obtained by entering the chi-square value into **Calc>Probability Distributions>Chisquare** as the **Cumulative probability** with df = 1 and entering the value as a constant (chi-square and the probability appear at the foot of the table). The *p*-value for the test is given by subtracting this from unity.

Table 6.1   Association analysis of two nominal variables (Minitab).

Crosstabulation with chi-square		
Data		
Gender	**Low fat**	**Frequency**
1	1	35
1	2	15
2	1	45
2	2	5

Rows: gender (1 = male, 2 = female)
Columns: low-fat (1 = yes, 2 = no)

	1	2	All
1	35	15	50
2	45	5	50
All	80	20	100

Chi-square = 6.250, with df = 1
Cell contents–count
MTB >
MTB > CDF 6.25;
SUBC > Chi-square 1
6.2500   0.9876   0.0124

For this example, the association is significant (p-value $= 1 - 0.987581 = 0.0124$) and the variables are not independent. The propensity for low-fat foods depends on gender. Examination of the table shows that there is strong tendency for female consumers to be in favour of low-fat foods compared with males. Using the output from the analysis and the formula above, Cramér's $V =$ SQRT$(6.25/(100 * (2 - 1))) = 0.25$, so the effect, although significant, is not particularly marked based on these data.

The Minitab analysis does not include the '***Yates's continuity correction***' to compensate for the use of a continuous distribution for nominal data frequencies. Without it the analysis can overestimate chi-square. This procedure is recommended by Upton and Cook (2001) for '2 × 2' tests, although there is still controversy over its use. In the example above, redoing the analysis by software that does include Yates's correction leads to the same conclusion (e.g. SPSS gives p-value $= 0.024$).

Contingency tables and their analysis by Pearson's chi-square are very common in research reports. Øygard (2000) used this test to compare proportions of consumers ($n = 703$) in categories of gender, education, income and social position with food taste categories including 'healthy' and 'exotic', and Kolodinsky *et al.* (2003) compared level of concern on genetically modified foods by gender, age and nationality. Both studies found that gender had a significant association with the other variables.

6.3   Correlation

Relationships in the form of correlation require that the variables are above nominal in level of measurement and that they are continuous and normally distributed. A measure of the correlation is given by calculation of a correlation coefficient. For ordinal data or metric data where there are doubts about the quality of the measurement, ***Spearman's rho*** (ρ) coefficient is appropriate. For an interval or ratio data, where it can be shown or assumed that the data are normally distributed, ***Pearson's (product-moment) correlation coefficient (r)*** is the choice.

6.3.1   *Main features of correlation*

Correlation coefficients are calculated such that the numerical values lie between -1 and $+1$. A magnitude of 1 indicates maximum correlation, and 0 indicates minimum correlation. These relationships are best viewed in graphical form as when the variables are plotted on a ***scatter graph*** (Section 3.2.4). In idealised form, maximum correlations are obvious (Fig. 6.1). Zero correlation in shown by a cloud of points with no obvious trend (diagram (a)). In actual research graphs, the correlation may not be so clear. When the variables increase in a positive direction together, there is a positive correlation (diagram (b)). When one increases and the other decreases, a negative correlation occurs (diagram (c)). The coefficients can be calculated by calculators and most statistical software. There are alternatives

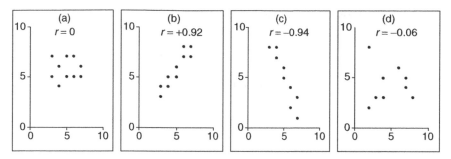

Fig. 6.1   Scatter diagrams of different levels of correlation (Excel).

such as *Kendall's coefficient of concordance*, which is a non-parametric method (available in MegaStat for Excel).

To achieve an adequate assessment of the effect in correlation studies, certain features must be present:

- Sufficient range of values
- Independent readings
- Linear relationship

If the relationship is not linear then the coefficient will not be valid. This may be indicated by a *curvilinear* appearance on the scatter diagram, although sampling variation can make this difficult. The data can be transformed to a linear format (typically, by taking the logarithm) or a curve-fitting procedure can be used. Hedonic measures can be curvilinear if measured over a wide enough range (see Fig. 6.1d), and this can mislead in correlation with sensory and instrumental data (Section 10.3).

The scatter graph is also useful for spotting *outlying points* as shown in Fig. 6.1d. These can give a false impression of the magnitude of the coefficient. Outliers should be considered for deletion on the basis that they may have been caused by a gross error (Section 3.3.4). Alternatively, using the non-parametric approach by ranking as with Spearman's coefficient can avoid the distortion that they cause.

The magnitude of the coefficient does not indicate significance (nor does it imply causation). Significance determination requires another calculation (see Table 6.2) along with the stepwise significance testing procedure (Chapter 4). Sample size (the number of pairs of readings) along with the effect size has a bearing on significance. As with all statistical significance, only the practitioner can decide whether or not such significance for a coefficient is *important*. Examination of effect size, significance level and sample size should be taken into account as well as the relevance to the particular topic under scrutiny.

An associated measure is the **coefficient of determination (R^2)** that is obtained by squaring the correlation coefficient (or directly from the data by the Excel function RSQ()). This gives an estimation of the variation in one variable explained by variation in the other (see the example below).

Table 6.2 Correlation analysis (Excel).

	A	B	C
1		**Correlation analysis**	
2			
3	**Sample data ($n = 6$)**		
4	X	Y	
5	Concentration	Panel	
6	0.01	6	
7	0.05	23	
8	0.25	34	
9	0.5	53	
10	0.75	79	
11	1	87	
12			
13	Result		Formula
14	Count (n)	6	$= \text{COUNT(A6:A11)}$
15	Correlation coefficient	0.98	$= \text{CORREL(A6:A11,B6:B11)}$
16	t	10.78	$= \text{ABS(B15)} * \text{SQRT}((n-2)/(1\text{-B15}^{\wedge 2}))$
17	p-value	0.0004	$= \text{T.DIST.2T}^{a}\text{(B16, }n\text{-2)}$
18	Coefficient of determination	96.67	$= \text{B15}^{\wedge 2} * 100$

[a]TDIST(B16,n-2,2) Excel 2003.

6.3.2 Correlation analysis

How to carry out a correlation analysis

Correlation analysis has several distinct stages:

1. Gather data (two measures on at least three sample units)
2. Draw scatter graph
3. Calculate the correlation coefficient
4. Calculate the significance of the coefficient

The sample size can be of great importance in deciding the importance of a correlation effect. Considerations of power and significance level apply if significance testing is included. Small samples (5–30) are possible in food laboratory studies where sensory and instrumental measures are being examined. For consumer studies, larger samples (100+) are necessary as relationships can be of lower magnitude and unit-to-unit (person) variability is often relatively high. As stated previously, each measure should be independent, but replication can be included to allow improved confidence in the estimate.

Ideally, measures should be on the same sampling unit, which is possible with consumer respondents. In the case of food materials where destructive measures are used or for hygienic reasons as in sensory tasting, convenience, etc., different subsamples may have to be used. This can introduce error causing the coefficient to be high as researched by MacFie and Hedderley (1993), who give a procedure to correct for this. In addition, for sensory versus instrumental studies the panel mean should be used.

Assumptions for significance testing are as given in Chapter 4 (Section 4.3) with the addition of a bivariate normal population. This is more difficult to assess, but approximate normality in the two measures is a compromise check. Otherwise a non-parametric method such as Spearman's coefficient can be used.

A typical application of correlation is in the establishing and utilisation of a relationship between sensory and instrumental data.

Example 6.3-1 *Correlation analysis (sensory and instrumental data)*
During training of a sensory panel, 'colour intensity' was measured on a series of samples (6) containing different concentrations of a red food dye. A graphic line scale was used and the readings were transformed to millimetres (0–100). Based on a correlation analysis what do the data reveal concerning any relationship and the performance of the panel?

The panel means were calculated and the data were entered into Excel as displayed in Table 6.2. The choice of X and Y is arbitrary in correlation analysis but requires consideration in regression (see below). For this example, the red dye concentration was taken as the X variate and the sensory intensity as the Y variate.

A scatter graph (see Example 3.2-4) was drawn that shows good linearity (Fig. 6.2). The coefficient was calculated using the CORREL() function, and it is significant using the formula as shown. For significance testing, a t-value is required based on the degrees of freedom of the test – obtained by the sample count (n) minus 2. The null hypothesis is that the coefficient $= 0$ and the alternative is that it is a non-zero value. The coefficient is high on the positive side and is significant even with only six points (objects) being examined.

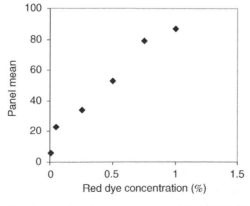

Fig. 6.2   Scatter plot of sensory versus instrumental data (Excel).

Based on the above analysis, there is evidence of a very strong relationship between the instrumental colour concentration and the panel mean colour intensity values. The significance test also shows significance and the null hypothesis is rejected – the coefficient is positive and highly significant. In this respect, the panel performed well and was accurate relative to the instrumental values.

The Pearson correlation coefficient calculation assumes that both measures are metric and that they come from a normal distribution. The coefficient of variation is obtained by squaring the correlation coefficient. For this example, 97% of the variation in X is explained by Y. Thus, Y accounts for almost all the variation in X.

Excel's **Data Analysis ToolPak** has a correlation tool, which can carry out the above analysis. It can incorporate more than two variables and an example in this form is given in Chapter 8 (Table 8.8).

Correlation analysis can be performed on ordinal data or a mixture of metric and ordinal. Spearman's coefficient of rank correlation is appropriate and it is available in MegaStat with correction for tied ranks. Minitab does not supply Spearman's test, but an alternative method is to analyse the ranks using Pearson's method as shown in Table 6.2. This is approximate, unless ties are present in which case Pearson's coefficient gives the correct value (Rees 2001). On this basis, an approximate p-value for Spearman's can be calculated using the above Excel formulae.

6.3.3   Correlation application

How are correlation studies applied?

Correlation studies are used widely in food research in examination of relations between instrumental, sensory and consumer measures of food and in nutritional studies comparing food intake with health and disease incidence. The above example has illustrated how the correlation results were used in checking panel performance.

The main point to be aware of is that it is incorrect to imply cause and effect automatically, when this may not be the case.

Correlation is also used as a starting point for the predictive process of *regression*, a very powerful technique. This method is a major application of both correlation and ANOVA and is dealt with in the next section. Other applications include assessment of reliability, which can establish whether or not two different methods of measurement correlate. This is a quick way of checking on that two data sets show similar changes, but it should not be confused with precision. Correlation can be high, but precision can be low.

The magnitude of correlation coefficients and their practical and statistical significance can be best appreciated by comparing results obtained in research. Rimal *et al.* (2001) gathered a range of consumer measures ($n = 236$) on food safety issues and examined correlation. Coefficients of greater than 0.4 were significant. This illustrates that increasing the sample size allows significance to be obtained with lower magnitudes for the coefficient. Thus, Unusan (2006), who studied a larger sample (684 students) for food preferences in childhood and in later life, found that coefficients of 0.3 were all considered important ('meaningfully large') because of the sample size, and magnitudes of $r = 0.33$ were found to be significant.

These relatively low coefficients contrast with those obtained by Zamora and Guirao (2004), who used correlation for comparing performance of a trained and an expert panel on wines ($n = 9$). Several sensory attributes were measured and coefficients of ±0.6–0.9 were produced, all significant. This demonstrates to an extent the contrasting nature of subjective consumer data with the more objective 'laboratory' data.

Another use of correlation coefficients lies with many of the measures given in Chapter 3 regarding the quality of survey instruments. An illustration of the use of correlation for assessing reliability of a consumer survey experiment is given in Eertmans *et al.* (2006), who reported on a satisfactory test–retest reliability for the food choice questionnaire over 2 and 3 weeks. Coefficients were relatively high (>0.7) with a sample of 358 respondents.

Such correlation studies, where large numbers of respondents are recruited, result in scatter graphs with many more points than the previous examples in this chapter (see Section 8.4.3 for an example of a scatter graph with 'many' points, and more use of correlation coefficients with survey data).

6.4 Regression

Regression is a natural extension of the results of a correlation analysis. The scatter diagram displays points that were determined in the investigation, but intermediate values can be judged. Thus, the possibility of *prediction* of one variable via the other exists.

6.4.1 *Main features of regression*

Regression analysis calculates a mathematical expression of this predictive ability and allows an accurate plot of a line on the graph. In regression, a clear distinction must be made on which variable is predicting the other. The **predictor variable** (or *explanatory variable*) is designated as the **independent variable** and the other as the **dependent** or response variable. In **simple linear regression**, there is one predictor and one response variable. More predictors are possible in *multiple regression* (Section 12.3.1).

Mathematically, regression is expressed as one variable being a function of the other. This gives a **regression equation** of the form $Y = bX + a$. In regression terminology, X is the predictor or explanatory variable and Y is the dependent variable. There are two *coefficients* in the equation: a is the **intercept** (where the line cuts the Y-axis) and b represents the **slope** of the line. These coefficients are determined during the regression analysis along with the plotting of the **best-fit line** using the method of **least squares**. This ensures that the experimental points are at a minimum distance from the line.

6.4.2   *Regression analysis*

Regression analysis follows similar stages to correlation analysis as above, plus a regression stage. Excel provides a full regression analysis. Using the example above, which has demonstrated evidence of strong correlation, a regression analysis was performed.

Allocation of X and Y demands some consideration. In this case, it is assumed that the concentration of the dye is causing the panel perception of intensity to increase. Logically, this is reasonable on the basis of chemical content on colour intensity. In other cases, it may not be so obvious and deeper reasoning may be required. Ideally, the X variable should be the one with least error (the regression model assumes this) and its values are fixed as accurately as possible for the experiment. Cohen (1988) points out that the horizontal axis (abscissa X) has less error on horizontal positioning of points in the visual display compared with the vertical (ordinate Y). Thus, the horizontal axis (X) is always the independent variable.

The data are in two columns (Table 6.2) and the regression option in the **Data Analysis ToolPak** menu provides the output (Table 6.3).

An overall measure of the strength of the regression is given by **R square**. The closer this is to 1, the stronger the effect (adjusted R^2 recalculates for the circumstances of the experiment so it can be compared with others). The coefficients are shown along with their individual significance. Thus, a (the intercept on the Y-axis) is 13.24 and b (the slope or gradient of the line) is 79.13, a large value indicating that the line is steep. As X (the concentration) changes, the panel mean increases rapidly and both these components are significant ($p < 0.05$). The ANOVA section of the table analyses the overall significance of the regression and

Table 6.3   Regression analysis (Excel).

Summary output

Regression statistics

Multiple R	0.983
R square	0.967
Adjusted R square	0.958
Standard error	6.503
Observations	6

ANOVA

	df	SS	MS	F	Significance F
Regression	1	4916.87	4916.87	116.28	0.0004
Residual	4	169.13	42.28		
Total	5	5086.00			

	Coefficients	Standard error	t-Stat	p	Lower 95%	Upper 95%
Intercept	13.24	4.10	3.23	0.03	1.84	24.64
X variable 1	79.13	7.34	10.78	0.00	58.75	99.50

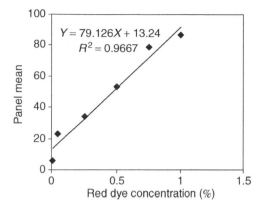

Fig. 6.3   Regression line and equation (Excel).

this is also significant. The null hypotheses for these tests are that the t coefficients are zero in effect.

Excel will also fit a line to the data and this is the line of best fit (determined mathematically). Right-click on one point of the graph in Fig. 6.2 and this will initiate a sub-menu. Select **Add Trendline** from this menu and fit a linear line, then in **Options** select to have the equation and R^2 on the graph (also accessed in Excel 2010 by clicking on the chart then, in **Chart Tools**, select **Layout/Trendline**; in Excel 2003 click chart then **Chart/Add Trendline**). The scatter graph should appear with the line fitted, and the equation is shown in Fig. 6.3.

The equation allows sensory measures to be predicted for any given concen tration of the dye. Predicted values can be obtained for all the experimental concentrations plus unknown ones. Thus, for a dye concentration of 0.1%:

Sensory colour intensity $= 79.126 \times 0.1 + 13.24 = 7.9126 + 13.24 = 21$ (result rounded to nearest integer).

This procedure is known as ***interpolation*** where predictions are made within the range of X values covered by the experiment. Taking prediction outwith this range is ***extrapolation***, which would assume that the relationship existed beyond the limits of the experiment. This can lead to invalid predictions especially with concentration effects on sensory perception. The above relationship was linear within the range of concentrations used, but according to psychophysical laws such as Fechner's law (Meilgaard *et al.* 1991), over wider ranges a non-linear effect occurs. Typically, as the stimulus increase, the response will level off, i.e. it will become non-linear and prediction at such levels would be incorrect.

6.4.3   *Regression assumptions*

Regression analysis has similar assumptions to those for correlation, but in addi-tion, no errors are assumed in the X values. Also, variability around each point is

Table 6.4   Calculation of residuals (Excel).

	Residual output	
Observation	Predicted Y	Residuals
1	14.03	−8.03
2	17.20	5.80
3	33.02	0.98
4	52.80	0.20
5	72.58	6.42
6	92.37	−5.37
		Sum = 0.00

assumed to be normal and to centre on zero. Some of these issues are covered in more detail in Section 9.6.2. Accepting that in the example above, the concentrations of the dye were made up very accurately, then the scatter around each point can be examined by calculation of ***residual values*** that are the 'left over' variation between the experimental values and the predicted ones:

Residual = Observed value − predicted value

Excel has options in the **Regression** analysis menu for calculation of residuals as above and provision of a *residual plot*.

The predicted values have been subtracted from the observed ones (Table 6.4) to obtain the residuals. For the 'perfect fit' regression equation, the residuals would be zero, but because of error they vary in magnitude from 0.2 to 8.03. One quick check is to see if they sum to zero – this has been done at the bottom of the table and this is so, providing some evidence for the spread centring on zero.

The ***residual plot*** (Fig. 6.4) shows the scatter more effectively. For compliance with assumptions, there should be a random pattern, but equal magnitudes on either side of the zero line. While not perfect, this is roughly so for the graph and it can be taken that assumptions are met.

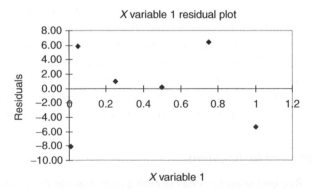

Fig. 6.4   Residual plot for regression analysis (Excel).

6.4.4 Regression application

How are regression studies applied?

One main application of regression is in the facility of replacing lengthy sensory assessment with a more rapid instrumental method. High linearity must be demonstrated along with good predictive power from instrument to sensory quality.

Ultimately, this can mean that routine chemical or physical analysis can predict sensory properties of varying raw material intakes, so these can be rapidly incorporated into production. One difficulty here is that the instrumental measures may not be a valid substitute (Piggot 1998). For instance, assuming that an instrumental texture measure of 'stickiness' is equivalent to the sensory sensation may be false. Other uses include calibration aids for sensory training and prediction of consumer preference via sensory and instrumental measures.

Regression studies abound in the literature, but there are fewer of the simple regression type. Homer *et al.* (2000) found high correlation and a linear relationship between raw and cooked fat content in part of a study on the acceptability of low-fat sausages. The use of low-fat and conventional fat content products enabled a wider range of fat values to be examined. The regression equation was able to predict level of fat loss on cooking. Other applications include prediction of the level of a food constituent required to achieve a particular sensory or acceptance target. Bower and Boyd (2003) used a simple linear regression model to predict the concentration of sucrose to produce 'ideal' and 'most-liked' sweetness. A more advanced application involves multiple regression (see Section 12.3.1) where more independent variables are studied. One example in consumer research is that of Rimal *et al.* (2001), who collected a range of consumer measures ($n = 236$) on food safety issues and the effects on consumption habits. Regression equations were formulated on the basis of demographic and attitudinal measures. Of the many variables studied, consumer's view of the importance of food labelling in concern over food safety produced the largest coefficient, i.e. it had the largest influence.

References

Bower, J. A. and Boyd, R. (2003) Effect of health concern measures and consumption patterns on measures of sweetness by hedonic and just-about-right scales. *Journal of Sensory Studies*, **18**(3), 235–248.

Carlberg, C. (2011) *Statistical Analysis: Microsoft® Excel 2010*. Pearson Education, QUE Publishing, Indianapolis, IN, pp. 133–139.

Cohen, S. S. (1988) *Practical Statistics*. Edward Arnold, London, pp. 62–85.

Eertmans, A., Victoir, A., Notelaers, G., Vansant, G. and Van Den Berg, O. (2006) The Food Choice Questionnaire: factorial invariant over western urban populations. *Food Quality and Preference*, **17**, 344–352.

Homer, D. B., Matthews, K. R. and Warkup, C. C. (2000) The acceptability of fat free sausages. *Nutrition and Food Science*, **30**(2), 67–72.

Kolodinsky, J., DeSisto, T. P. and Labrecque, J. (2003) Understanding the factors related to concerns over genetically engineered foods products: are national differences real? *International Journal of Consumer Studies*, **27**(4), 266–276.

MacFie, H. J. H. and Hedderley, D. (1993) Current practice in relating sensory perception to instrumental measurements. *Food Quality and Preference*, **4**(1), 41–49.

Meilgaard, M., Civille, G. V. and Carr, B. T. (1991) *Sensory Evaluation Techniques*, 2nd edn. CRC Press, Boca Raton, FL, p. 48.

Middleton, R. M. (2004) *Data Analysis using Microsoft Excel*. Thomson Brooks/Cole Learning, Belmont, CA, pp. 49–60.

Øygard, L. (2000) Studying food tastes among young adults using Bourdieu's theory. *Journal of Consumer Studies and Home economics*, **24**(3), 160–169.

Piggot, J. (1998) Relating sensory and instrumental data: why is it still a problem? Abstract – 4th Sensometrics Meeting, Copenhagen, 1998.

Rees, D. G. (2001) *Essential Statistics*, 4th edn. Chapman & Hall, London, p. 211.

Rimal, A., Fletcher, S. M., McWatters, K. H., Misra, S. K. and Deodhar, S. (2001) Perception of food safety and changes in food consumption habits: a consumer analysis. *International Journal of Consumer Studies*, **25**(1), 43–52.

Unusan, N. (2006) University students – food preference and practice now and during childhood. *Food Quality and Preference*, **17**, 362–368.

Upton, G. and Cook, I. (2001) *Introductory Statistics*, 2nd edn. Oxford University Press, New York, p. 381.

Upton, G. and Cook, I. (2002) *A Dictionary of Statistics*. Oxford University Press, New York, p. 89.

Zamora, M. C. and Guirao, M. (2004) Performance comparison between trained assessors and wine experts using specific sensory attributes. *Journal of Sensory Studies*, **19**, 530–545.

Chapter 7
Experimental design

7.1 Introduction

All scientific investigations involve experiments. To ensure that results are valid and trustworthy, the organisation, planning and procedures of the experiment must be of adequate quality. Design processes bring together all such aspects including specific statistical analysis techniques. The ultimate purpose is to maximise the precision (preciseness) of the experiment, i.e. the ability of the design to discriminate between effects. To achieve this, effects must be quantified efficiently. Additional benefits also accrue in longer-term use of design. Reduction of experimental time and cost is possible, as design can accelerate the whole process. Generally, all stages are optimised giving increased efficiency and improved transfer of research to practical applications. The experimenter should also take a 'step back' and examine the intended work from all points of view in terms of its feasibility, robustness and cost as described by MacFie (1986). Design features have been touched on in previous chapters and these will now be formalised. Most designs given herein involve testing by analysis of variance (ANOVA). This chapter gives a descriptive account and the application chapters that follow illustrate selected design types in detail.

7.2 Terminology and general procedure

Several key terms are used in describing experimental design (Table 7.1).

7.2.1 Experiments, studies and investigations

Several *experiments* have been described throughout earlier chapters. Typically, they are studies of the effect of variation on a product, process or consumer subject to obtain new facts or to confirm or deny a hypothesis. The types of

Statistical Methods for Food Science: Introductory Procedures for the Food Practitioner, Second Edition. John A. Bower.

Table 7.1   Terminology in experimental design.

Term	Meaning
Experiment	A planned inquiry or a controlled investigation
Variable	Any characteristic of the experiment that can vary (known or unknown; contributing; under scrutiny)
Sampling unit	Individual units sampled from the population
Experimental unit	Sample unit(s) that are allocated in the experiment to receive treatment
Factor	Independent variable(s) which are under study
Response(s)	Dependent variable(s) that are under study
Level(s)	The 'settings' for individual factors
Control(s) and base line(s)	Reference points
Treatment	The application of the combined 'settings' of the factor variables
Effect	The outcome of the treatment on the response
Main effect	The outcome of a single factor effect
Interaction	The outcome of a combined factor effect

experiment can vary from exploratory studies, where data are gathered to gain facts, such as characterising products or balloting consumer views on foods. Other investigations come in the form of comparative and confirmatory experiments or as relationship studies, where the focus is on specified factors and effects. An account and explanation of some pertinent features of the above terms follows below.

7.2.2   *Experimental units and sampling units*

Experimental units are the individual amounts or lots of material or numbers of objects (or subjects) to which the treatments are applied and which are assessed at the termination of treatments. They may be represented by one *sampling unit* (Section 2.3.1) or more depending on the design (Bender *et al.* 1988). For instance, in the case of consumer survey trials, the experimental unit and the sampling unit are often one and the same, i.e. the consumer respondent or subject, but this is not always applicable. For an experiment on the effect of label information on preference, one experimental unit could be the control treatment with no label consisting of 50 subjects and a second experimental unit could be the test treatment (another 50 subjects). Treatments are applied to the experimental units. These need to be identified in terms of the number of individual amounts or items sufficient to cover the requirements of the experiment.

7.2.3   *Variables, factors, levels and treatments*

Variables are identified and set up as independent and dependent. An *independent variable* is a *factor*, which is set at specific (fixed) or randomly selected *levels*. The factor levels constitute the specific *treatments* of the experiment, i.e. 'what the experimental units are subjected to'. A treatment description will specify and define the conditions of the experimental factors and levels; thus, one treatment could be factor A at level 1 combined with factor B at level 2.

A *dependent variable* is a variable that may or may not change according to the effect of the treatments. It is described as a *response* – the term used to describe the way(s) in which the experimental units react to the application of the treatment conditions. One or more responses are measured to assess the effect of the treatments.

Experimental factors can be of two types:

Quantitative – different **numerical** levels
Qualitative – different **identity** levels

Hence, with a *quantitative factor*, the factor is in the form of a variable that can be set at different numerical levels, e.g. the quantity of an ingredient, or process temperature. This contrasts with a **qualitative factor** which is nominal or categorical in nature and which is set at different category or identity levels, e.g. 'flavour additive' with three different types of flavour. Note that the quantity may not be the same.

In food study experiments, many factors can be used depending on the focus of the experiment. For example, in consumer testing, consumer age may be a factor of interest; in sensory panel training the panel and the panellists as factors may be under scrutiny; for inter-laboratory trials, the laboratories are the factor. Each factor is an independent variable, as it is controlled by the experimenter.

The *levels* are the settings that are assigned to the factors – this can be done on a fixed or random basis. Each factor plus level combination constitutes a *treatment* condition. The more factors and levels, the greater the number of treatments. If replication is included, this will increase the number of treatments again. The number of factors starts at one and there is theoretically no upper limit. At least two levels are required or a single level with a reference value. As factors and levels are added, the experiment becomes larger and the organisation and interpretation become more complex. *Blocking* (see below) may be required to spread the work – this should be done so that variation is minimised.

Beyond these guidelines, there is no restriction on types and numbers of factors and levels; thus, ingredients, process types and conditions, etc., can all be included within the one experiment. In addition, the types of responses can include sensory, consumer and instrumental in more advanced experiments, e.g. Baardseth *et al.* (1992), who included a number of qualitative and quantitative factors at 2–5 levels and included 15 sensory plus 2 instrumental response measures.

7.2.4   *Controls and base lines*

A fundamental characteristic of a 'good' design could be that it includes a control measure. This ensures that there is a reference point ('baseline') from which to gauge the magnitude of the response(s). Thus, control treatments provide a bench-mark to compare treatment effects more validly. This can be achieved by providing experimental units that are of *known nature* and characteristics (chemical, physical, psychological, behavioural, etc.).

7.2.5   *Responses and effects*

The response(s) are the measures that are taken after the experimental units have been treated. Thus, they depend on 'what happens' and they are the **dependent** variable(s). Responses can be one or more measures of instrumental, sensory, etc.

Main effects are due to the effect of the *individual factors* on the response(s). **Interactive effects** are caused by *two or more factors* interacting in some way to produce an effect specific to each factor involved; hence, a significant 'two-way' interaction involves two factors acting together to modify the response in a way that is different from the effect of each single factor. The importance of interactions can outweigh the individual main effects.

Effect size

At this point, it is pertinent to consider the **effect size**, a concept raised in Chapter 4. The question will arise: what is an important effect size? Essentially, this comes down to the ultimate consequences of any difference or relationship that is found. Thus, if there is an 'appreciable' improvement or lack of improvement in a response such as the percentage yield from a process, or the number of sales for a food product, then the treatment that produces such an effect can be adopted or rejected, respectively. Assigning quantification to 'appreciable' requires input from the subject specialist(s). In some food-related subjects, there are clear guidelines, e.g. in dietary studies an important effect is a weight change of 10%. Some forms of chemical analysis for statutory purposes ascertain whether a compound such as an additive is within the legal maximum or not. Effects that exceed this even by small amounts are technically infringing the requirement, e.g. an additive must be present at no more than 200 mg/kg, but on analysis samples values average out at 205 mg/kg, after adjustment for uncertainty. With sensory measures, the decisions of important effects sizes can be based on significance of results such as in discrimination tests. Changes in product formulation can be monitored here, but changing production settings based on such tests should take into consideration power levels (Section 8.5.4). With hedonic measures, appreciable effects can be judged by the overall rating or score achieved, e.g. Pyne (2000) quotes a score of 6.0 or over on the nine-point scale as one qualification for a successful product. Another way to view such data was taken by Baird *et al.* (1988), who studied hedonic scores for fish products. They calculated mean values to gauge overall position and all products were liked, but they also calculated the percentage of consumers who scored on the like side of the scale. A cut-off point of '70% of consumers liking' was identified as an indicator for a viable market product.

Identification and specification of many of the above features are part of the design process. Additional details are also needed, such as methods of sample preparation, treatment application and the selection of the statistical design being applied for set-up and eventual analysis of effects. Familiarity with these terms is essential to enable a procedural plan to be drawn up.

7.2.6 Stages in the design procedure

Clear objectives must be set at each stage of the overall design plan, with formulating a rationale, selecting an appropriate design and drawing up detailed specifications. In the rationale, the experimenter must:

- Decide on the questions for which the experiments will provide answers
- Decide on the aims and objectives
- Formulate the specific hypotheses to be tested
- Consider the cause and effect structure being applied

One of the objectives of many experiments is to establish 'cause and effect' to enable greater understanding of reactions and changes. It requires identification of the independent variable and the dependent variable. Causality should be argued logically with consideration of extraneous variation, and the need for association should be established first, etc.

Harper and Wanninger (1969) introduce the concept of cause and effect in food experiments. They describe the procedure as one of understanding, and eventually quantifying, the ways in which causes (input variables in the form of ingredients and process parameters) result in effects (the output in form of sensory, instrumental, responses, etc.). Causal analysis requires experimental control so that the validity of any modelling of reactions can survive scrutiny, etc.

For design selection, the nature of the experiment, number of groups and methods of analysis must be taken into account. Statistical knowledge is required from the beginning of this process. Specifications will decide on the scope and extent of the experiment(s) (preliminary to critical) and will detail all aspects of methodology: statistical, experimental units, factor treatments and levels, controls, assessment techniques (sensory, chemical, instrumental, etc.).

7.3 Sources of experimental error and its reduction

Many error sources can act against the validity of an experiment. Some of these were described in Section 3.4. Other possibilities can occur and relate to inadequacy or omission at some stage of the whole process (Box 7.1). Experimental error refers specifically to variation occurring between experimental units that have received the same treatment. Ideally, these should be similar before treatment, but a number of influences act against this.

> **Box 7.1   Source of experimental error.**
>
> Inherent and other sources of variation between experimental units within treatments
> Lack of uniformity during conduct of experiments
> Lack of calibration of processes and assessment techniques
> Inadequate statistical design and planning

Foods are recognised as being variable in chemical composition and physical structure to various degrees. This is ***inherent variation*** and it means that each experimental unit will differ even before treatments are applied. It cannot be removed, but it can be reduced by size reduction of the food material if the product nature or experimental circumstances permit. In addition, there may be other sources of variation between experimental units when large amounts of material are required. Different sources, different storage history, etc., will have inherent variation plus other possible effects. The *relative amount of information* gained in experimentation is a function of these sources: as the degree of variation increases, the information decreases as the estimates are 'swamped' by the large variation (Steel and Torrie 1981).

Conduct of experiments must be as constant as possible over the time of the experiments. Lack of calibration of processes and assessment techniques and inadequate or absent measures of accuracy and precision, all contribute to weakness in an experiment. It is not always possible to ensure that these sources are covered.

A lack of an appreciation of statistical aspects of design can serious undermine experimental results; this applies from sampling through to the end analysis. Inadequate control of variables during the experiments and response assessment stages give a methodology that is weak and does not appreciate all sources of variation – all the above factors can contribute to *error*, which if not recognised and quantified can invalidate an experiment.

7.3.1   Ways of reducing error

Statistical appreciation

Errors due to inadequacy of design planning and appreciation of variability, etc., can be avoided by consultation of statistical texts in selection of appropriate designs. This may not provide full answers, and the experience of the scientist will be called upon in decisions regarding the number of factors and levels, and the settings of the levels (the treatments). The whole experiment can be undermined by treatment levels that are outwith the regions which matter. For example, choosing levels of a process factor that results in products that are non-viable or un-assessable by the response measurement technique employed. Preliminary work and screening designs can guide selection here.

Variation between experimental units

Replication

A single application of each treatment gives no estimate of error; depending on the nature of the experiment, observed differences between treatments could be due to inherent variation. Replication (two or more) gives *quantification of error variation*, some of which will be due to inherent and other differences in the experimental units. The experiment will have more precision, confidence intervals will have decreased width and statistical tests will have more power. As the number of replicates increases, so does the amount of information gained for a constant level of variability (Steel and Torrie 1981).

In more complex designs, where the experimenter has the experience and guidance to provide an assurance that certain assumptions hold, specific replication of individual treatments can be omitted, thereby reducing the work involved. Hence, in certain designs with no replication, an estimate of error is still provided, but this is distinct from *'pure error'* which only replication provides. The number of replicates required depends on the degree of inherent variability present, as assessed by prior knowledge or determined by preliminary work, and the degree of precision required (Section 3.4).

The relative gain in experimental precision falls as the number of replicates rises, i.e. there is more of a gain by increasing from two to four replicates than is obtained by increasing from four to eight.

NB: Replicate treatments differ from replicate end determinations on one treatment, i.e. the whole treatment must be repeated using another experimental unit.

Manipulation and dilution

If data are available on any other unwanted source of variation then **analysis of covariance** can be performed to take this into account, e.g. material with known differences in storage time.

While Excel and basic Minitab can calculate covariance for two variables, analysis of the style just mentioned (available in more advanced packages and SPSS) is not so easily attained, although Carlberg (2011) describes its implementation with Excel.

When no specific data are available, it is possible to 'manipulate' inherent variation, i.e. utilise any variable characteristic within the experimental material to contribute as little as possible to treatment differences. Thus, with food materials, any obvious variation should be equally represented in each experimental unit, e.g. size distribution. It may also be possible to subdivide items and allocate fractions to the treatment lots – ideal where there is a compositional or quality variation within the structure of the food items (Fig. 7.1).

Other aspects may also be improved in this way, such as pairing and 'matching' of samples in order to eliminate unwanted variation. Division of food material that occurs in large masses such as root vegetables and meat cuts, etc., can be done to suit the experimental requirements. The resulting pieces now originate from a

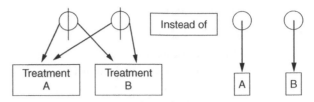

Fig. 7.1   Subdivision of units to reduce variation within treatments.

common source and variation in chemical and physical features between separate original material will be distributed throughout the treatments.

With consumer surveys, respondents can be allocated to treatment groups so that variation is evened out – characteristics such as gender, age, socioeconomic, nationality, region, etc. It is important that beyond these requirements, respondents be *randomly allocated* to treatments.

Procedural techniques

Control of error resulting from procedures can be achieved by *uniformity* of the tasks in all details. While this is desirable in one sense, an excessively systematic approach to the *order* of tasks can introduce error and *randomisation* is universally applied. Despite these solutions, the size of the experiment may itself cause unwanted variation, necessitating additional design techniques.

Blocking

In many instances, it is not possible to do all treatments at the one time and an experiment has to be divided into a series of separate stages. This introduces another possible source of uncontrolled variation and error. The technique of **blocking** involves organisation of experimental stages such that *all treatments occur within each stage*, e.g. four treatments (A–D) with three replicates:

Unblocked design:

Stage 1 – Do A on three experimental units
Stage 2 – Do B on ”   ”   ”
Stage 3 – Do C on ”   ”   ” etc.
Stage 4 – Do D on ”   ”   ” etc.

Blocked design:

Stage 1 – Do A, B, C, and D on one experimental unit each
Stage 2 – Do ”   ”   ”   ”   ”   ”
Stage 3 – Do ”   ”   ”   ”   ”   ”

The consequences of blocking are that any circumstance or event, which causes error or variation within a stage, will now *affect all treatments*. It is possible to have replicate blocks and there are a number of block designs such as complete block, split block and incomplete block (ICB), etc. (see below). In sensory evaluation, blocking can have special significance if the assessors are allocated as the blocks. Any variation introduced by the panel can now be quantified and removed if desired from subsequent statistical analysis.

Randomisation

This has been proven to reduce error and bias that is caused by taking a systematic, non-random approach to organisation of experimental stages. The technique involves *randomisation of the order* in which stages and individual treatments are performed. For example, a completely randomised design (CRD; no blocks) of five treatments and four replicates would have 20 random numbers in the range 1–20 assigned to individual treatment replicates; these numbers would specify the order in which the treatment replicates were done – instead of doing treatment 1, replicate 1 first, then treatment 1, replicate 2 second, etc. In sensory evaluation, samples are presented to each judge in a random order to avoid bias and '*the order effect*' (see later for more on the specific problems for sensory evaluation).

Coding and blinding

If the designer, co-worker and participants (consumers, if included) in the experiment are aware of all the details, they may be *biased* in their actions and decisions. As far as possible, the exposure of knowledge of treatment details should be minimised. Techniques to achieve this include concealing treatment and replicate identity by labelling with **random codes** and using **blind** or **double-blind** treatment allocation methods. 'Blinding' refers to a design feature where the participants are unaware which treatment is which (due to coding or concealment of identity). 'Double-blinding' is where both the participants and the organiser of the experiment are unaware of the exact identity of the treatments being allocated. In this case, a third party must code treatments.

Use of controls or base lines

Without use of a reference point, conclusions from an experiment, no matter how well designed in other respects, could be invalid. Thus, controls should be included whenever possible. Examples could be a control group of consumers who received no information before assessing foods, or addition of flavourless powder in a formulation experiment. This use of units with a 'zero' setting for the treatment (null-effect or 'dummy' treatments) is similar to the **placebo effect** used in drug-testing trials. They provide the **baseline** for measuring treatment effects, which can be of critical importance in some applications such as cosmetic testing.

Other possibilities are units that contain a presence of the response being measured. For instance, a raw sample of food for comparison with processed, or reference standards of known composition in chemical analysis. In the latter, the standards would constituent a 'positive control' (as distinct from those with 'zero' settings sometimes referred to as 'negative controls'). Responses are taken for the control units, but their identity must be concealed by coding and they must be dealt with as for all other experimental units.

Measurement error

Efficient design can be limited by inadequate control and assessment of the response measures. There should be measures in place or previous data available, regarding accuracy and precision of the techniques employed. Refinement of the methods can be used to improve any weaknesses, e.g. more panel training in sensory experiments.

Pilot studies in surveys are invaluable in this respect. Even with these steps, response measures in isolation could lead to faulty interpretation and some form of measurement quality checks should be performed.

Suitability of the response measurement

The effectiveness of experiments in achieving their target depends also on the 'suitability' of the response(s), i.e. does the response truly reflect the nature of the feature under study? Are the response data coming from a population that has a normal distribution and is the response linear? With instrumental or objective measures, there may be little doubt. In such circumstances, the response variable is at ratio level and would be assumed to be normally distributed. More problems arise when using sensory evaluation tests for optimisation of 'sensory quality', as such techniques may assume the above conditions.

7.4   Types of design

NB: It is assumed that the responses to be measured in these designs are at least interval in nature. Designs for non-parametric responses are indicated where available, but generally they are less accessible than their parametric counterparts. Some other designs are detailed in the application chapters.

There are a great number of possible designs available for the food practitioner. These range from simple one-factor designs, to more complex forms, which have evolved to allow more precision and to overcome limitations of the simpler types. Readers requiring more details should consult texts such as Cochran and Cox (1957) and Box *et al.* (1978).

The simplest type of design is the *single factor* case (Table 7.2). Such an experiment can be analysed by one-way analysis of variance (ANOVA). This can be done for two or more levels. These are all classed as *one-factor designs* or 'one variable at a time' (OVAT) designs (sometimes OFAT – one factor at a time).

Table 7.2   Types of experimental design.

Design	Number of factors	Number of levels	Analysis method	Example
One factor				
(i) Two groups	One	Two or more	*t*-Test; Mann–Whitney; binomial; chi-square	Sections 5.3–5.5
Completely randomised design (CRD) more than two groups	One	Two or more	One-way ANOVA; Kruskall–Wallis	Sections 9.5.2 and 5.4
(ii) Paired	One pair		*t*-Test; Wilcoxon	Sections 5.4 and 5.5
Randomised complete block (RCB)	One	Two or more	Two-way ANOVA; repeated measures ANOVA; Friedman's	Section 8.5.5
Two or more factors				
(i) Factorial	Two or more	Two or more	Two-way/three-way ANOVA, etc.	Section 10.4
(ii) Fractional factorials	Two or more	Two	Special ANOVA two-way/three-way, etc.	Section 10.5
(iii) Latin square	Two or more	Two or more	Three-way ANOVA	—
(iv) Split plot	Two or more	Two or more	Special	
(v) Incomplete block	Two or more	Two or more	Special	Section 8.4.2
Changeover	Two	Two	Special	—
Optimisation	Two or more	Two or more	Special	—

They have the advantages of simplicity of operation and of interpretation, e.g. one factor at three levels – compare the three levels – an OVAT design, i.e. only *one* factor is varied; all other controllable variables are kept constant – simple design features, relatively easy to control. The disadvantages of OVAT designs are that the procedure needs to be repeated separately for any additional factors and because such designs cannot detect *interaction* of factors – hence, they may 'miss' optimum levels.

7.4.1   One-variable designs

Despite the disadvantages, OVAT designs appear regularly in research publications. They are used for simple experiments and for later stages of experimentation (Meilgaard *et al.* 1991). In all cases, one factor at two or more levels is examined. A number of forms are common.

Independent groups

Two-group designs (independent)
Terms appropriate for these designs were described in Section 4.5. They are simple designs appropriate for comparison of two groups. One factor at two levels (i.e. two treatments). The experimental units are randomly allocated to the two treatments. Analysis employs the ***independent t-test*** or the ***Mann–Whitney U test***.

Completely randomised design (more than two groups)

This is a three or more independent samples design equivalent of the two-group design above. The effect of one factor is examined, but three or more levels can be compared. Analysis in this case uses *one-way ANOVA* or the *Kruskall–Wallis Test*.

With these independent group designs, it is important that treatments lots are prepared such that there is similar variability within the units. Food materials have inherent variability especially for solid raw materials and mixing, milling, etc., at the sampling stages can reduce this. In some circumstances, this may be sufficient and the initial variation in instrumentally measured responses will be acceptable. Where the units are assessors in sensory panels and respondents in consumer trials then the variation is likely to be larger. The CRD is possible in such cases if circumstances demand that the same assessor or consumer cannot handle all treatments. Gacula (1993) gives some guidance on application, with requirements for uniformity between separate panels and increasing the sample size to 100 or more for consumers. The CRD can be used for two or more factors, but this demands more treatments.

Despite these possibilities, it may not be possible to reduce the variation sufficiently for the required number of units. Thus, independent group designs suffer in that differences between experimental units can be present before the effect of the treatments. For instance, four sets of meat samples are processed by four process settings and then analysed for texture tenderness by a shear device. Initially, the four sets could have been quite different in texture and thus any differences after processing would include this initial variation. It is possible to determine a control unit measure of the texture by testing a raw sample, but as the test is destructive, it is not possible to assess the raw texture of the original experimental units without destroying them. These designs do have advantages of simplicity of ease of use and they do not require equal numbers in each treatment.

A good example of the use of a CRD design is provided by Mumba *et al.* (2004) in a study on the effect of soaking time on trypsin inhibitor activity in soya beans. There were five soaking times (the factor) plus a positive control (raw) and a negative control (heat-treated) for the assessment of activity. Weighed lots of soya beans were randomly allocated to the seven treatments and it was found that activity decreased significantly with soaking time.

Related (dependent) groups

The variability between experimental units is reduced by *pairing* or *matching* when related designs are used. Formation of related groups is described in Section 4.5. Treatment effects are now not clouded to the same extent by the initial variation.

Certain sources of unwanted variation are evened out and quantified in some cases by application across all units. Thus, a process is applied to three treatment units using the same machine or a sensory test is performed by the same assessors for all units. Other possibilities for blocking are analytical methods, laboratories, technicians, sessions, days and replicate sets, etc.

One of the key features of this type of design is the order of application of treatments for the single related set. Thus, if machine X processes food products under conditions a, b and c – which order is used? If order 'a–b–c' is followed then 'a' may be biased because it is first, etc. Similarly, when comparing analytical methods or assessing sensory properties, the order of procedures is often critical. The procedure of *randomisation* and *balancing* of order is the design method to overcome such bias (Section 8.3). Often the related sets are referred to as *blocks* (see above) that are a second factor. As explained by Bower (1997), there are a number of different terms used to describe related designs depending on the application. All come under the heading of *two-factor designs*, but the second factor is usually not of direct interest and the term *one-factor repeated measures* is more appropriate. Other terms are possible for sensory designs of this type (see Chapter 8).

Paired designs (related)
Pairs of experimental unit are treated and measured. The difference between the response measures is analysed using the *paired t-test* or the *Wilcoxon signed pairs test*.

Randomised complete block (more than two related groups)
Randomised complete block (RCB) designs are analysed as *two-way ANOVA*. In the simplest form, the block-by-treatment interaction would not be assessed and replication is not required. If interaction is within the data, it cannot be quantified due to lack of degrees of freedom (no replication). This interaction will be contained in the ANOVA error term resulting in loss of precision in assessing the treatment effect. For RCB use where replication is difficult, e.g. in consumer tests, a non-parametric test is recommended (O'Mahony 1986).

The RCB is a very common and popular design because of its improved precision. It is the equivalent of the paired design for three or more treatments. The limitations of the CRD are overcome by inclusion and assessment of the *block effect*. Variation inherent to the units is assessed and removed. The blocking method can be applied to any factor that is identified as causing unwanted variation. The block design is therefore a type of two-way ANOVA. The second factor (block) variation is calculated and removed from the residual error, thus improving the focus on the main factor of interest. Lee and Inglett (2006) used an RCB design in examination of physical properties of cookies with varying levels of oat bran replacing the shortening. The main factor was the level of bran and the blocks were replicate sample lots. ANOVA followed by Duncan's multiple range tests revealed significant differences in dough spread, water activity and colour due to bran level.

Repeated measures design
This term is sometimes used to describe design conditions where treatments are judged by the same assessor on the same samples in sensory experiments during a single session. In this circumstance, the repetition refers to the fact that the same assessor is testing repeat samples within a short space of time. Extending this concept, *repeated measures* applies to any measurement repeated over time, such as in storage studies and long-term use of products. Essentially, this is an RCB-type, which can be analysed by two-way ANOVA as described above. This analysis is simpler, but it ignores possible effects such as correlation between

the responses, and for this inclusion *repeated measures ANOVA* should be used instead (Schlich 1993). If different groups of assessors are used then a split plot type of analysis is more appropriate (see below).

Latin square

This design extends the block effect principle and includes two blocks of unwanted variation. It can be analysed as a three-way ANOVA. While giving more precision, it does require that treatment numbers equal block size. Order of application can be balanced out using Latin squares. For instance, with three treatments where one block factor is 'assessor' and a second is 'order of assessment':

Assessor	Treatment order		
1	a	b	c
2	b	c	a
3	c	a	b

This simple example shows that the assessment order has been balanced out to an extent – treatments have been applied first, second and third. However, each treatment has not been assessed after every other one.

Three treatments can be applied in six orders (abc, bca, cba, ...). The number of orders is given by $n!$ ('n factorial'); thus, $3 \times 2 \times 1 = 6$. If the designs were to incorporate all six orders then the square would need to be replicated. Numbers of orders increase dramatically with treatments, e.g. five treatments $5! = 120$ orders.

In sensory tests, where several samples are assessed, bias can occur is several ways for order, such as *first treatment effect*, when treatments with 'noticeable 'features occur before a 'less noticeable' one. This can be balanced out using *Williams Latin squares*, which implement all possible orders with balanced designs for a range of treatment and blocks (MacFie and Bratchell 1989; Wakeling and MacFie 1995). These give the ultimate balance, as every treatment occurs before every other. As mentioned above, the number of presentation orders increases dramatically with the number of treatments, and such balance is only achieved in sensory work where the numbers can be spread across the assessor (consumer) panel.

All the above designs can result in large numbers of treatments (see below) when the basic design is used. The main treatment effect can be estimated, but other aspects such as interaction are not unless there is replication.

7.4.2 Factorial designs

The RCB-type of design and Latin square are able to assess the main treatment effect plus one or two block effects. These can be viewed as additional factors. Where more than one factor is of interest, the design become factorial in nature.

The disadvantages of OVAT designs are overcome by use of *factorial* designs where all effects are evaluated.

Factorial designs

Examining the effect of factors in combination has more advantages. In addition to all the **main effects**, **interaction effects** can be calculated. This requires that the treatments are replicated and the experiment becomes larger. In addition, interpretation becomes more difficult when significant interactions are found (Section 10.4). One other feature of design is the manner in which factors are defined. They can be *fixed*, i.e. specifically selected, then the results are only applicable to the levels of the fixed factor(s). It is possible to choose the factors randomly, just as with the sample. This would give a **random effects model** for the design and would allow generalisation to the population from which the factor and levels were chosen. Combinations of these choices as **mixed models** are also possible. It is important that the nature of the factor in this respect is taken into account in ANOVA – different error terms are used in the calculations and therefore different results are obtained (Section 8.5.5).

Full factorials

With replication, both simple and interactive effectives are assessed, e.g. a three-factor experiment (A, B and C) would assess all main and all interactive effects:

> Treatment effects: A, B, C (simple, i.e. main effects)
> Treatment effects: AB, BC, AC, ABC (interactive)

There are 3 two-way interactions and 1 three-way (ABC), i.e. all possible combinations.

Such designs can be analysed by ANOVA (two-way → three-way, etc.) depending on the number of factors. Hence, two factors using two different levels of each – two factors (A, B) plus two levels (1, 2). The treatments required for this design as a *full factorial* design would be:

Treatment	Combination
1	A1 + B1
2	A1 + B2
3	A2 + B1
4	A2 + B2

This is equivalent to 2^2, i.e. the total number of treatment combinations for a factorial experiment is given by the product of the number of levels for each factor.

If the number of levels is increased to 3 then $3^2 = 9$ treatments are required and this is *without replication*. Thus, five factors at three levels each $= 3^5 = 243$ done in duplicate $= 486$ treatments to be made up and tested – much more difficult to organise and control, plus a more complex analysis. Baardseth *et al.* (1992) used a full factorial design to study ingredient and processing effects on sausage properties. There were four factors at various levels (5, 3, 2 and 2) that gave 60 treatments, plus replication of 3 treatments and one reference formulation, resulting in a total of 64 production runs. Replicating all treatment formulations would have meant at least doubling the main treatments (i.e. 120 runs) so the researchers used partial replication instead to give a measure of pure error, presumably by random selection of three treatments.

Changeover (crossover) designs

This procedure in a design refers to the manipulation of order variation to eliminate bias (Gacula 1993). This has been explained above with respect to the Latin square. Treatments order is varied so that carry-over from one treatment affects every other treatment and not just a few. This is especially important for sensory experiments involving several samples where strong flavoured samples affect weaker flavoured ones, depending on order. The problem arises because unlike non-sentient material, human assessors can remember a previous treatment effect. The same point applies to other experiments where biological material can retain a residual effect from a previous treatment.

Similar circumstances can occur in consumer studies. If the experiment uses one group to measure as a control condition followed by the same group for the test condition, the subjects will already have experience of the measurement procedure and they could be biased by this carry-over effect. The crossover technique would divide the main group into two and vary the order. Examples of this could be home-use trial for food products (Gacula 1993) or assessing food with and without nutritional information. The crossover effect can be assessed in the analysis, although this requires that the two subgroups be compared. If there is no significant effect then the main treatment effect can be determined with confidence. If there is significant carry-over then it may be that only the first exposure of the treatment can be compared.

In sensory experiments, the carry-over effect can be determined as another factor provided that there is sufficient replication, but in some experiments, it is assumed to be cancelled out by use of randomisation of order or balanced order techniques such as the Williams Latin square (Wakeling and MacFie 1995). Schlich (1993) describes use of changeover design in sensory work.

7.4.3   Optimisation designs

For optimisation, the statistical analysis of the experimental data allows calculation of a mathematical equation or *model*, which describes the relationship between the response, the treatments and error effects (Harper and Wanninger 1970). This can then be used to predict conditions that would give the most desired response; in fact, the model can be used to generate or 'map out' a *response surface*, which can show in two or three dimensions how the response is affected by the treatments.

Two-level factorials

Factorial experiments done at two levels can be used for this purpose, but the linear first-order models produced may be inadequate and provide a poor description of the geometric shape of the response surface. One reason is that two-level factorials, even if full, cannot completely assess any *non-linear effects* caused by treatments; i.e. three levels at least would be required. Thus, to elucidate fully and eventually to optimise a system, more information is needed.

Three-level factorials

These allow generation of a second-order model, which will give better predictive power. There is still one drawback: fitted responses from three-level factorials do not predict with equal power in all directions, i.e. such designs are not *rotatable*. This demands inclusion of extra design points such as in the *central composite design* (see Section 10.5).

7.4.4   Designs to reduce the number of treatments and experimentation

A common problem with experimentation and the use of the above designs is that the number of treatments can become so large that it becomes difficult to run in one session and certain requirements may be difficult to achieve. Additionally, in sensory work, this can lead to assessor fatigue. Even simple designs such as the RCB can have this drawback as each block has to have one of each treatment. Certain designs can overcome such difficulties by a number of modifications.

Split plot

The *split plot* design is a type of incomplete block. Essentially, a two-factor factorial structure, but one factor is designated as the 'main plot' (from agricultural terminology) and the second is divided into subplots within the main one. This contrasts with a full factorial where each factor is treated in a similar manner. The design has application in food science and technology experiments and some types of consumer studies.

To illustrate the principle, consider a study on the effect of flavour content in a product (four levels), subjected to different freezing methods (two forms). A full factorial would require at least a 4×2 design $= 8$ treatments, and there would be eight freezing runs done in random order. For some runs, the freezing equipment would need to be cleaned and reset, etc.

A split plot can do this as in two runs, one for each freezing method. Thus, for four flavours (W, X, Y and Z) and two freezer types (T and F):

Full factorial of eight random runs – Fy, Fx, Tz, Fw, Tw, Ty, Fz, Tx
Split plot of two randomised runs – Run 1 (T freezer) y + w + z + x
Run 2 (F freezer) x + z + w + y

For the split plot, the four flavours are randomly allocated to the machines with respect to the order of preparation and the order in which there are placed or stacked in the freezer compartments. As the freezers are taking a unit of each flavoured product, the amounts may require to be reduced. The order of the freezing runs can also be randomised, but in this case, they can be run together.

In this experiment, the **main plots** are the freezers and the product flavours are the **subplots**. It may appear on first glance to be similar to be the same as a full factorial, but there are subtle differences. The subplots are complete blocks, but the main plot can be viewed as incomplete because there is not a unique occurrence of each factorial combination. A consequence is that on analysis the subplot effect is determined with more precision than that of the main plot.

There are a number of other features of the split plot that make it one of the more complex designs. Its advantages are that it has more precision in one factor than in the other. Thus, treatments known to produce small effects should be allocated to the subplot. An alternative use for this feature is that it allows inclusion of a second source of variation with more precision than just 'unknown'. Additionally, subplot units can be smaller (as above) that may save on materials; e.g. a processing machine that requires a minimum load of units to function, or for when large amounts are needed for one factor, but not the other.

Processing experiments are an ideal application whenever there is machinery, with complicated, lengthy procedures for change of setting, etc. This is particularly so when only one machine is available. Seker (2004) used a split plot design for this purpose when looking at the effect of screw speed and configuration on residence time in extruders. Only one extruder was available and using a full factorial would have meant changing configuration (two types of screw) for certain runs. The split plot method allocated the extruder configuration as the main plot and allowed different speed to be applied (randomly) as subplots.

Another name used in some disciplines for the split plot is **mixed ANOVA** (Huck 2004). This refers to experiments that have '**between subjects**' and '**within subjects**' comparisons within the same experiment. This is similar to the example above: the between subject factor is the main plot and the within subject is the subplot one. This can be applied in consumer studies, e.g. an investigation of the effect of 'information' and 'price level' on purchase intention for home-grown food commodities. If the 'information' factor (two levels – 'information given', 'no information') is the 'between subject' factor (main plot) then the participants for level 1 ('information given') will not take part in the 'no information' setting.

The within subject factor ('price level') is exposed to all participants. Bower and Saadat (1998) used a similar *'between and within design'* where two groups of consumers assessed liking of spreading fats, one without label information (the control group) and the other with information (the test group). Thus, all subjects assessed the fat samples (the within group factor), but only half received labelled samples (the between group factor).

In sensory work (Lawless and Heymann 1998), the sub-units can be different groups of assessors, i.e. one or more main factors have full plots, but the third (assessor) is split into two or more groups. Stone and Sidel (1993) describe its use in storage studies, but the use of the split plot for trained assessors, where the panel is divided, is viewed as rare by Hunter (1996).

Fractional factorials

It is possible to reduce the work involved in full factorial designs, but still gain most of the important information – this is done by use of *fractional factorial* designs. Care is required in their application as certain assumptions are made regarding interactive effects, namely that they are *relatively insignificant*; if this is so then the main effects of many factors can be evaluated with much fewer trials. Two levels, which are estimated to 'bracket' the optimum region, are used for each factor, e.g. a 1/4 fraction of a 2^5 full factorial (32 trials) can be performed in 8 trials. There are savings in cost and time plus increased efficiency as more variables can be incorporated. Typically, they are used as **screening trials** to identify, quickly and efficiently, the critical factors before more detailed fuller factorial experiments. The data from these latter experiments are analysed to obtain the optimum combination in terms of the response(s) measured (see Section 10.5 for an example of a fractional factorial).

Incomplete block designs

Another way to reduce experimental time and resources is to use *ICB* experiments. When there are large numbers of treatments that would cause separation of assessment into separate time periods, it is possible to handle all treatments within one run by reducing the number of treatments per block. ICB designs are particularly useful in sensory experiments where for reasons of sensory fatigue or strongly flavoured samples, assessors cannot deal with all samples. In the balanced ICB application (BICB), a reduced number of treatments are assessed and the allocation of these is balanced so that each treatment sample receives the same number of assessors and degree of replication. Published designs of this type are available in sensory texts and Cochran and Cox (1957). The analysis of ICB data is more complex and requires use of BICB ANOVA (Section 8.4.2).

7.5   Analysis methods and issues

Except in cases where non-parametric methods are readily available, most of the above designs are analysed by parametric ANOVA in one of a number of forms. As indicated, some designs require more advanced analysis, which is usually outwith the ability of basic software. Using Excel, the application of many of the above designs is limited. Paired, two-group and one-way CRD, two-way RCB analysis is possible. Basic versions of Minitab can perform all these plus three or more factors, but designs must be full and balanced. Analysis of Latin square and split plot designs is not directly possible with Excel, Megastat and Minitab, and users must turn to more advanced versions or packages such as SPSS which can do all these and ICB designs. In addition, the more advanced assumption checking such as those associated with repeated measures designs (e.g. homogeneity of covariance; Lawless and Heymann 1998) will not be available in Excel and basic Minitab.

7.5.1   *Identification of design effects*

What is the basis of design precision?

ANOVA was devised by Fisher (1966) to separate or partition sources of variation so that they could be quantified, individually and interactively, as to the effect on treatments. ANOVA does this by comparing variability within a treatment with variability between different treatments. For the full analysis, this requires that a treatment (experimental unit) be repeated, i.e. there are at least two replications. The extent to which these units, which have received the same treatment, constitutes the **within-variation**. The **between variation** is due to any effect of different treatments (i.e. the differing factor and level combinations). If these have little effect then between-treatment variation will be low – if at least two different treatments cause a large response difference then it will be high. There are two sources of variation, expressed as **variance** (Section 3.3), which are expressed as a quotient – the **F ratio**:

F = Variability between treatments/variability within[1]

F is the statistic produced when ANOVA is performed and its magnitude varies according to the variance values. Efficient experimental design attempts to minimise the within variance so that any between variance can be detected. The larger the *F* value, the larger the *treatment effect*.

In some designs, ANOVA cannot assess each effect distinctly due to **confounding** that refers to the confusion of one effect with others (also referred to as '*aliasing*'). For designs with absent or limited replication, the interaction effect

---

[1] Within variation is usually referred to as 'left over' **residual variation** or **error variation**.

Table 7.3   ANOVA assumptions.

Error term	Nature
General	Normally distributed, random nature, uncorrelated, variance homogeneity
Treatment effect	Independent of level
Factor effect	Additive
Repeated measures	Uncorrelated

can be confounded with main effects. In such situations, it is not possible to assess any interaction and it may or may not be present.

This is a design decision that requires some knowledge of likely interaction effects.

If new circumstances are being examined then replication would be required to assess them, but it depends on the type of interaction. Higher order interactions are less likely to be of significance, but second-order ones may be more marked in effect. In some designs it is possible to manipulate the confounding effect to advantage, to cut down on experimentation (Section 10.5).

Additional assumptions for experimental design and ANOVA

Assumptions for ANOVA were mentioned in Chapter 4. While these give the basic picture, there are more complex details (Table 7.3).

Thus, in addition to the sample data coming from a normal population, parametric ANOVA requires that the errors be distributed in this way. Additionally, they must be randomly 'scattered' on either side of the mean and must be independent or uncorrelated with one another. The effect on treatments must not depend on the particular magnitude of the response in the experiment. That is, an effect identified as significant between 5 and 10 mg/100 g of a substance should apply to lower and higher concentration ranges for the response substance (this can be infringed in some chemical analysis cases (Section 9.6)).

ANOVA calculates a *model* in the form of an equation that quantifies the effect of factors on the response. Each factor effect is assumed to 'add-on' to the effect in a linear manner. This is the concept of **additivity** of the ANOVA model (hence, a 'linear effect model'). Thus:

$$\text{Effect} = \text{effect factor A} + \text{effect factor B} + \cdots + \text{error}$$

Each factor will have a *coefficient* calculated relative to the strength of the effect.

How can these additional assumptions be checked?

Methods of analysis are available to determine some of these error aspects as described in Sections 3.5 and 4.4. Examination of the sample data should be done first by graphs and tables. Non-additivity arises when there is interaction between two or more factors. Replication in the design allows this to be quantified (see previous section on the RCB design and example in Section 10.4), otherwise examination of the data for patterns which suggest that changes are not occurring 'in parallel' is the only alternative. In sensory rating and scoring tests, one or more assessors may assign values 'out of sequence' with other assessors. For instance, one assessor scores three samples as 5, 2 and 8, but another scores them as 3, 9 and 1, so the middle sample is causing a divergence. This can manifest itself as interaction if sufficiently prevalent. If the software is capable, visual examination of the residuals from ANOVA using graphs within diagnostic techniques is possible in a similar manner to the procedure used for regression (Section 6.4).

7.6   Applicability of designs

What type of design is best?

Selection of the design should be guided by the purpose and the circumstances of the experiment, but there can be choices. Box *et al.* (1978) give an account of the role of experimental design with examples of the types of scrutiny that is required for effective design. The *sensitivity* (Bender *et al.* 1988) of a design is its ability to 'pick out' differences, its preciseness and how it 'reacts' to changes in the response. This can be improved by lowering the within-variation (residual error) as explained above, increasing sample size, replication, etc. Such techniques lower the standard error (Section 3.3.2) of the estimates in an experiment and hence calculated statistics will be larger. Advance knowledge of the degree of within-variation likely (e.g. in a preliminary or pilot experiment) can be used to calculate efficient sample size and replication for a design.

There are arguments to support use of smaller, simpler designs in that they are easier to manage and interpret. This is acceptable if key design essentials are included (balancing, randomisation, controls, coding, etc.). What is lost in such a case is the level of precision for detecting effects – so A, B and C – analyse as one-way ANOVA is less precise than RCB by two-way ANOVA or repeated measures. Related designs have advantages over independent because of the more precise focus on effects without the interference of the variation between groups.

References

Baardseth, P., Næs, T., Mielnik, J., Skrede, G., Hølland, S. and Eide, O. (1992) Dairy in-gredient effects on sausage sensory properties studied by principal component analysis. *Journal of Food Science*, **57**(4), 822–828.

Baird, P. D., Bennet, R. and Hamilton, M. (1988) The consumer acceptability of some underutilized fish species. In *Food Acceptability* by D. M. H. Thomson (eds). Elsevier Applied Science, Barking Essex, pp. 431–442.

Bender, F. E., Douglass, L. W. and Kramer, A. (eds) (1988) *Statistics for Food and Agriculture*. Food Products Press, New York, pp. 79–86.

Bower, J. A. (1997) Statistics for food science V. Part A: comparison of many groups. *Nutrition and Food science*, **97**(2), 78–84.

Bower, J. A. and Saadat, M. A. (1998) Consumer preference for retail fat spreads: an olive oil based product compared with market dominant brands. *Food Quality and Preference*, **9**, 367–376.

Box, G. E. P., Hunter, W. G. and Hunter, J. S. (1978) *Statistics for Experimenters*. John Wiley & Sons, Inc., New York.

Carlberg, C. (2011) *Statistical Analysis: Microsoft® Excel 2010*, Pearson Education Inc., QUE Publishing, Indianapolis, IN, pp. 361–372.

Cochran, W. G. and Cox, G. M. (1957) *Experimental Designs*, 2nd edn. John Wiley & Sons, Inc., New York.

Fisher, R. A. (1966) *The Design of Experiments*, 8th edn. Hafner, New York.

Gacula, M. C. (1993) *Design and Analysis of Sensory Optimization*. Food & Nutrition Press, Trumbull, CT, pp. 29–34.

Harper, J. M. and Wanninger, L. A., Jr. (1969) Process modelling and optimization 1. Experimental design. *Food Technology*, **23**(9), 49–51.

Harper, J. M. and Wanninger, L. A., Jr. (1970) Process modelling and optimization 2. Modelling. *Food Technology*, **24**(3), 60–62.

Huck, S. W. (2004) *Reading Statistics and Research*, 4th edn. International Edition, Pearson Education, New York, pp. 378–389.

Hunter, E. A. (1996) Experimental design. In *Multivariate Analysis of Data in Sensory Science* by T. Næs and E. Risvik (eds). Elsevier Science BV, Amsterdam, pp. 37–70.

Lawless, H. T. and Heymann, H. (1998) *Sensory Evaluation of Food: Principles and Practices*. Chapman & Hall, New York, 717pp.

Lee, S. and Inglett, G. E. (2006) Rheological and physical evaluation of jet-cooked oat bran in low calorie cookies. *International Journal of Food Science and Technology*, **41**, 553–559.

MacFie, H. F. H. (1986) Aspects of experimental design. In *Statistical Procedures in Food Research* by J. R. Piggot (ed.). Elsevier Applied Science, London, pp. 1–18.

MacFie, H. and Bratchell, N. (1989) Designs to balance the effect of presentation and first-order carry-over effects in hall tests. *Journal of Sensory Studies*, **4**, 129–148.

Meilgaard, M., Civille, G. V. and Carr, B. T. (1991) *Sensory Evaluation Techniques*, 2nd edn. CRC Press, Boca Raton, FL, pp. 292–294.

Mumba, P. P., Chilera, F. and Alinafe, G. (2004) The effect of length of soaking time on trypsin inhibitor, crude protein and phosphorus contents of soybeans (*Glycine max*). *International Journal of Consumer Studies*, **28**(1), 49–54.

O'Mahony, M. (1986) *Sensory Evaluation of Food – Statistical Methods and Procedures*. Marcel Dekker, New York, pp. 203–204.

Pyne, A. W. (2000) Innovative new products: Technical development in the laboratory. In *Developing New Food Products for a Changing Marketplace* by A. L. Brody and J. B. Lord (eds). Technomic Publishing, Lancaster, PA, pp. 259–276.

Schlich, P. (1993) Use of change-over designs and repeated measurements in sensory and consumer studies. *Food Quality and Preference*, **4**, 223–235.

Seker, M. (2004) Distribution of the residence time in a single-screw extruder with differing numbers of mixing elements. *International Journal of Food Science and Technology*, **39**, 1053–1060.

Steel, R. G. D. and Torrie, J. H. (1981) *Principles and Procedures of Statistics – A Biometrical Approach*, 2nd edn. McGraw-Hill International Book Company, Singapore, pp. 122–136.

Stone, H. and Sidel, J. L. (1993) *Sensory Evaluation Practices*, 2nd edn. Academic Press, San Diego, CA; London, pp. 106–144.

Wakeling, I. N. and MacFie, H. J. H. (1995) Designing consumer trials balanced for first and higher orders of carry-over effect when only a subset of k samples from t may be tested. *Food Quality and Preference*, **6**, 299–308.

Part II

Applications

Part I provided an account of the nature of data and their collection and analysis by a variety of methods. Part II builds upon this foundation with some applications. These cannot be viewed as comprehensive as they include selected topics only. Each chapter describes the nature of the data source and then outlines experimental design approaches followed by a selection of analysis illustrations and summary methods.

Presentation of all tests in detail with full statement of hypotheses, etc., is not possible, but the first test in Chapter 8 is given as an example of the full procedure. Beyond this, frequent reference to methods and fuller examples covered in previous chapters is made.

Chapter 8
Sensory and consumer data

8.1 Introduction

In this chapter, much of the basic statistical information of Part I is applied to sensory and consumer data. The data sources have a number of features in common which are considered together before examination of specific detail.

8.2 The quality and nature of sensory and consumer data

These data are generated by groups of people who are consumers, food company staff or others recruited to participate in food trials for varying lengths of time. Such participants are viewed in different ways depending on the application; if they are assessing food samples then they can be referred to as *judges*, *assessors* or *panellists*. The data from this type of source will provide *sensory data* by various sensory evaluation tests. Within this type there is another division: data generated by selected, trained panellists (BSI 1993) are viewed as **trained panel sensory data** and that from consumers (untrained or lay respondents) as **consumer sensory data**. This is similar in some respects to the two types of sensory evaluation described by O'Mahony (1988). Separating the data sources in this way is by no means clear-cut as many practitioners use consumer panels for tests that others limit to trained panel work; some sensory tests (such as difference tests) are commonly used by both forms of panel. Additionally, when time permits, training consumers can also be incorporated in some experiments. There are a wide variety of statistical applications to sensory data, with identification of product preference and differences being common objectives.

The second main data form is that originating from consumers in *surveys* where views and perceptions on food issues, concepts, etc., are recorded and these are referred to as **consumer survey data** (Fig. 8.1). Here, the participants can

Statistical Methods for Food Science: Introductory Procedures for the Food Practitioner,
Second Edition. John A. Bower.
© 2013 John Wiley & Sons, Ltd. Published 2013 by John Wiley & Sons, Ltd.

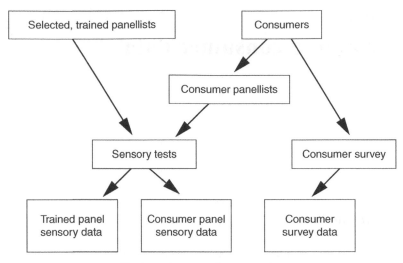

Fig. 8.1   Sensory and consumer data.

be viewed as **respondents** for general questionnaire trials or as **subjects** when differing conditions are administered to groups, e.g. half of the subjects receive brand details of products in the questionnaire and the other half do not.

Measurement systems based on human participants in the above formats result in data that are 'noisy' in terms of error, when compared with instrumental data. People are prone to a large number of physiological and psychological influences, which affect their performance in such exercises. These are described in detail in sensory texts (e.g. Stone and Sidel 1993), but they also affect performance of consumers in surveys. Although both forms of data are subject to relatively large error due to their origin, the extent of effects tends to be higher for consumer sources. Additionally, consumer data (sensory and survey) have much more variability due to individual characteristics in terms of likes and dislikes, opinions, beliefs, etc.

Trained panel sensory data suffer less from error due to the training and constant monitoring procedures that are employed. Methods of evaluation of error sources are an important application to these latter data. The error determination methods in Section 3.4.3 are not used to the same extent, although this topic is of ongoing interest in sensory methodology (Calviño *et al.* 2005).

Error appreciation, evaluation and control are crucial when one considers the considerable investment made by food companies in sensory and consumer evaluations, and the consequent critical nature of decisions based on the data. Accepting the divisions above, there are three types of data that are relatively distinct in a number of ways (Table 8.1).

Trained panel data have a lower subjectivity as their nature approaches objectivity like an instrument, due to training and use of standards, etc. Again, this separation based on error level is not sacrosanct – some research points towards

Table 8.1 Sensory and consumer data: relative differences in nature.

Sources	Subjective nature	Error level	Analysis methods
Trained panel sensory	Low	Low	Parametric
Consumer sensory	High	High	Parametric, non-parametric
Consumer survey	High	High	Non-parametric, parametric

little difference – and consumer data should not be viewed as 'inferior' in this respect.

8.3   Experimental design issues

Given the large number of possible sources of unwanted variation in such data (sensory and consumer), scientists must use experimental design to combat them. For experimentation with food units, all procedures in presentation of the samples must be controlled, in terms of unit size, shape, container, etc. The extent of this does depend on the characteristics of individual foods. The use of specially designed ***sensory booths*** is a feature of such control. These ensure minimisation of distraction and provide a calm environment for the assessment. More advanced versions include positive pressure conditions (to prevent ingress of foreign odours) and computer control of test instructions and data collection to obviate the errors that can occur during manual data collection. Variables such as *order of presen tation* can be randomised or balanced to overcome carry-over and 'first sample' effects. Designs based on the ***Williams Latin square*** meet this need (Wakeling and MacFie 1995).

Related or independent designs are possible examining two or more samples. When participants assess all samples, related tests are used with the *randomised complete block* (RCB) being common. For consumer sensory trials, it is not usual for repeat sessions to be held, but for trained panels this is the norm. In the latter case, additional blocking can be used for repeat sessions. More specifics on design types are given below and other examples are provided int Section 7.4.

8.4   Consumer data (sensory and survey)

8.4.1   *Sampling issues*

Food sampling for consumer sensory tests

Sampling for consumer testing involves both the consumers and any food samples to be assessed. When food samples are to be presented, a separate sampling procedure is required for the food material. This is similar to the procedures described in Chapter 2 for instrumental and sensory measures (also below), but much greater quantities are required as the consumer panel will be much larger (typically, 100–500) than a sensory panel (8–12). Although recommended, replication of both sample and subsample (the determination or observation) is not feasible with large consumer panels. Complication arises due to the large number of subsamples required and sample amounts must take this into account (Section 2.3.2). Homogeneity between samples of the same material is important as well as representativeness. The number of observations and the complexity of the task are usually limited with consumer sensory evaluation (typically, six to eight subsamples). This is restricted by fatigue factors, although some practitioners (Moskowitz 2003) argue that this can be exceeded when short recovery periods are included within a trial. Usually, with consumer sensory tests such as hedonic rating the data are analysed directly, with all values per sample included, unlike trained panellist data that may be averaged first to give a ***panel mean***. Measures of precision are not possible without replication and all that can be calculated is the variability of the whole consumer group. Often this is very high and %CVs of 50% or more are common. Accuracy measures for hedonics are difficult, as this would require a standard for affective measures (Gacula 2003) which to date has not been established. The nearest would be a comparable rating for a current product that was recognised as being 'not liked' or 'well liked', such as, in the latter case, a brand leader.

Consumer sampling (sensory evaluation and survey)

Also of consideration is the *sampling of consumers* for sensory work and when the intention is to gauge consumer population attitudes, opinions, beliefs, etc., in surveys. Here, the sample units are the individual consumers. Selection can be random, but more commonly, *convenience methods* are used in shopping malls, etc. Sometimes both forms of test (sensory and survey) will take place and there are a number of possible locations for the gathering of the data.

Consumer data come from tests and trials performed in a wide variety of settings including in-house (food company), central location, shopping malls and home-use. Control of experimentation does apply in some of the above, but home-use is less controlled, although it has the advantage of being more realistic in reflection of the real food consumption situation. There are arguments for less 'laboratory'-type control in consumer tests demarcating these methods from trained sensory panels (Lawless and Claassen 1993).

8.4.2 Analysis of consumer sensory tests

The data are viewed as being more subjective and more 'noisy' in term of error in nature compared with trained sensory panels and the focus is on tests of an *affective* nature dealing with *preference* and *acceptability*. A number of possibilities exist here and the food scientist will be familiar with these tests (Table 8.2) or if not the reader can consult one of a number of texts such as Meilgaard *et al.* (1991).

Table 8.2　Consumer sensory tests.

Test	Basic form/instruction
Preference tests	
Paired	Two samples – choose the one preferred
Ranking	Three to six samples rank in order of preference
Acceptance	
Hedonic	Score on scale dislike to like degree
Diagnostic	
Just right	Select on scale for level of attribute relative to ideal

The data generated are in the form of incidence, ratings, scores, etc. Perhaps the simplest sensory test is where two samples are compared for preference.

Analysis of preference tests

Paired preference test
How to analyse a paired preference test

In its basic form, the ***paired preference test*** gives the incidence or count of the selections for two samples (A and B) by untrained consumers numbering 30 or more (BSI 2005). A very simple results table is produced: number of consumers choosing sample A and the number of consumers choosing sample B. The tests can identify a difference in preference, but they cannot establish the magnitude of difference.

There are several methods of analysis:

- Binomial test
- Binomial tables
- Chi-square test

For the example below, the ***binomial test***, which is the commonest method, is used. The basis of the test is described in Section 5.3.2. The number of *trials* refers to the number of assessors or judges. The choice results in two groups: those selecting A and B, respectively. One of these, the larger if there is predominance, is designated as the number of *agreements* (referred to as *successes* for one-tailed tests in Section 8.5).

The test statistic can be calculated from the binomial terms as described by Bower (1996), but software is used here. *Cumulative probabilities* must be calculated, i.e. all the events including and above the ones which qualify as an 'agreement' (or 'success').

In Excel, the BINOM.DIST() function (BINOMDIST() Excel 2003) is used. In the basic form of the function, probabilities are calculated for exact numbers of agreements ('*successes*' in Excel) and for '*at most*' levels. To get the '*number of agreements (successes) or more*' format requires slight modification of the formula. This can be done in a number of ways, but in the examples subtracting one from the number of agreements gives the probability of getting 'up to one less' than the number of agreements. This probability can then be subtracted from unity to give 'number of agreements or more'. The enumerated data are entered as the number of judges selecting the higher or equal count minus one (**Number_s-1**), and as the total number of **Trials** (number of judges). **Probability_s** is the chance probability and '**Cumulative**' is set to '**TRUE**' so that individual terms are summed. This gives the *one-tailed probability* of getting 'more or equal than' for one sample dominating the other, i.e. A > B. To get the two-tailed (A>B and B>A), this probability is doubled (Box 8.1). This analysis is a significance test and the hypothesis testing approach is assumed.

Box 8.1   Worked example of paired preference test.

Paired preference test
Objective: To establish whether two products differ in terms **of preference**
Question: *'Is there a significant difference between A and B in terms of preference?'*
Experiment design: Paired design – 27 consumers presented with two subsamples of two products A and B, in balanced presentation order, coded with random three-digit numbers

Result:	Sample	Number preferring
	A	20
	B	7

Thus, there are 27 trials and 20 agreements
Statistical hypotheses:
H_0 Proportion$_A$ = Proportion$_B$ (numbers preferring A = numbers preferring B)
H_1 Proportion$_A$ ≠ Proportion$_B$ (numbers preferring A or B are not equal; two-tailed)
Significant level α: 5% (0.05)
Selection of test:
Level of measurement is nominal; discrete data (counts); number of groups = 2, independent. Therefore non-parametric test for two group proportions – binomial test.
Calculation (Excel):
Probability (two-tailed) due to chance:
= (1 – BINOM.DIST[a] (number of agreements – 1, number of trials, 0.5, True)) × 2
= (1 – BINOM.DIST (19, 27, 0.5, True)) × 2
$p = 0.0096 \times 2 = 0.0192$ (***p*** < 0.05)
Conclusion: H_0 is rejected and H_1 is accepted – there is a difference in preference: product sample A is preferred over B.

[a]BINOMDIST() Excel 2003.

Example 8.4-1 *Analysis of paired preference test by binomial method*
Initial prototypes (A and B) of a new product were examined by a small panel
of consumers for preference. A paired preference test was used. Twenty-seven
participants completed the test and 20 out of the 27 chose the A product sample.
Does this indicate preference of A over B?

The analysis of this test is presented in a full worked format (Box 8.1), illustrating aspects of sensory protocol and statistical significance testing.

The conclusion has found a significant difference in preference. Published tables
agree with this result, but they also show that if there was one less agreement then
the test would not be significant ($p > 0.05$). This underlies the nature of such
tests, which are affected by the number of trials and the proportion of agreements.
With ten trials, 90% or more agreements are required, whereas with 50 trials
approximately 60% give $p < 0.05$. Useful information can be gained by considering
'close to significance' results and ascertaining how many choice changes it would
take to 'turn' the results. The consequence lies in the power of the test. Tests with
small sample numbers (number of judges) with marginal numbers of agreements
will have lower power for a given difference than larger sample numbers (see
Section 4.2.2).

A confidence interval (CI) can be calculated for the results of the paired preference test (and for difference tests) using the formula given in Section 3.4.3
and Table 3.7 for proportions. This is based on the number of agreements. In
the example above, the sample size is small, but to illustrate the uncertainty, the
calculation was performed. The estimate of the proportion of consumers in the
population preferring sample A is 74% (20/27), thus 0.74 and 0.26 are substituted
in Table 3.7 for proportion yes and no, respectively. The 95% CI is 58–91% which
is very wide (±17%) and a larger sample size would narrow the CI. Thus, for
a similar exercise with 150 consumers, 111 agreed on A. This gives the same
average estimate, but the interval is 67–81% (±7%).

Published tables may be limited in terms of larger numbers of trials. Once the
number of trials gets larger (>50), the scatter of the other outcomes around the
50:50 division becomes more symmetrical and approximates a normal distribution
(Upton and Cook 2001) and probabilities can be calculated on the basis of an
approximation. This is not required by the Excel function, which can deal with
large numbers of trials, sufficient to cover the needs of most practitioners.

The chi-square test can also be used for this analysis using the procedure in
Chapter 5 (Table 5.3). The observed frequencies, 20 and 7, are entered and the
expected values are set to one half of the total count of 27 (13.5). It reaches a
similar conclusion in this case, but there are differences in that the chi-square in
its uncorrected form is more powerful (Section 5.3.1).

Preference ranking
How to analyse a sensory ranking test
Ranked data can be analysed using ***Friedman's test*** (BSI 2006), although ***rank sum***
tables (Section 5.4.4) are an alternative. A sample size of two is a special case of

ranking and such a test can be analysed as a paired preference test above. Typically, ranking will involve three to six samples. There is an example of Friedman's test on ranking data in Section 5.4.1 and another example of the test appears below on hedonic data, including follow-up multiple comparisons.

Acceptance test – hedonic rating

Hedonic rating data can also be analysed using the Friedman's test above. This is still the case in some research reports, but due to the evidence concerning this scale, it is treated as being interval (Section 2.2.3). This is assumed, but rarely checked, in many publications. The example below includes checks and analysis by ANOVA. Use of the hedonic rating scale in an acceptance test does not involve a ranking procedure as such, but doubts may exist about its use in an interval manner. The conservative approach is to treat the data as ordinal. Samples are assigned ratings from a category scale with 5–9 anchors (Section 2.2.1), or using a line scale with anchors on the extreme points and the central neutral point. Friedman's analysis converts the data to ranks. This is, of course, a loss of information and power and is a consequence of opting for the conservative approach of a non-parametric test.

Example 8.4-2 *Analysis of hedonic data by Friedman's test*
A hedonic rating test was carried out by a panel of assessors on three product samples. The nine-point hedonic scale was used with category box anchors ('dislike extremely' (1) to 'like extremely' (9)). Subsamples were presented to ten participants in randomised order. The objective was to answer the question: 'Is there a significant difference in liking?'

The design used is an *RCB* with three product samples and ten assessors (*blocks*). For this test, the null hypothesis is that all three medians are the same, the alternative is that at least two differ. The data are entered in columns in Minitab and analysed using Friedman's test (Table 8.3).

The *p*-value is less than 0.05, thus H_0 is rejected and H_1 accepted – there is a difference in liking.

Subsequent to a significant result by ANOVA, a *pair-wise comparison* must be done to identify where the differences are located. As seen in the table the highest rank (most liked in this case) is sample 3 (c) – but is this significantly different from all the others? Or just one? A paired comparison procedure is required as explained in Section 5.5. For Friedman's test, there are several possibilities, but the recommended one is given in BS ISO (2006) (Section 5.4.3). Others include more the conservative sign test and Wilcoxon's test. These latter tests, along with the more powerful *t*-test, were performed by Minitab (Table 8.4).

As seen, the sign test (least powerful) cannot locate any significant differences, but Wilcoxon's test locates two. However, if a ***Bonferroni adjustment*** (Table 5.16) is made to the probability value for the multiple testing as 0.05/3 = 0.017, then there is but one significant difference – sample b versus sample c. Megastat for Excel gives similar results to Minitab, but the multiple comparison tests did not detect any significant differences as with use of the sign test above.

Table 8.3   Hedonic rating test (Minitab).

		Sample	
Assessor	**a**	**b**	**c**
1	5	1	5
2	3	6	9
3	6	4	8
4	5	4	7
5	4	4	4
6	6	4	8
7	4	5	6
8	6	5	5
9	5	2	6
10	2	6	5

Friedman test of rating by sample blocked by consumer
$$S = 5.85, \text{df} = 2, p = 0.054$$
$$S = 6.88, \text{df} = 2, p = 0.032 \text{ (adjusted for ties)}$$

Sample	N	Estimated median	Sum of ranks
1(a)	10	4.500	18.5
2(b)	10	3.500	15.5
3(c)	10	6.500	26.0

Grand median $= 4.833$

Hedonic rating treated as interval data (RCB design)

Checking of the hedonic data above with Excel using the procedures detailed in Sections 3.5 and 5.5 (not shown here) revealed non-zero, but acceptable values for skew and kurtosis. Histograms, normal plot and the Ryan–Joiner test (Minitab) show that the distribution of the sample data is skew for all, but n-plots and

Table 8.4   Pair-wise comparisons after a significant Friedman's ANOVA (Minitab).

Sign test of median $= 0.00000$ versus N.E. 0.00000					
	N	**Below**	**Equal**	**Above**	**p-value**	**Median**
Sign a:b	10	3	1	6	0.5078	1.000
Sign a:c	10	7	2	1	0.0703	−2.000
Sign b:c	10	7	2	1	0.0703	−3.000

N for Wilcoxon					Estimated
	N	Test	Statistic	p-value	median
Sign a:b	10	9	28.0	0.554	0.5000
Sign a:c	10	8	1.5	0.025	−1.500
Sign b:c	10	8	1.5	0.025	−2.000

1 sample t on signs					
	N	Mean	sd	se mean	T	p-value
Sign a:b	10	0.500	2.550	0.806	0.62	0.55
Sign a:c	10	−1.700	1.947	0.616	−2.76	0.022
Sign b:c	10	−2.200	1.989	0.629	−3.50	0.0068

Table 8.5   Analysis of variance on hedonic rating data (Excel).

ANOVA: Two-factor without replication

Summary	Count	Sum	Average	Variance		
Sample a	10	46	4.6	1.82		
Sample b	10	41	4.1	2.54		
Sample c	10	63	6.3	2.68		

ANOVA

Source of variation	SS	df	MS	F	p-value	F-critical
Rows	20.67	9	2.30	0.97	0.50	2.46
Columns	26.60	2	13.30	5.60	0.01	3.55
Error	42.73	18	2.37			
Total	90	29				

the significance test are acceptable. If an assumption is made that these slight deviations from normal are acceptable then a parametric analysis can be proceed.

This analysis was performed by Excel using the **Anova: Two-Factor Without Replication** option in the **Data Analysis Tools** menu. Excel does not give an RCB design directly, but this is the equivalent. The data are analysed in columns as above (Table 8.3) with the samples as columns and the rows as assessor blocks. The complete set of data, including labels and the assessor column, is entered into Excel as the **Input Range**.

In the output (Table 8.5), there is significance for the sample effect as the p-value for the 'Columns' F ratio is 0.01. Thus, at least two samples differ in degree of liking (similar results were found using Minitab and Megastat). In this analysis, the assessor blocks (rows) have not exerted a significant effect (see below).

Follow-up t-tests with *alpha* specified as 0.017 (Bonferroni adjustment) with Excel (not shown here) give a significant difference for samples b and c. Further analysis in Minitab using one-way ANOVA with multiple comparisons gives two significances for Fisher's least significant difference, but Tukey's test gives one, as above (see Section 5.5.2 for details of how to perform ANOVA and multiple comparisons with Excel and Minitab). Megastat provides a randomised block ANOVA option with Tukey and t-tests for multiple comparison – these also give the same results as Minitab. Essentially, in this example, the non-parametric and parametric tests have given similar results.

Just-about-right tests

Just-about-right (JAR) scales (Section 2.2.1) are a type of ***attribute diagnosis*** tool. They can be constructed with line scales, but three or more category boxes are more usual (usually of the style 'too low', 'slightly low', 'just right', 'slightly high' and 'too high'). They cannot be assumed normal and continuous and should initially be examined by graphical means (Lawless and Heymann 1998). Analysis can be done simply by frequencies or percentages of the choices. It is possible to use chi-square methods for the proportional responses for each category by sample in a contingency table (Bi 2006), where the data are viewed as 'ordered categories'. This demands that there is a different consumer group for each product. When consumers assess all samples, then the cell counts are not independent, and individual chi-square tests must be used for each sample. These can then be compared with a separate independent set of standard proportions based on existing brands, etc. (Meilgaard *et al.* 1991). Another way is to 'fold-over' the upper part of the scale so that the ideal point is the maximum. This makes it more convenient for interpretation and for comparison with degree of liking data (Bower and Boyd 2003), but loses some information.

Example 8.4-3 *Analysis of JAR data*
A series of JAR tests was carried out on four biscuit products by 60 consumers as part of a product development study. The attribute 'crunchiness' was considered particularly important for acceptability. One of the objectives was to identify a suitable level (i.e. ideal) of 'crunchiness' that would be a characteristic to incorporate in future new products. Samples of the products were rated on a five-point JAR scale.

The first stage in analysis of these data was to calculate the frequencies of each JAR category for each product. This can be done by Excel (Example 3.2-1) or more conveniently by Minitab using the **Stat>Tables>Tally** function set to **count**. The data were arranged in columns according to the choices of 60 consumers and summarised (Minitab), then a series of individual chi-square analysis was carried out on the data for each product. This is possible using Minitab or Excel (Example 5.3-1). The latter method was used (Excel) by extension of Table 5.3 (five rows in total; all expected frequencies = 12; Table 8.6).

The chi-square statistics are very large and highly significant ($p < 0.001$) due to the presence of some high and very low counts for some of the frequencies,

Table 8.6 Analysis of JAR scale (Minitab and Excel).

		Sample				
		1	2	3	4	All
Too low	1	2	2	8	14	26
Slightly low	2	2	20	30	24	76
Just right	3	10	32	16	18	76
Slightly high	4	20	4	2	2	28
Too high	5	26	2	4	2	34
	All	60	60	60	60	240
Chi-square test (Excel):						
		s1	s2	s3	s4	
Chi		38.67	60.67	43.33	32.00	
df		4	4	4	4	
p-value two-tailed		0.000	0.000	0.000	0.000	

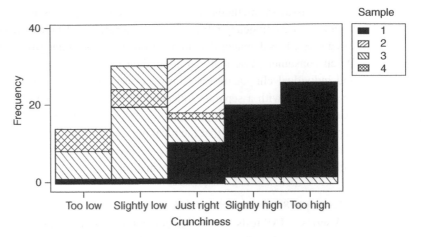

Fig. 8.2   Stacked histogram with sample by JAR rating (Minitab).

so there is a large effect. The data have also been graphed (Fig. 8.2) in a stacked chart in the form of a histogram. The responses show that sample 1 has received more of the 'too crunchy' category, and sample 2 the most of 'just right'. Based on these samples, the product with the 'ideal' degree of 'crunchiness' is number 2.

JAR scale data can be analysed by a variety of other methods as described by Lesniauskas and Carr (2004) and other authors, and more recently by Gacula *et al.* (2008). These methods include conversion of the data to percentages of consumers who express an attribute as 'just right' (ideal) and 'not just right'. Thus, in Table 8.6, 10 out of 60 or approximately 17% of the consumer panel found that the 'crunchiness' of sample 1 was 'just right'. For large samples of consumers, the central limit theorem allows these percentages to be analysed by parametric ANOVA to determine differences in the ideal totals across different products. 'Just right' responses are coded as '1' and 'not just right' as '0'. The columns of this 'incidence metric' for each sample can be compared by one-way ANOVA.

Another possibility is that the number of assessors on either side of the ideal point can be compared using a binomial test (Section 8.5.5) to decide if one non-ideal condition is significantly (p-value for sample is <0.05) prevalent over the other. This information guides decisions by the product developer on how to modify the product attribute(s).

Scales similar to Likert items such as 'level of importance' and 'purchase intent' (Section 2.2) are often a focus of consumer surveys and research. They can be treated in a similar manner to the JAR analysis and reported as frequencies.

Balanced incomplete block design

An important application in sensory assessment of food is the use of ***balanced incomplete blocks*** (BICB; Section 7.4.4). One application of the BICB design

Table 8.7   Balanced incomplete block structure, data and analysis (SPSS).

Consumer	Sample				
	1	2	3	4	5
1	2	5	—	—	—
2	—	1	5	—	—
3	—	—	3	—	6
4	4	—	5	—	—
5	—	2	—	8	—
6	2	—	—	7	—
7	—	3	—	—	4
8	4	—	—	—	3
9	—	—	3	9	—
10	—	—	—	7	5
Mean (adjusted)	2.9	2.5	3.7	8.3	5.1

Dependent variable: hedonic

Source	Sum of squares	df	Mean square	F	Significance
Total	478.000	14			
Consumer	7.500	9	0.833	0.278	0.958
Sample	55.000	4	13.750	4.583	0.049
Error	18.000	6	3.000		
Total	496.000	20			

could be for 'fiery' and 'frosty' foods as described by Allison and Work (2004). An example details an experiment where five samples were assessed, but due to their extremely strong flavour, assessors could tackle only two. As the assessors (consumers) were available for a single session only, a BICB design was used (Cochran and Cox 1957). It should be noted that unlike an RCB design, the number of assessors is fixed once the number of treatments and replications are decided (Gacula and Singh 1984). The example (Table 8.7) is one of the simpler BICB designs and it is chosen with the maximum reduction in the number of 'tastings' (i.e. the minimum is two).

The adjusted mean values for the samples show large differences. The ANOVA is significant (p-value for sample is <0.05), thus multiple comparisons can be performed. Tukey's test gives significance for sample 4 compared with 1 and 2, whereas the least significant difference test finds these plus an additional one between 1 and 4. If the full 'multiple testing rule' is employed by using the Bonferroni test then none of the comparisons are significant! This is equivalent to specifying the significance level for ten comparisons ($=0.05/10 = 0.005$). Differences of the above nature with a full block design would be more convincingly significant. Part of the difficulty is that, for the example, the number of assessors is small (10) and the p-value is just significant ($p = 0.049$). As consumers were involved more would usually be available, and in such a case, the BICB structure above could be repeated to suit.

8.4.3   Analysis of consumer survey data

Questionnaires for testing of consumer food habits, views and opinions can include a wide variety of levels of measurement and scale forms. A few of the many possibilities, and data of interest in research, are listed below:

- Types, forms, brands of food commodities and food products consumed
- Frequency of consumption
- Views, opinions, attitudes, beliefs on food issues, concepts, etc.
- Food-purchasing habits and intentions
- Tests of food knowledge and awareness
- Rating of factors important in food choice
- Reasons for likes and dislikes
- Demographic questions of importance in food habits

Statistical analysis can employ many of the methods covered in previous chapters. Only a selection of examples is given on the basis of typical questions for different levels of measurement, with reference to previous sections for the method.

Nominal data questions

Survey questions of the type with a nominal reply can be analysed by:

- Frequency summary by table or chart (Section 3.2)
- Proportional significance (Section 5.3)

These methods apply to both dichotomous and multiple-choice questions. Pooling of responses into a smaller number of categories and reduction of higher levels of measurement (to give a more robust measure) are also possible in some cases.

Ordinal data questions

These include ranking tests and arbitrary scales on views and opinions, such as 'importance'. These data can be summarised by calculation of a summary value (median) for comparison with other measures, plus a display of the frequency for each point on the scale when it is in the form of a category scale. In related designs, Friedman's test can be used as indicated above. Scales similar to JAR, e.g. 'purchase intent', where there is more emphasis on the categories and there are doubts about the continuity of the scale, should be analysed as per the JAR data above.

Likert scales and items

Likert items and scales are very common in consumer surveys for eliciting opinions on statements concerning food and nutrition. Individual items can be viewed in the above manner as ordered categories and analysed in a similar manner by a frequency count and chi-square. When a multi-component Likert scale is used to give a summated score or average, the data are more continuous and parametric analysis is possible. Thus, the following statements (s1–s5) could be rated together as a full Likert scale (level of agreement) or treated as individual items:

---

s1 : 'I would buy this product on a regular basis'
s2: 'This product represents value for money'
s3: 'This product would be suitable as an inexpensive snack'
s4: 'I would buy more than one item of this product'
s5: 'This product would be at the top of my snack list'

Taken together, all the statements are measuring the same underlying concept or **construct** – the *commitment to purchase*. Consumers who score high on each item are more likely to buy. Data generated by the above questions can be analysed in a number of ways:

- Expression of the average values of the construct in the sample
- Comparison of the average scores obtained by subgroups
- Correlation analysis on the scores for checks on reliability and validity

Analysis of Likert data was illustrated in Section 5.4.2. Individual items can be analysed as per the JAR scale example above and the other forms of analysis are discussed below.

Metric data questions

Survey questions on variables attaining interval and ratio status can receive treatment by ANOVA and regression analysis. This may be preceded by division of the sample into smaller parts or *subgroups*, often a main objective in consumer surveys.

Subgroup analysis

Analysis by subgroups contained in the consumer sample can be done in a number of ways. One is by comparisons of choices or ratings according to gender, age, socioeconomic group, etc. These characteristics have been shown to influence food habits, such as choice of 'healthy' foods, etc. In addition, consumers of different gender and age group can occupy distinct market groups for certain products.

Another way is to create subgroups based on a feature of interest such as 'educational level' or 'nutritional knowledge' or 'environmental awareness'. The division of the whole sample into two or more groups is achieved via a particular set of questions(s) that measure or at least record the respondents' status with respect to the feature of interest.

For example, consumers could select their 'educational level' from a nominal list. A more detailed example could be a set of five questions to test their level of awareness of 'environmental issues' related to foods. For nominal data, the subgroups can be created on the basis of two or more levels of education, e.g. 'school diploma', 'degree' and 'higher degree'. With features measured on ordinal or higher scales then measures of location such as the median and quartiles can be used to divide the sample into two or more subgroups. The ***median-split*** is the simplest way of doing this – the median is calculated, then two subgroups arise – one below the median (the 'low' group) and the other above (the 'high' group). Consumer respondents with values at the median can be analysed as a third group if numbers are comparable with the low and high groups, otherwise they are considered 'middle of the road' in attitude and are not compared. The median-split with two subgroups will usually be more effective at revealing any possible difference than comparing several groups.

All the above subgroups are independent, and thus, independent group tests are required for testing of differences in further measures of interest. Thus, taking the example above, assume that a median-split on 'environmental awareness' was used to create two subgroups: 'low awareness' and 'high awareness'. Another question measured the 'acceptance of animal foods' on a scale. The groups could be compared using a Mann–Whitney U test (ordinal; Section 5.4.3) or an independent *t*-test (metric; Section 5.5.1) to test the research hypothesis that '*high environmental awareness prompts low acceptance of animal foods*'.

A common subgroup comparison is that of *gender*. For instance, a consumer researcher poses the question '*does purchase intention for a new "teen" soft drink vary according to gender?*' The null hypothesis is that males and females do not differ in their intention to purchase and the alternative is that they do. A purchase intent scale of a more continuous nature is used: five categories from 'will not buy, might buy, probably buy, buy and definitely buy'. Assuming that this scale is *ordinal* then an appropriate non-parametric test with two independent samples is the Mann–Whitney U test. A sample of 50 males and 50 females of ages 15–18 years are surveyed and asked to rate purchase intention for the declared new beverage. The median values are 3 and 2, respectively, for male and female. The test gives a *p*-value of 0.75. The null hypothesis is retained – there is no difference in purchase intention between male and female teenagers for the proposed new drink.

Comparison with published consumer characteristic

Consumer surveys are performed regularly on food consumption habits so there are likely to be published figures available. As habits change and new influences appear, researchers often wish to compare the existing figures with new survey data.

One such consumer survey was interested in confectionery consumption. An official diet report (year 2005) on 'food facts' stated that on average, people in the country consumed the equivalent of twelve 50 g bars of chocolate in a year. The new survey included a question on this point so that the published value could be compared with new (2007) test data to establish whether or not the situation had changed. Thus, '*do people currently consume a different amount of chocolate (bars) compared with the national average for the year 2005?'* The researchers argued logically that on the basis of increasing leisure time and increased obesity levels chocolate consumption might be higher. Countered against this was the level of increased nutritional awareness, which could have a lowering effect. They took the conservative view and posed the research hypothesis: 'People currently consume a different amount of chocolate that the national average for the year 2005'. Null and alternative hypotheses were stated on this basis and data were gathered from 110 consumers in a quota sample similar to that of the previous research. The previous population average chocolate consumption was the equivalent of 12 bars. The new data enumerated to 14 bars. As the scale is ratio, a ***one-sample t-test*** (Section 5.5.1) was used to compare the test sample mean against the published mean to see if there was a significant difference. On analysis, the difference was significant ($p = 0.007$) and the null hypothesis was rejected – there was a small, but significant increase in chocolate consumption.

Consumer survey checks on reliability and validity

Survey questionnaires often include checks on the instruments used to gather data. Some mention of these was given in Section 6.3 and correlation analysis is a feature of such procedures. *Internal reliability* is analysed by examining the correlation between the items in a survey instrument. Taking the example of the Likert scale (five items) above, correlation analysis of 145 consumer ratings for these items produced Table 8.8.

This analysis is accessed in Excel via the **Toolpak** version of correlation in the **Data Analysis** menu. It allows data from two or more variables to be analysed for inter-correlation. In this case, data for five variables (items s1 – s5) were entered as a block of 5 columns by 145 rows. Each item has a positive correlation with each other item. Coefficients are significant ($p < 0.05$; analysis not shown), although the magnitudes vary. On this basis, the Likert scale construct has ***internal reliability*** or consistency. More comprehensive measures are available for this aspect of reliability such as ***Cronbach's alpha*** (available with some software packages) that produces a single overall value. Coefficients above 0.7 are interpreted as a satisfactory level of internal consistency, although any above 0.5 can qualify. The above data are well within this as the coefficient for Cronbach's alpha is 0.90 (SPSS).

Table 8.8   Inter-correlation between Likert items (Excel).

	s1	s2	s3	s4	s5
s1	1				
s2	0.82	1			
s3	0.68	0.66	1		
s4	0.58	0.54	0.57	1	
s5	0.54	0.70	0.61	0.81	1

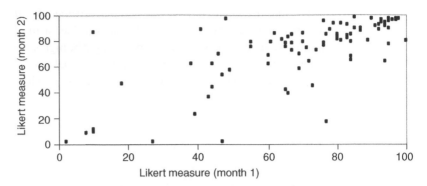

Fig. 8.3   Scatter plot of test–retest survey (Excel).

One month later, the survey was delivered again to the same respondents, but return was lower than the original count ($n = 86$). The average scores for the construct for each month were plotted in a scatter graph (Fig. 8.3).

This shows evidence of a positive relationship. The **test–retest reliability coefficient** (calculated by the formulae in Table 6.2) is 0.72 ($p < 0.001$) that is highly significant. Thus, the instrument can be viewed as reliable in that a similar result was obtained when it was repeated.

Other checks include the ability of the survey instrument to distinguish consumers with various traits. To illustrate this form of analysis, a follow-up study examined *actual purchase* of the product type after a period. Based on these data and the previous commitment to buy scale, the researchers wished to know if the construct showed *predictive validity*, by its ability to classify 'low' and 'high' 'purchasers' correctly. To calculate this form of validity, both the original 'purchase commitment' score and the current 'actual purchase level' were divided into two categories (median-split) as 'low' and 'high' groups. These were then compared using a crosstabulation analysis as described in Table 6.1.

The analysis (Table 8.9) shows that the original construct has predictive validity in that those consumers identified as 'low purchase commitment' had low incidence in the new measure of actual purchases ($p < 0.05$). 'Actual purchase' is not independent – it depends on which group the consumers were in, based on the original 'commitment' measure. Therefore, the construct has predictive ability in that it could be used to predict future sales. Validity can also be calculated using correlation coefficients – 'low-commitment' consumers would be expected to show a negative relationship with actual scales and vice versa for 'high commitment'.

Similar analyses to the above appear in consumer research reports. Tepper *et al.* (1997) used reliability and validity checks on a 'restrained eating questionnaire'. The researchers developed a shorter version of the original, established questionnaire instrument. The test–retest reliability was significant and the new version was able to classify 274 consumers as having 'low' or 'high restraint', with a '90% success'.

Table 8.9   Association analysis as a predictive validity check (Minitab).

Crosstabulation with chi-square		
Data		
Purcom.	**Puract.**	**Freq**
1(lo)	1(lo)	25
1(lo)	2(hi)	13
2(hi)	1(lo)	15
2 (hi)	2(hi)	23

Rows: purcom	Columns: puract		
	1	2	All
1	25	13	38
2	15	23	38
All	40	36	76

Chi-square = 5.278 with df = 1
Cell contents – count
MTB >
MTB > CDF 5.278;
SUBC > Chi-square 1.
5.2780   0.9784   0.022

Eertmans *et al.* (2006) reported on a previous work where a satisfactory test–retest reliability analysis for the 'food choice questionnaire' was performed over 2- and 3-week periods. Coefficients were relatively high (>0.7) with a sample of 358 consumers ($r = 0.71$, 0.83, respectively).

8.5   Trained panel sensory data

Data generated by trained panels are usually more objective. Here, the measurement system consists of a small group of selected, trained participants. In this sense, they can be viewed as an objective 'instrument'. Generally, data are more valid and reliable. For some tests, levels of measurement become more interval in nature and distributions more normal. Ultimately, the quality of data can become sufficiently high to warrant accreditation of sensory laboratories just as with other testing laboratories (EAL 1995; Elortondo *et al.* 2007).

Data comprise two main types – those from **difference tests** and those from **descriptive analysis methods**. With the former, product differences and similarity are the focus. In the latter, average values and descriptive statistics for product differences are more common objectives, but the data can also provide diagnostic information on panel and panellist performance and other measures of data quality. Some of these features are described by Wolters and Allchurch (1994), who give a summary list of criteria for panel performance, including reliability over replications, convergence (validity) and concordance between panellists, etc.

8.5.1   Samples for sensory assessment by trained panels

The assessors themselves are selected by a screening process and not by a sampling procedure, per se.[1] Thus, the samples referred to here are the food materials. The panel will require enough of the food material for the sensory test and it is subject to the sample selection procedures described above for consumer sensory samples (Section 8.4.1).

In this situation, the sample unit will have to be divided up into smaller portions, one or more per panellist, thus giving many more observations. Ultimately, for intensity-scaling tests, the scores are averaged to give a single result for the samples (the panel mean) before further analysis.

With trained assessors, replication is more likely. The term 'replicate' refers to a repeat determination by the same assessor on two or more unique sample units. The arguments above regarding numbers of end determinations apply to sensory determination also. In the sensory situation, more replicates mean more 'load' on the panellists and higher risk of sensory fatigue. However, just as with instrumental data, single determinations cannot detect error. In addition, several checks on panellist performance rely on at least duplicate determinations being performed.

Numbers of replicates can be approximately four to six for routine profiling work and ten or more during training with difference tests (Stone and Sidel 1993). Blocking of sessions can be used to spread out this load (Section 7.3.1). This has the advantage that replicates are not viewed together, thus reducing bias and improving independence of repeat measures.

8.5.2   Quality of trained panel sensory data

Precision

For metric data in descriptive tests, precision (consistency) can be calculated by use of standard deviations and %CV provided that a measure is repeated. Values will usually be larger than instrumental measures and %CVs can be of the order of 5–10% (panel and panellist) depending on the extent of training. CIs can be calculated and the CI *width* can be used as an indicator of precision and uncertainty. The width of the interval does depend on how replicates are treated in the analysis. Using different panellists as replicates gives narrow interval widths as they are likely to be more numerous than true replicates, e.g. a CI based on $n = 10$ (panellist) will usually be narrower than one based on two replicates ($n = 2$). BSI (1993) describes a more comprehensive consistency check for selected panellists on the basis of scoring intensities of an attribute for six samples in triplicate. In the BSI test, the value of interest is the *error variation* calculated as the **residual standard**

---

[1] This is one of the arguments put forward by practitioners who are of the view that trained panellists should be treated as a fixed effect for analysis. Despite this, most sensory scientists tend to take the opposite view (random effect).

deviation, which is assessed by ANOVA. An example of this type of analysis is detailed below.

Accuracy

Specific accuracy measures are possible only when some estimate of the true measure is known, e.g. as when trained panels score intensities of standard reference samples with known concentrations of compounds that affect sensory properties. There may be some certified standards available, e.g. colour plates. Standard sensory scales are available, but are likely to be more variable. These include the 'food-hardness scale' (Muñoz 1986) and 'spectrum scales', with reference scores (Meilgaard *et al.* 1991). Such calibration tools can also include one or more food samples of known chemical concentration, which will produce a range of sensory stimuli on the scale (Ishii and O'Mahony 1991).

Ultimately, descriptive tests produce some summary measure and CIs can be calculated, although this is not often reported in publications. When CIs are calculated they indicate, in addition to precision, the degree of uncertainty, in that they give a probability of repeated similar samples producing an interval containing the true value. Narrow intervals indicate less uncertainty.

Accuracy is of use for monitoring trained sensory panel performance in this way using reference samples, and in other tests such as ranking of intensities in the correct order. Accuracy in the *qualitative* sense for descriptive sensory tests is measured by the comprehensiveness of profile measures – does the profile include all the pertinent characteristics of the product? For instance, does the attribute 'rancidity' really measure what was intended? This is more difficult to gauge, although the sensory scientist can consult previous work on similar products for comparison and can use chemical and physical standards to 'fine-tune' definitions.

Accuracy can also be judged in difference tests when the result can be predicted, i.e. a true or 'correct' result is known, as when comparing samples with a known formulation difference. The level of accuracy can be judged on the ba sis of the number of correct responses on repeated testing. Some sensory texts give guidance on this with levels of approximately 50–75% quoted. Such figures will depend on the difficulty of the test and as yet there does not appear to be a standard test, although there are several 'selection criteria' tests described (e.g. BSI 1993).

8.5.3 Sources and test types

A wide variety of procedures and tests are currently in operation worldwide for sensory testing food by trained panels. These tests are applied in quality control, product development and sensory research work. The data produced range from simple univariate values to large databanks of multivariate values. Univariate and some bivariate application are dealt with in this chapter. Multivariate methods are dealt with in Chapter 12. Most, if not all, trained panel work is laboratory-based under controlled conditions. A selection of test types is listed in

Table 8.10 Sensory tests, data types and level of measurement.

Test	Data	Level of measurement
Difference	Univariate counts	Nominal
Descriptive	Univariate/multivariate scores	Ordinal/interval
Magnitude estimation	Univariate score relative to reference	Ratio
Time intensity	Univariate score over time	Interval/ratio
Psychophysical	Score intensities over controlled stimuli, e.g. (just-noticeable-difference)	Ratio

Table 8.10, but it is not possible to include examples of all the methods in this chapter.

The objective of the analysis must be borne in mind when selecting some of the above tests as analysis can be applied to panel and panellist performance checks as well as to difference in products, etc.

8.5.4 Experimental design issues

In addition to the general points applicable to both consumer and trained panel experiments, a feature of the latter for descriptive analysis tests is designs to enable assessment not just of product differences, but assessor affects and interactions of assessor by sample, carry-over effects, etc. Such designs are basically factorial in nature, and demand increased replication in relation to the number of factors included. Analysis methods are based on parametric ANOVA.

Lea *et al.* (1998) give a detailed discourse on the application of ANOVA in treatment of sensory descriptive data. They describe a variety of replication structures that result in different error terms for significance assessment. They recommend *at least three full replications* where a single large lot of the food material supplies all assessors, where individual units can be given to each assessor then *at least two units per assessor* is appropriate. These authors also view selected trained assessors as a *random effect* (as appears to be the opinion of most sensory practitioners). Steinsholt (1998) discusses this issue and the case for the *fixed effect* viewpoint. He argues for use of the split plot design in analysis of panel data and explains how different conclusions can be reached depending on these differing approaches. Use of *random effect* factors requires software that can accommodate such terms (basic Excel cannot, but some forms of Minitab can; it is also possible to manipulate the error term in Excel ANOVA to achieve the necessary analysis).

Research experiments may concentrate on such effects, but ultimately, the main focus for other practitioners is the *product* or *treatment effect.* Again, *randomised block structures* are common along with *repeated measures designs*.

The other main types of tests used by trained panels are difference or discrimination tests. These can be of the paired design type as described above for consumer panels.

Table 8.11 Chance probability of some common difference tests and minimum numbers (one-tailed) for significance (5% level) with 12 trials.

Test	Chance probability (%)	Number trials	Number for significance
Paired	50	12	10
Duo/Trio	50	12	10
Triangle	33	12	8
'2 out of 5'	10	12	4

8.5.5   Analysis of trained panel sensory tests

Discrimination tests (difference testing)

How to analyse difference tests

Difference tests produce counts of occurrences. Assessors make choices and these are enumerated. These data can be analysed by the binomial method or the chi-square test as described above for the paired preference test (Box 8.1).

For some tests there is a 'correct' result in that the panel organiser knows in advance that there is a real difference present, e.g. as when one sample has more of an ingredient than the other. In such tests, the outcome can be predicted if the null hypothesis is false, and therefore, they are *directional* (one-tailed). Correct results are expressed as the number of 'successes'. Some important differences arise in respect of the chance probabilities. This can give significant (non-chance) results with smaller panel sizes, at the expense of a more complex test. Thus, the well-known **triangle test** has a chance probability of 33% (Table 8.11).

All these tests can be analysed by binomial probabilities (Table 8.12) where for one-tailed tests the *number of successes* are used and for two-tailed tests the *number of agreements*.

The chi-square test can also be used (Table 5.3; test of proportions) by entering the frequency counts of the outcome of the test. Chi-square is more powerful and

Table 8.12 Binomial formulae for difference test analysis using Excel.

Test	H_1	Binomial Formula (Excel)
Paired difference	Two-tailed	= (1- (BINOM.DIST[a] (Number of successes-1, number of trials, 0.5, True)) * 2
Paired difference (directional)	One-tailed	= 1- (BINOM.DIST (Number of successes-1, number of trials, 0.5, True)
Paired difference (preference)	Two-tailed	= (1- (BINOM.DIST (Number of agreements-1, number of trials, 0.5, True)) * 2
Duo-Trio	One-tailed	= 1- BINOM.DIST (Number of successes-1, number of trials, 0.5, True)
Triangle	One-tailed	= 1- BINOM.DIST (Number of successes-1, number of trials, 1/3, True)
'2 out of 5'	One-tailed	= 1- BINOM.DIST (Number of successes-1, number of trials, 0.1, True)

[a]BINOMDIST() Excel 2003.

gives slightly lower (more significant) p-values.[2] To illustrate one of the above formulae, the ***triangle test*** in its basic form is used below.

How to analyse a triangle test

Example 8.5-1 *Analysis of a triangle test by the binomial method*
A laboratory-based sensory experiment with trained assessors compared two products (A and B) for evidence of a difference. Coded samples (A, B) were presented in balanced order, for a triangle test (where A is the odd sample) to 15 assessors. Eight chose A as the odd sample and seven chose B. On the basis of these data, do the products differ?

The objective was to establish whether or not two products, A and B, differ. The null hypothesis is that the proportion of assessors choosing A equals the proportion choosing B. The alternative is that proportion choosing A (the correct one) is greater than the proportion choosing B (one-tailed $\alpha = 5\%$). These data were entered into the Excel function (Table 8.12) and a p-value of 0.088 was obtained (not significant): 1- BINOM.DIST (8-1, 15, 1/3, TRUE) = 0.088.

The conclusion is that H_0 is retained: there is not enough evidence for a difference between the samples. (NB: This should not be interpreted as 'proof' that the samples are the same.)

Results for the other difference tests are calculated in similar manner using the binomial formulae. The *A not A test* and the ***same–different*** test are analysed using chi-square (see below). CIs can be calculated for difference tests as described by Smith (1981) on the basis of the normal approximation (see Section 8.4.2 for paired preference). Macrae (1995) gives more advanced procedures.

The power of the above paired and triangle tests has been shown to be lower than other equivalents due to various factors such as not specifying the nature of the difference (Ennis 1990). Forms of the test that do specify this have the same choice probabilities, but have higher power. They are referred to as ***m-alternative forced choice*** procedures. An attribute is specified as the criterion for difference detection (Bi 2006). Thus, the ***2-AFC*** is a directional paired difference test, e.g. which of the two samples is sweeter? Rather than 'are the two samples the same or different?' The ***3-AFC*** is performed as for the triangle test, but specifies an attribute; e.g. there are three samples, two are the same, one is different in saltiness–identify the one which is more 'salty'.

Similarity testing

In some circumstances, the sensory practitioner may wish to establish whether two products are the *same* or not. This can occur when a formulation requires a slight

---

[2] Expected frequencies are 50:50 for paired tests, 1/3:2/3; for the triangle test and 10:90 for 'two out of five'; the chi-square probability must be adjusted to one-tailed by halving the two-tailed value.

change in an ingredient. The manufacturer does not want consumers to notice the change and ideally a test that concludes 'no difference' is required.

At first sight, the logical route would be to use a difference test, but this is not so. These tests can tell the sensory scientist that products exhibit a difference as specified by the alternative hypothesis, which applies if the null hypothesis is rejected. If the null is retained, this does not mean that products do not differ (i.e. the null is not proved by non-significance – rather there is not enough evident to reject it) (see Section 4.2.2).

If the scientist wishes to establish that this latter situation is present then a variation called ***similarity testing*** can be used. The basis of this type of testing requires examination of the *alpha* and *beta* risks; with the usual discrimination tests, the *alpha* risk is protected by the level of significance – the chance of declaring a false difference is *alpha*. In the case of similarity tests, the *beta* risk is of more interest – the analysis must ensure that the chance of *missing a true difference* must be kept low. This probability depends on a number of factors as discussed in Section 4.2.2.

One factor is the *size of the difference* present. This can be visualised in more than one way in the context of a sensory trial. In similarity testing, the difference is viewed as the *degree above chance probability*. Thus, a 50% detection above the chance probability would mean a large difference and 25% above would indicate a smaller difference, etc. (these values are based roughly on the publications of Schlich (1993) and Lawless and Heymann (1998)). The concept can be restated as the number or *proportion of assessors* (trained or consumer) likely to detect the difference if it is present. Ideally, the manufacturer would wish this to be 0%, but it should be remembered that in difference testing when assessors cannot tell a difference they are forced to make a chance choice. This means that a certain percentage above zero will occur by chance even when products are the same. Thus, a maximum acceptable proportion is chosen. The manufacturer wants this to be as low as possible, but typically, 20% is used.

On this basis, the power can be determined for various degrees of difference and samples size (Table 8.13). As seen, small differences assessed with high power require many more assessors. Hence, detection of a small difference with *alpha* at its usual setting of 5%, and with *beta* at the same value requires 150 assessors for the test. Preliminary testing can give an estimate of the difference to enable selection of assessor numbers, but large numbers of assessors could cause problems for in-house panels and consumer assessors are usually required.

The requirement can depend on other factors as discussed by Schlich (1993) and Macrae (1995), who provide tables and charts for the required power and assessor numbers. Worked examples are given in BS ISO 2004. In the similarity test, the *fewer* the number of assessors who select correctly, the *lower* the probability that such a difference would be detected by the consumer. Hence, published tables indicate the *maximum number of correct responses* allowable for a *similarity* to be concluded. Unfortunately, such low levels are required to be at or below chance levels, which tends to undermine the confidence in the results of similarity tests based on difference testing (Piggot *et al.* 1998).

Table 8.13   Approximate power for triangle test at varying levels of sensory difference, *alpha*, *beta*, proportion of discriminators and number of assessors[a].

Number of assessors	Proportion of discriminators in population (%)	Sensory difference	Alpha (%)	Beta (%)	Power (%)
20	20	Medium	5	20	80
20	20	Small	5	45	55
40	20	Medium	5	5	95
150	20	Small	5	5	95
120	20	Small	10	5	95
210	20	Small	5	1	99
120	20	Small	5	10	90
600	10	Small	5	5	95
70	30	Small	5	5	95
60	20	Small	5	10	90

[a]Approximations from Schlich (1993) and BSI (2004).

How to perform a similarity test using a triangle test

Only a limited example of the procedure is given here with approximate numbers of assessors and these are based on the publications above. Essentially, decisions need to be taken on the settings for *alpha*, *beta* and the maximum proportion of discriminators and the size of the difference. Assuming that these are 5%, 5% (power is 95%) and 20%, respectively, then the assessor number figure in Table 8.13 would provide a suitable sample size, i.e. 150 consumers. Changing any of the settings will change this number. Usually, it should not be necessary to use a similarly test for larger differences, thus the '*small difference*' setting can be retained. *Alpha* is of less important and it can be increased, e.g. setting *alpha* to 10% reduces the number of assessors to 90. Power is high with *beta* at 5% – the number of assessors is directly related to power. Thus, increasing power to 99% means about 210 assessors are required, reducing it to 90% requires about 120. The largest influence appears to be the proportion of discriminators in the consumer population. Cutting this to 10% brings the assessor count up to near 600! Using 30% discriminators approximately halves the number. Thus, the number of assessors for a ***similarity triangle test*** is higher than that for a ***difference triangle test***. BS ISO (2004) recommends that not less than 30 should be recruited (e.g. *alpha* at 20% and power at 80% would require 40 assessors for 20% discriminators).

Using the guidelines above, a sample of consumers can be collected and presented with the samples as for a triangle test protocol. The difference is in the analysis where the decision on a *significant similarity* depends on the *maximum* numbers of correct responses. Figures are given in published tables for this, but a useful check is given by the formula in the BS ISO document that calculates the proportion of discriminators for experimental numbers of assessors, correct responses and the settings above. The formula has been entered into Excel (Table 8.14) for the settings in the last row of Table 8.13. The terms necessary for the formula are named by abbreviations, which simplify its display. The

Table 8.14 Similarity test calculator for triangle test (Excel).

	A	B	C	D	E
1			Similarity test		
2		Proportion of discriminators	0.20		
3		*alpha*	0.05		
4		*beta(b)*	0.10	Formula	
5		*Z beta(z)*	1.28	$= ABS(NORM.S.INV^{a}(b))$	
6		Number of assessors (n)	60		
7		Number of correct responses (cr)	28		
8		Result			
9		Calculated proportion	0.324	$=$ Main formula	
10		Correct responses below (%)	33%	$= INT(C9 * 100 + 1.5)$	
11					
12					
13			Main formula^{b}		
14	$= ((1.5 + (cr/n) - 0.5) + 1.5 * z * SQRT((n * cr - cr^{\wedge}2)/n^{\wedge}3)$				
15			Correct responses for other results		
16	-5	-4	-3	-2	-1
17	0.196	0.222	0.247	0.273	0.333
18	20%	22%	25%	28%	31%

[a] NORMSINV() Excel 2003.
[b] BSI (2004).

formula requires to be entered for the calculated proportion (cell C9) and for the five cells below (A17–E17) which calculate other proportions for 'correct' results below the one found. Data required are *beta*, the number of assessors and the number of correct responses. *Alpha* and the number of discriminators are also in the table for information, but are not required for the calculation (*alpha* affects the total number of assessors used). The calculated proportion is adjusted slightly to take into account the approximation being used, and expressed as a percentage.

A similarity test was run with 60 consumer assessors and a result of 28 correct responses was obtained.

The calculated proportion is 33%, which is well above the specified 20%, i.e. if the changed product goes out in the tested form, an estimated 33% of the consumer population will detect a difference. To get the maximum number for a similarity requires entry of figures below the test result. This has been done for up to 5 below the number attained in the original test. As the number of correct responses falls, the number of discriminators also reduces, but it is not until 23 that 20% (before adjustment) is achieved (22 reduces the count to 18%). Thus, if no more than 23 out of 60 declare a difference then the samples can be deemed similar at 90% power. In the example, 28 declare a difference so H_0 is retained and there is

Table 8.15   A not A test (Minitab).

Identified	Presented	Frequency	
A	A	11	
A	Not A	28	
Not A	A	49	
Not A	Not A	32	
Identified	Presented	Frequency	
1	1	11	
1	2	28	
2	1	49	
2	2	32	
Rows: Identified		Columns: Presented	
	1	2	All
1	11	28	39
2	49	32	81
All	60	60	120

Chi-square = 10.978 with df = 1
Cell contents – count
MTB ≫ CDF 10.978
SUBC > Chi-square 1.
10.978 0.9991 0.001
MTB >

insufficient evidence for the samples being similar. Bi (2007) describes a method for similarity testing with the paired comparison test.

A not A

'A not A' is a more involved difference test and is analysed by chi-square in a '2 × 2' crosstabulation as illustrated in Table 6.1 using Minitab. In the example, ten assessors received three samples of 'A' and three of 'not A'. The frequency values for the test were recorded in a crosstabulation as displayed at the top of Table 8.15. Samples *identified* as 'A' or 'not A' are compared with samples that were *presented* as 'A' or 'not A'.

In Minitab, the 'A' or 'not A' terms need to be coded as 1 and 2, respectively.

The analysis proceeds as described for Table 6.1. The p-value is less than 0.05 and the null hypothesis is rejected: there is evidence of a difference. More details of this test are given in BSI (1988).

'Same–different' test

The *'Same–different' test* (also referred to as the 'simple difference test'; Meilgaard *et al.* 1991) involves presenting samples with a known 'non-difference', i.e. they are the same, along with the two samples that are different. Assessors can receive one pair or two and the occurrence of the four combinations are distributed evenly. Data from such test can be analysed by chi-square in the same way as 'A not A'. Frequencies are entered as explained above for 20 assessors who received a single pair of samples (Table 8.16).

In this test, the result is not significant: there is no evidence of a difference.

Table 8.16   'Same-different' test (Minitab).

Identified	Presented	Frequency	
Same	Same (AA or BB)	9	
Same	Different (AB or BA)	5	
Different	Same (AA or BB)	11	
Different	Different (AB or BA)	15	
Rows: Identified	Columns: Presented		
	1	2	All
1	9	5	14
2	11	15	26
All	20	20	40

Chi-square = 1.758 with df = 1
Cell contents – count
MTB > CDF 1.758;
SUBC > Chi-square 1.
1.758   0.8152   0.185
MTB >

Descriptive analysis data

How to analyse descriptive analysis panel data and panel performance
Descriptive tests

There are a number of different descriptive tests often called *profiling methods*, typified by the ***quantitative descriptive analysis*** method of Stone and Sidel (1993). They are characterised by using intensity scales to score a number of sensory attributes in an objective manner, with replication and by avoiding reference to liking or preference. Other forms of ***quantitative sensory profiling*** techniques such as ***conventional profiling*** (Lyon 1992) have developed and possibly broadened from the above. Analysis of profiling data is more complex because of the increased number of variables generated. If several samples are assessed, it becomes difficult to detect possible trends by visual examination of the data. Display methods such as spider plots or radar charts (Fig. 3.9) go some way to improving visual detection. Only individual attribute analysis is dealt with below (see Chapter 12 for a multivariate example).

As stated above, a trained panel can be envisaged as an objective instrument. Calibration is possible using prepared model samples or by manufacturing products with varying sensory properties. Thus, accuracy and reliability can be improved, but there may still be a requirement for procedures to even out the differences between panellists. Certain types of design can do this such as the RCB, which is common in profiling experiments (also referred to as ***treatment by subjects*** (Stone and Sidel 1993) where each sample or product is assessed by each assessor. Thus, it is a ***repeated measures*** type of design – more fully with replications over time.

A number of (unwanted) effects arise when using panels as the measuring instrument. The intention is to identify and quantify product differences, i.e. the ***sample effect*** (also known as the *product or treatment effect*). However, although this effect is of primary interest, an ***assessor effect*** can arise due to different panellists using different parts of the scale. For example, assessor 5 scores all products high on the scale whereas assessor 8 uses a lower section of the scale. Additionally, assessors may score products in a different rank order, e.g. assessor 1 scores product sample A higher than B, but other assessors score B as higher than A, etc. This is known as the ***assessor by sample (or treatment) interaction effect***. If these effects are large, they can undermine the analysis in several ways (Table 7.4). Hunter (1996) gives a more detailed account of these effects.

Sensory data sets can be large as trained panel size can number 10–12 individuals. Each sample has number of measures for each attribute plus any replications. It is assumed that the data are from an interval scale, that they conform to normal and that variances between groups (samples) are similar. Some of these assumptions are checked as part of the analysis. Usually panel results are in the form of a summary of means followed by ANOVA of each attribute in an RCB design. In the examples below, all factors (including assessors) are treated as *fixed effects* unless otherwise indicated.

The sample effect (identifying product (treatment) differences)

In the simpler forms of analysis, profiling data can provide estimates of the product sample mean values and these can be compared for significant differences.

Example 8.5-2 *Analysis of descriptive analysis data for the sample effect*
A trained panel is involved in development of fruit drink product and one attribute of interest is 'sourness'. The objective was to establish whether differences in 'sourness' exist in four retail fruit juices (p1–p4) or not, hence '*is there a significant difference in sourness level between the beverages?*' The experiment involved ten trained assessors, scoring on a graphic line scale for intensity of sourness (0–100), with balanced presentation order.

The appropriate design for these data is an 'RCB without replication', with assessors as blocks and analysis by ANOVA.

The null hypothesis was that there was no difference in sourness between the samples and the alternative was that at least two samples differed. Based on the assumptions above, a parametric test for four groups was required (ANOVA): analysis was by two-way ANOVA using Excel's **Data Analysis Toolpak** (the '**Two-Factor Without Replication**' option). Data were entered as columns and identified as a block in the ANOVA menu (see Table 8.5 details for guidance). A summary of the product scores (Table 8.17) was produced (the summary of the assessor scores has been removed).

It shows that there are differences in the mean values and ANOVA assesses these differences for significance. The *sample effect* is given by the 'columns' section and it shows a p-value of 0.11 which is not significant ($p > 0.05$). Therefore, H_0 is retained: there is insufficient evidence for a real difference in sourness.

Table 8.17   ANOVA of profile data (Excel).

Assessor Block	p1	p2	p3	p4
1	44	12	60	58
2	78	79	12	19
3	29	14	66	82
4	33	47	13	63
5	19	1	6	70
6	71	56	85	63
7	58	15	79	68
8	34	26	19	58
9	50	13	12	21
10	93	84	32	96

ANOVA: Two-factor without replication

Summary	Count	Sum	Average	Variance
p1	10	509	50.9	568.1
p2	10	347	34.7	890.2
p3	10	384	38.4	950.5
p4	10	598	59.8	574.6

ANOVA

Source of variation	SS	df	MS	F	p-value	F-critical
Rows	10715.90	9	1190.66	1.99	0.08	2.25
Columns	3998.90	3	1332.97	2.23	0.11	2.96
Error	16135.10	27	597.60			
Total	30849.90	39				

The assessor effect

An assessor effect is given by the 'rows' section in the analysis: it shows a p-value of 0.08, which is also not significant, although in the 'inconclusive' region. Even if the assessor effect were to show high significance, this would not undermine the analysis because the RCB method has removed the assessor variation.

Assume that a replicate session was included with the session above (Table 8.17). These data are combined with replicate session 1 and are analysed in a different manner: the **Two-way ANOVA with Replication** option in Excel. For this, the data require to be rearranged so that replicate two data occur as consecutive rows for each product, i.e. samples are in rows to ensure that the samples are identified as such in the output. In the table, the data are displayed in similar manner to replicate one for reasons of space. The data are selected as before (include label headings, and then enter '2' for the number of replications; Table 8.18).

The output is large as a summary is given for each assessor. This has been removed and an overall mean value for each product calculated. In the analysis, both the sample and the assessor effect are significant (sample and columns, respectively; if data are analysed as above, these effects will be switched round). Again the RCB design has extracted the variation due to the large assessor effect allowing a more precise view of the treatment (sample) differences.

Table 8.18   ANOVA for assessor (fixed effect) by sample interaction effect[a] (Excel).

	ANOVA: Two-factor with replication					
Assessor	p1	p2	p3	p4		
a1	44	12	60	58		
	61	17	82	85	D	
a2	78	79	12	19		
	73	65	43	77	A	
a3	29	14	66	82		
	66	13	47	69	T	
a4	33	47	13	63		
	39	52	64	66	A	
a5	19	1	6	70		
	73	17	87	81		
a6	71	56	85	63		
	62	71	61	77		
a7	58	15	79	68		
	82	28	75	74		
a8	34	26	19	58		
	35	65	54	83		
a9	50	13	12	21		
	36	15	56	47		
a10	93	84	32	96		
	98	77	94	99		
Mean	56.70	38.35	52.35	67.80		

ANOVA			Sourness			
Source of variation	SS	df	MS	F	p-value	F-critical
Sample	8904.30	3	2968.10	7.31	0.0005	2.84
Columns	15167.30	9	1685.26	4.15	0.0008	2.12
Interaction	16083.20	27	595.67	1.47	0.13	1.77
Within	16242.00	40	406.05			
			Res. sd			
Total	56396.8	79	20.151			

[a]Analysis of the data with the assessor effect assigned as 'random' produces the same overall conclusion.

The assessor by sample interaction

The interaction term has also been analysed: it is not significant. This indicates that although the panel are using the scale in different ways, panel members do not appear to be scoring the sample set in a significantly 'crossed' manner in terms of interaction. This is desirable as a significant assessor-by-sample interaction is a more serious error in terms of panel performance.

Significant product differences have been found by ANOVA and multiple comparisons are now required to locate these. Ideally, a package that includes the necessary procedures after this form of ANOVA should be used. MegaStat has this facility and analysis by Tukey's test locates products 1 and 2 and 2 and 4 as significantly different; the more powerful t-test procedures (MegaStat) agree with these findings but products 3 and 4 and 2 and 3 are also different. If the Bonferroni adjustment (significance levels at 0.008) is applied to these latter results, then they are rendered not significant.

Overall, the products displayed a range of sourness levels, with product 4 having the highest, with a significantly different step down to the lowest value (product 2).

Panel diagnostics, training and selection

Training of selected assessors entails monitoring and feedback to improve performance. Such checks are also used as aids for selection of panellists and of sensory attributes. In eventual testing of samples, the data are often scanned for confirmation that panel and panellist ability is at the standard attained during training.

Analysis can be done by a number of methods and in a number of styles:

- Summary methods of accuracy and precision
- Correlation methods
- Sensory cf. sensory
- Panel cf. panel
- Panellist cf. panel
- Sensory cf. instrumental

Precision, discrimination and accuracy of panel and panellist

The summary methods of Section 3.4.3 can be used to calculate precision, accuracy and the discrimination ability of both the panel and panellists. This has been done with the measures on 'sourness' above along with a second attribute ('sweetness').

The methods can also be used to aid selection of attributes. Thus, individual attributes can be checked and those that show poor precision, etc., can be developed with more training, or be deleted from the profile if there are alternatives.

Panel precision

For *precision* (consistency), the panel means for replicates 1 and 2 are compared. A summary of the panel means sd and range for the means of the attribute 'sourness' is compared with sweetness (Table 8.19).

The precision of the panel is given by the overall %CV for each attribute. In this case, the panel is more precise with measurement of 'sweetness' as it has a lower %CV than that of 'sourness'. The overall range of the means across products gives a guide to the level of *discrimination* of the attribute. With 'sourness', the range is lower than that of 'sweetness' indicating that sweetness provide a more effective discriminator between products than does 'sourness'. That is, the closer the panel means are for different products, the less that attribute can distinguish between them. Another measure of panel precision can be obtained by ANOVA – here the magnitude of the error term (the **'Within MS'** (mean square) term in Table 8.18) is a measure of agreement between replications. Taking the square root of this gives the *residual standard deviation*. A comparison of the ANOVA tables for 'sour' and 'sweet' shows the latter's 'within' error term is lower

Table 8.19   Measures of panel precision and discrimination (Excel).

	p1	p2	p3	p4	Range	Overall mean
			'Sour' Data			
Replicate 1	50.9	34.7	38.4	59.8		
Replicate 2	62.5	42	66.3	75.8		
Mean	56.7	38.35	52.35	67.8	29.45	53.8
sd	8.20	5.16	19.73	11.31		
			Av. sd	11.10		
			%CV	20.64		
			'Sweet' data			
Replicate 1	20.6	67.5	32.6	11.5		
Replicate 2	27.3	61.5	40.1	14.7		
Mean	23.95	64.5	36.35	13.1	51.4	34.5
sd	4.74	4.24	5.30	2.26		
			Av. sd	4.14		
			%CV	12.00		

(Tables 8.18 and 8.20) confirming that this attribute is determined more consistently than the other.

A third possibility, when there are many products, is to run a correlation analysis of replicate 1 versus replicate 2. For high precision, there should be a high positive and significant correlation.

Panel accuracy

If instrumental data are available for the product sample set then *accuracy* can be assessed by performing a correlation analysis on the mean values for the sensory versus instrumental values across the products. In terms of calibration, scores on known concentrations of products made up with a range of flavour/texture/colour ingredients, etc., should show good agreement with high (>0.7) correlation coefficients. In the case of sourness, assume that figures are available for the added amount of citric acid in the beverages.

The correlation coefficient has been calculated using Excel's CORREL() function and is displayed on the chart along with a fitted line (Fig. 8.4). The number of points is limited in this example because only four products were examined, but a linear trend is apparent. Two points are 'off the line', but the acid content values in the data are within a short range. The correlation coefficient is 0.87, which is high

Table 8.20   ANOVA of 'sweetness' attribute (Excel).

Source of variation	SS	df	MS	F	p-value	F-critical
Sample	29453.65	3	9817.88	40.08	0.00	2.84
Columns	3423.70	9	380.41	1.55	0.16	2.12
Interaction	7361.60	27	272.65	1.11	0.37	1.77
Within	9799.00	40	244.98			
			Res. sd			
Total	50037.95	79	15.65			

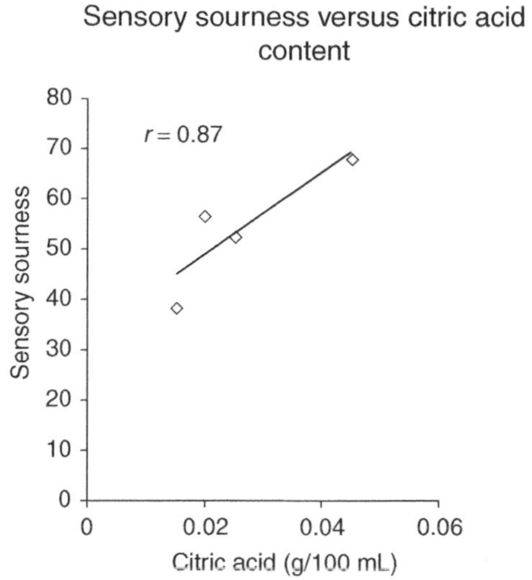

Sensory sourness versus citric acid
content

Fig. 8.4   Scatter plot of sensory sourness and acid content (Excel).

in that >50% of the variation in the sensory data is explained by the chemical data. The significance of the coefficient was calculated using the equation given in Table 6.2. It is not significant and this is the same even if the data are analysed using the rank coefficient (Spearman's). Although the panel has performed 'reasonably well', usually at least perfect *rank correlation* should be achievable with more training. Other tests such as *Kendall's rank concordance* (Megastat) and the *Page test* can be used to check that the order of ranking corresponds to the accurate (instrumental) rank.

Panellist precision and accuracy
Similar analysis can be done for individual panellists. Each section of the data can be treated as above. The various measures for each panellist can be graphed for comparison and review.

Other panel checks
Correlation of panellist with panel

Ideally, trained panellists should demonstrate concurrence with the panel in that each member's scores for an attribute should correlate well with the panel. This was done for the mean scores for panellists 2 and 5 plotted against the panel means (Fig. 8.5).

Panellist 5 has a positive high correlation ($r = 0.99$) and agrees with the panel. Panellist 2 has strong divergence from the panel with a low negative coefficient (−0.32) and would benefit from additional training.

Additional correlation checks can be done using all attributes to see which ones correlate with others. This can reveal useful information on the patterns that are

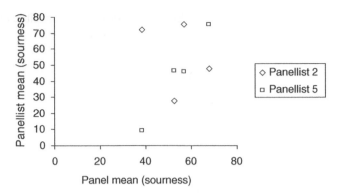

Fig. 8.5   Scatter plot of panel and panellist means for sensory sourness (Excel).

building up in the data in a multivariate sense (see Chapter 12), but which can also be used to eliminate some attributes at early stages of vocabulary selection.

8.6   Analysis of relationships

Ultimately, a major application of sensory data is in the identification of which sensory attributes are important in deciding whether a product is liked and purchased, or not. Thus, the study of relationships between sensory profiles and consumer measures is of great interest to food companies. Some examples of these relationships are given in Chapters 10 and 12.

References

Allison, A. and Work, T. (2004) Fiery and frosty foods pose challenges in sensory evaluation. *Food Technology*, **58**(5), 32–37.

Bi, J. (2006) *Sensory Discrimination Tests and Measurements*. Blackwell Publishing Ltd., Oxford.

Bi, J. (2007) Similarity testing using paired comparison method. *Food Quality and Preference*, **18**, 500–507.

Bower, J. A. (1996) Statistics for food science III: sensory evaluation data. Part B – discrimination tests. *Nutrition and Food Science*, **96**(2), 16–22.

Bower, J. A. and Boyd, R. (2003) Effect of health concern measures and consumption patterns on measures of sweetness by hedonic and just-about-right scales. *Journal of Sensory Studies*, **18**(3), 235–248.

BSI (1988) BS 5929–5:1988 ISO 8588:1987. *Sensory Analysis of Food. Methodology. 'A'–'not A' Test*. British Standards Institution, London.

BSI (1993) BS 7667–1:1993 ISO 8586–1. *Assessors for Sensory Analysis. Guide to the Selection, Training and Monitoring of Selected Assessors*. British Standards Institution, London.

BSI (2004) BS ISO 4120:2004. *Sensory Analysis. Methodology. Triangle Test*. British Standards Institution, London.

BSI (2005) BS ISO 5495:2005. *Sensory Analysis. Methodology. Paired Comparison Test.* British Standards Institution, London.

BSI (2006) BS ISO 8587:2006. *Sensory Analysis. Methodology. Ranking.* British Standards Institution, London.

Calviño, A., Delia Garrido, D., Drunday, F. and Tamasi, O. (2005) A comparison of methods for monitoring individual performances in taste selection tests. *Journal of Sensory Studies*, **20**, 301–312.

Cochran, W. G. and Cox, G. M. (1957) *Experimental Designs*, 2nd edn. John Wiley & Sons, Inc., New York.

EAL (European cooperation for accreditation of laboratories) (1995) *EAL-G16: Accreditation for Sensory Laboratories.* United Kingdom Accreditation Service, Teddington.

Eertmans, A., Victoir, A., Notelaers, G., Vansant, G. and Van Den Berg, O. (2006) The Food Choice Questionnaire: factorial invariant over western urban populations. *Food Quality and Preference*, **17**, 344–352.

Elortondo, F. J. P., Ojeda, M., Albisu, M., Salmerón, J., Etayo, I. and Molina, M. (2007) Food quality certification: an approach for the development of accredited sensory evaluation methods. *Food Quality and Preference*, **18**, 425–439.

Ennis, D. M. (1990) Relative power of difference testing methods in sensory evaluation. *Food Technology*, **44**(4), 115–117.

Gacula, M. C., Jr. (2003) Validity and reliability in sensory science. In *Viewpoints and Controversies in Sensory Sciences and Consumer* by H. R. Moskowitz, A. M. Muñoz and M. C. Gacula, Jr. (eds). Blackwell Publishing Ltd., Oxford, pp. 100–101.

Gacula, M. C., Jr., Mohan, P., Faller, J., Pollack, L. and Moskowitz, H. R. (2008) Questionnaire practice: what happens when the JAR scale is placed between two 'overall' acceptance scales? *Journal of Sensory Studies*, **23**, 136–147.

Gacula, M. C. and Singh, J. (1984) *Statistical Methods in Food and Consumer Research.* Academic Press Inc., Orlando, FL, pp. 323–336.

Hunter, E. A. (1996) Experimental design. In *Multivariate Analysis of Data in Sensory Science* by T. Næs and E. Risvik (eds). Elsevier Science BV, Amsterdam, pp. 37–70.

Ishii, R. and O'Mahony, M. (1991) Use of multiple standards to define sensory characteristics for descriptive analysis: aspects of concept formation. *Journal of Food Science*, **56**(3), 838–842.

Lawless, H. T. and Claassen, M. R. (1993) Application of the central dogma in sensory evaluation. *Food Technology*, **47**(6), 139–146.

Lawless, H. T. and Heymann, H. (1998) *Sensory Evaluation of Food: Principles and Practices.* Chapman and Hall, New York.

Lea, P., Næs, T. and Rødbotten, M. (1998) *Analysis of Variance for Sensory Data.* John Wiley & Sons, Inc., New York, pp. 56–63.

Lesniauskas, R. O. and Carr, B. T. (2004) Workshop summary: Data analysis: getting the most out of just-about-right data. *Food Quality and Preference*, **15**, 891–899.

Lyon, D. H. (1992) *Guidelines for Sensory Analysis in Food Product Development and Quality Control.* Chapman & Hall, London.

Macrae, A. W. (1995) Confidence intervals for the triangle test can give reassurance that products are similar. *Food Quality and Preference*, **6**, 61–67.

Meilgaard, M., Civille, G. V. and Carr, B. T. (1991) *Sensory Evaluation Techniques*, 2nd edn. CRC Press, Boca Raton, FL.

Moskowitz, H. R. (2003) Sample issues in consumer testing. In *Viewpoints and Controversies in Sensory Science and Consumer* by H. R. Moskowitz, A. M. Muñoz and M. C. Gacula (eds). Blackwell Publishing Ltd., Oxford, pp. 128–131.

Muñoz, A. M. (1986) Development and application of texture reference scales. *Journal of Sensory Studies*, **1**, 55–83.

O'Mahony, M. (1988) Sensory difference and preference testing: the use of signal detection measures. In *Applied Sensory Analysis of Foods*, Vol. **1** by R. H. Moskowitz (ed.). CRC Press, Boca Raton, FL, pp. 145–176.

Piggot, J. R., Simpson, S. J. and Williams, S. A. R. (1998) Sensory analysis. *International Journal of Food Science and Technology*, **33**, 7–18.

Schlich, P. (1993) Risk tables for discrimination tests. *Food Quality and Preference*, **4**, 141–151.

Smith, G. L. (1981) Statistical properties of simple sensory difference tests: confidence limits and significance tests. *Journal of the Science of Food and Agriculture*, **32**, 513–520.

Steinsholt, K. (1998) Are assessors levels of a split plot factor in the analysis of variance of sensory profile experiments? *Food Quality and Preference*, **9**(3), 153–156.

Stone, H. and Sidel, J. L. (1993) *Sensory Evaluation Practices*, 2nd edn. Academic Press, San Diego, CA, London, pp. 99–106.

Tepper, B., Choi, Y. S. and Nayga, R., Jr. (1997) Understanding food choice in adult men: influence of nutrition knowledge, food beliefs and dietary restraint. *Food Quality and Preference*, **8**, 307–318.

Upton, G. and Cook, I. (2001) *Introductory Statistics*, 2nd edn. Oxford University Press, New York.

Wakeling, I. N. and Macfie, H. J. H. (1995) Designing consumer trials balanced for first and higher orders of carry-over effect when only a subset of k samples from t may be tested. *Food Quality and Preference*, **6**, 299–308.

Wolters, C. J. and Allchurch, E. M. (1994) Effect of training on the performance of descriptive panels. *Food Quality and Preference*, **5**, 203–214.

Chapter 9
Instrumental data

9.1 Introduction

The applications dealt with in this chapter apply mainly to chemical data generated in analytical chemistry experiments and investigations. This subject area has received much attention in recent years regarding data quality. The relative importance of the strict control requirements for levels of toxins, additives, contaminants and nutrients in food reflect this. With chemical data, there are legal compositional requirements and limits of constituents for legal and safety reasons. Various initiatives have developed from procedures set up by the Department of Trade and Industry's Valid Analytical Measurement scheme and by national standards such as NAMAS (National Measurement Accreditation Service). Currently, much new work is ongoing with Food Standards Agency projects, some of which are referred to below. There is also great importance in microbiology from the safety, quality and storage stability point of view and data from this discipline must also be subject to scrutiny (not covered in this chapter).

Other instrumental techniques used for food analysis may not have legal or other statutory limits imposed on the measurand, but they are important for different reasons. These comprise such factors as that of ensuring efficient and reliable data on which to base ongoing activities in food processing and research. Such measures include not only a host of physical, biochemical and biological measures of food, but also processing parameters.

9.2 Quality and nature of instrumental data

Data from instruments will usually conform to ratio level of measurement, originating from a normal population. Discrete and continuous forms are possible depending on individual instruments. The values are gathered by an instrumental measuring system during analytical experiments in many science fields. Food is

Statistical Methods for Food Science: Introductory Procedures for the Food Practitioner,
Second Edition. John A. Bower.
© 2013 John Wiley & Sons, Ltd. Published 2013 by John Wiley & Sons, Ltd.

Table 9.1   Some examples of instrumental measures for foods.

Method/measure
Chemical content
Nutrient: Fat, protein, carbohydrate, mineral, vitamin, fibre
Quality/legal: Moisture, ash, meat content, additives, peroxides
Contaminants: Heavy metals, pesticides
Food properties and analysis: Near-infrared, X-ray diffraction, gas chromatography, high-performance liquid chromatography, atomic absorption spectroscopy, plasma analysis
Physical properties
Mechanical properties: Firmness, cohesiveness, yield point
Fundamental: Viscometry, rheometry
Colour: CIELAB, colour matching
General: Humidity, a_w, specific gravity, weight, volume, yield, overrun
Biological
Microbial counts: Non-pathogenic, pathogenic
Nutritional measures: Biological value, net protein utilisation
Processing parameters
Air speed, screw speed, temperature, time, pressure

treated like any other material in respect of analysis and many such methods can be applied. A selection of these is indicated (Table 9.1).

All such methods will have an end determination where the analyte, property or count is measured. There may be a subsequent calculation involved, but at that point, the data are ratio in nature and are ready for summarising and possible further analysis.

9.2.1   Quality of instrumental data

Error

Error levels are lower for instrumental data than those from consumer and sensory sources. Calibration methods are available for most instruments such that systematic error is removed or minimised. Many modern instruments have also reduced the operator error by confining human intervention to sampling and taking of the final reading. This contrasts with many 'wet' methods of chemical analysis, which may involve several individual stages of intermediate measurement (Nielsen 2003). Thus, in addition to sampling and sample measurement, there can be intermediate measures of volume, weight, distance, temperature, etc., and measurement decision depends on the observations of the analyst. All such stages can contribute to overall error. For instance, compare the stages in the wet traditional Kjeldahl method with that of a nitrogen analyser based on the Dumas method. The former has stages of sample preparation, sample weighing, digestion, dilution, distillation, titration aliquot measure, burette reading and the calculation. A modern nitrogen analyser will include sample preparation and weighing, placing in the machine, followed by the final reading.

These latter descriptions apply to chemical analysis, but similar considerations can be highlighted for other instruments, which measure properties in food, such as colour and texture by non-chemical means. Mechanical testing machines can measure empirical and fundamental properties of food texture. They can be calibrated using strain gauges or materials of known mechanical properties. A major source of variability can arise with sample preparation for such devices. Food samples cannot be comminuted for some texture tests and standard-shaped pieces must be cut. Even if this is controlled, inherent variation can result in wide piece-to-piece variation. Individual pieces may deform and fracture in different ways. Reflectance instruments that measure colour are very rapid and involve minimum operator error, but surface variation on food will cause variability. Colour matching systems are more dependent on operator acuity and fatigue level to decide on match points, etc.

With processing trials, several parameters require to be set and measured during the runs over considerable periods. Thus, temperature, humidity, pressure, etc., require to be monitored with instruments that are calibrated.

As seen, a number of error sources will result in a lessening of confidence in the measure. Together, the combined error sources result in **uncertainty** of the measurement and in the case of analytical chemistry at least, it is referred to as a property of the data. This uncertainty can be decomposed into sample, method, operator error, etc. Data from any instrument can be assessed for accuracy and precision of the results, if certain requirements are met (such as an estimate of true value and replication, respectively). These can be determined by the methods given in Section 3.4.3. Some typical values are given in Table 9.2.

Similar measures occur in current research journals, some of which have been referred to in other chapters. Nielsen (2003) quotes a figure of 5% as an attainable %CV in chemical analysis of foods. There are examples of better levels of precision, but it is not always possible to ascertain whether they are based on a 'within-unit' or 'between-unit' calculation or not. Gao *et al.* (2006) appear to have extremely low %CVs (<1%) in their experiments on microbial survival (see below). This contrasts with the apparently high values (%CVs >25%) reported by

Table 9.2 Typical accuracy and precision levels of instrumental data.

Method	Precision within unit (duplicate) %CV[a]	Precision between unit (duplicate) %CV	Accuracy %REM[b]
Soxhlet fat	4	11	±11
Dumas nitrogen	4	12	±3
Brittleness	7	18	—
Luminance	1	5	—
Moisture	2	5	±10
Mineral	3	7	±1

[a] Percent coefficient of variation.
[b] Percent relative error of the mean.

Martín-Diana *et al.* (2007) on a study by other workers of firmness measurement on cantaloupes. The exact method of calculation in these studies was not apparent, although in the latter case a quoted '$n = 14$' implies a 'between-unit' measure.

'Within-unit' calculations assess *measurement variation*, as they are usually carried out on subsamples from a homogenised sample unit. 'Between precision' is based on unit-to-unit variation and is affected by variability within the food material, etc.

In addition, if the above measures are calculated with duplicate end determination and duplicate sample units – more replication would result in lower figures. Precision measures may also be quoted using the terms *repeatability* and *reproducibility* (see below).

Accuracy measures are less commonly published. In some cases, there may not be a reference standard for comparison and accuracy is reliant on use of calibration, as with some instances of colour and texture instrumentation. Chemical methods have a number of possible standards to calculate accuracy. With a certified reference material and modern instrumentation such as with mineral content determination, using atomic absorption spectroscopy, very high accuracy is possible of $\pm 1\%$ or less.

Beyond these summaries, measures that are more specific are used, primarily with chemical analysis data in the food field. Decisions on the levels of accuracy and precision that are 'acceptable' depend on specific requirements. Greenfield and Southgate (2003) point out that for chemical composition tables, attempts to achieve improvement of accuracy beyond $\pm 10\%$ can be offset by the additional work involved. In cases where legality is an issue, the highest possible levels of accuracy and precision may be required.

9.3   Sampling and replication

Chapter 2 described the detail of sample size and amount and how these affect the sample representativeness with respect to the population (Section 2.3.2). Often in analytical methods, the amount of material in the sample is decided by the method of analysis, but there may be a range of possible sizes, e.g. a traditional method such as Soxhlet for fat, maximum sample size could be limited to 10 g, whereas other methods use sample sizes of 50–100 g. Ideally, the larger sample should be taken, but this would depend on the amount of material available, etc. It also depends in turn on other points as indicated above such as the variability of the material. For more variable material, larger samples are more representative.

9.3.1   *Sample sizes in instrumental determinations*

In analytical methods (e.g. AOAC 1999), the number of determinations is often referred to as the *level of replication*. In this case, the term refers to how often the basic measurement was performed. In analytical terms, it means repeated measures by the same analysis system on subsamples of a unique sample unit. Typically, this is often as duplicates – 'duplicate end determination'. Sometimes one determination plus a reference sample may be the norm. These 'end determinations' or observations are carried out on subsamples.

The number of subsamples does not refer to the *sample size*, which is based on the number of distinct sample units (Section 2.3.2). The sample size in this respect depends on the scope of the experiment. For instance, in a quality control laboratory there may be 20 cans of corned beef selected for nitrite content analysis – the contents of each can be blended to even out any in-can variation, then duplicate end determinations are performed. Thus, the sample size is 20 and the number of end determinations is 2 per unit, giving 40 in total. Such levels could be typical of a day's laboratory work. Thus, sample sizes are relatively small in chemical analysis, as are the number of determinations per unit. This also tends to be the situation with other instrumental methods.

Are duplicate end determinations adequate?
Decisions on an acceptable level of replication need to take into account several factors:

- Statistical measures
- Cost of the analysis time and personnel
- The instrumental method itself

If the method involves a long timely procedure using expensive reagents and several personnel, etc., then this may limit increased levels of end determination. If the method is simple, rapid and low cost then more determinations can be incorporated. The method itself may be subject to high levels of variability (low precision would prompt increased numbers). Conversely, with a precise method and well-trained analysts a single determination may suffice, and duplicates are quite acceptable.

The effect of these variations in level of replication on the accuracy of the estimate is also important. Performing a single determination has one big disadvantage: there is no reference point for detection of measurement error. Duplicates provide this and will provide a more accurate estimation of the analyte for the sampling unit, but confidence intervals based on duplicates can be very wide. Increasing the number of replicates will increase the accuracy, but at a decreasing rate, i.e. the gain is greater by increasing from two to four replicates than from four to eight.

This point is illustrated in Section 3.4.3. Thus, duplicates are a compromise for routine analyses with established techniques where a reliable estimate of a standard deviation is available. Other factors can influence the decision within a laboratory, such as costs in terms of time and finance, plus the experience of the analyst and criticality of decisions based on results. The advantage of increased replication is that there is increased confidence – the estimate is more accurate and 'on target' (assuming no systematic error) and confidence intervals (CI) will be narrower:

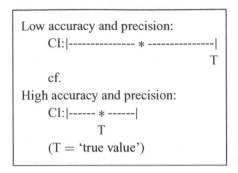

Note that increasing subsample numbers can only increase the confidence in the estimate for a single sampling unit. The only way to increase confidence in the treatment effect is to replicate the full sampling unit.

With other forms of instrumental analysis, the same points apply, but in some forms cost of additional replicates on both sampling units and subsamples may be minimal as in the case of physical texture and colour, but even here the time cost for subsample preparation prior to testing may outweigh any gain.

9.4   Experimental design issues

Just as with sensory properties, instrumental properties can be ascertained for products, samples and prototypes, etc. Ongoing quality control, production development and research programs continually establish and monitor various forms of instrumental properties in foods. Interest lies with chemical levels of nutrients and non-nutrients, as well as instrumental measures of texture, colour and more fundamental aspects of food's structure and chemical make-up. Such properties can be determined simply and monitored, or experiments can study changes that occur during formulation, processing and storage, etc.

All forms of experimental design are used in the above work. Simple one-variable designs are routine, e.g. establishing the nutrient content in new and existing commodities, or routine analysis of additive levels, is common in chemical analysis studies. At the other extreme are multifactor experiments, studying the effect of a process factor on yield of a product, e.g. the effect of air speed, temperature, pressure and time on %yield of dried milk. Often with large experiments, both sensory and instrumental measures are employed.

Such investigations are organised by a number of designs, but in most cases, analysis is by parametric ANOVA, which predominates in published accounts, where differences in products and processes are determined. Designs of the completely randomised design (CRD) and randomised complete block form are very common, with factorial structure and optimisation techniques (Section 7.4). The former designs can all be dealt with using Excel's Toolpak ANOVA with up to two factors with replication being possible. Basic Minitab can accommodate up to three factors. All statistical software can deal with single factor experiments with many levels in the CRD format and a worked example of this type of analysis is illustrated below.

In the literature, food investigations that include instrumental measures range across many applications. Often several analyses are performed on treatments in formulation and process experiments. Abdullah (2007) included moisture, fat, protein, thiobarbituric acid, peroxide value, pH and percent acidity in canned meat formulations with different levels of meat. Many analyses were required to confirm that the formulations complied with national criteria. On the processing side, there is great interest in the study of process machinery and the various settings for effects on critical factors. Dehydration studies such as that by Severini *et al.* (2005) on potato quality looked at effects of drying rate using three drying methods combined with five blanching treatments. Fifteen process runs were required and best quality was found with microwave blanching and belt drying. A relatively simple experiment by Villavicencio *et al.* (2007) compared quality in cereal bars that had been subjected to varying level of irradiation to control insect contamination. Eating quality was maintained with dose levels up to 2.0 kGy and 3.0 kGy, depending on the product flavour stability.

Therdthai *et al.* (2002) carried out a 3^5 partial factorial when examining temperature zone effects on bread crust colour along with other dependent variables. Colour was measured instrumentally, and they quoted a 'consumer acceptable range' for a CIELAB colour parameter (L* lightness) which provided a target response level. They used the data to model the process and identified optimum temperature zoning conditions to give best quality with the crust colour within the acceptable range.

There are also many examples of larger industrial level processes and instrumental investigations that follow the experimental design procedures in proscribed sequences (as in Section 7.2). Yann *et al.* (2005) provide a good example of this in their use of design in attempts to reduce browning in hard cheeses. They discussed the problems of industrial scale experiments, particularly with a product such as cheese, which required long storage times before development of the browning defect. It was, therefore, essential to use design procedures to reduce experimentation. Five critical factors were identified and then subjected to a half fraction of a two-level factorial design (5^2 treatments = 32, half fraction = 16). Also included were *centre points* giving 19 treatment runs. Colour was measured by instrumental means, and after 6 and 12 months, stored sample colour measurements were taken. They identified 'added acid' and 'temperature of starter culture' as critical in effect. These factors along with 'rennet concentration', which was included as it was 'technologically essential', were then incorporated in an optimisation design. This was a full three-level factorial with partial replication and additional points to aid the predictive power of the model (a 2^3 structure = 8 + 6 + 2 replicates = a further 16 runs). The researchers were able to reduce browning by lowering the three factor effects within constraints of the cheese making essentials. A simple example of the use of a fractional design is given in Chapter 10.

Obviously, reduction of experimentation is important and May and Chappell (2002) describe use of a **_Taguchi design_** (a type of incomplete factorial) for finding critical variables in food canning of pasta. They examined the effect of 11 variables using only 12 experiments, including fill weights, ingredient concentration and headspace. The dependent variable was the F_0 value and they identified the concentration of tomato puree as the most critical variable.

A very important area of study in processing is *microbial effects*. The *survival count* is the dependent variable in such experiments. Gao *et al.* (2006) describe inactivation of *Bacillus subtilis* by high pressure and heat treatments. They also used fractionation to identify critical factors, followed by an optimisation structure. Three factors were studied: pressure, temperature and holding time. These were optimised to give maximum microbial destruction. These workers also had a target level of survival identified as a maximum number of spoilage organisms for 'healthy people'.

Measurement of a dependent variable as retention or percent *retention* is often used in instrumental studies of vitamin and mineral losses on processing. Moreira *et al.* (2006) explored the effects of heat shocks in fresh lettuce. Ascorbic acid retention fell slightly (70–80%), but effects on microbial growth and sensory acceptability were mixed. This study provides an example of the reasoning process in cause-and-effect explanations.

9.5   Statistical analysis of instrumental data

9.5.1   Summary methods

Instrumental data can be summarised by a variety of table and graph types. Typically, for routine work, a table of the average values is listed (e.g. Table 3.6). For larger experiments, or where changes with time are involved, graphical displays can give a better interpretation of data (Section 3.2.3). Equally important as the results themselves are the *levels of confidence* in the values. These aspects, in form of standard deviations or standard errors, are commonly presented in tables or graphs with average values of instrumental data.

Example 9.5-1 Analysis of instrumental data in a processing experiment
An experiment was conducted into the effects of drying methods with seven types of drier to establish the extent of effects on reconstitution of dried fruit when dried to constant weight. High reconstitution was taken as a desirable characteristic of the dried food, as it gave a product that compared well with the fresh control material. A CRD was used with one factor, at seven levels (qualitative), plus a raw fresh control. Each treatment was replicated three times:

Factor (1): drying method
Levels (7): tray, fluid-bed, freeze-dry, microwave vacuum, microwave pulsed and fluid pulsed
Control: raw material lot

Table 9.3   Run numbers for CRD experiment.

	Tray	Fluid	Microwave	Vacuum	Freeze	Pulsed microwave	Pulsed fluid
Replicate 1	2	10	20	5	17	11	9
Replicate 2	14	19	3	12	8	1	16
Replicate 3	21	7	13	4	18	15	6

A large batch of the fruit was randomly sampled for 24 lots (1000 g each) of prepared, thinly sliced material. Each drying method was carried out three times in a CRD. Treatments (21) were allocated randomly to a run number and drying performed. At the termination of the drying process, the product was vacuumed packed and samples were taken for reconstitution testing. This was taken as the water absorbed in 90 s, and the measure was expressed as g H_2O per 100 g dried material. All analyses were performed on duplicate subsamples taken from the dried material from each treatment. Duplicates were averaged, rounded to the nearest gram and reported for each replicate treatment. Is there evidence that the different drying methods produce products with different reconstitution properties?

Run number allocation is shown in Table 9.3. Thus, the first run was the 'pulsed microwave' (replicate 2) and the last was the 'tray drier' (replicate 3), etc.

The reconstitution value data are listed in Table 9.4 followed by single factor ANOVA using Excel (see Example 5.5-2 for procedure).

The reconstitution values are also plotted using an error bar graph (Fig. 9.1). The standard deviations are similar (approximately 5–6 g) except for the 'microwave' triplicate. A variance check (F test, Section 4.4) for the microwave data and the treatment with the smallest variance (fluid-bed) showed no significance and equality of variance was assumed. The sd values are equivalent to %CV of approximately 15% (35% for 'microwave'), which is reflective of variation between separate units (true replicates).

The ANOVA p-value is less than 0.05; therefore, at least two treatments differ. Perusal of the summary mean values reveals that the 'pulsed fluid-bed' drying method has the highest reconstitution value. Multiple comparisons could be performed for all possible pairs (21 comparisons), but as the p-value of the result is not markedly low (0.02), it is probable that not all pairs will show a significant difference.

MegaStat t-test comparisons locate six significant differences with α at 0.05, but with Bonferroni adjustment (significance level at 0.00238), only Microwave cf. Pulsed fluid are significantly difference. The more conservative Tukey's comparison agrees with this latter difference but also includes Microwave cf. Freeze.

Overall, the most marked difference was for the 'pulsed fluid-bed' dryer (the highest reconstitution level of the dried food), which was significantly higher than the 'microwave' method.

Table 9.4   Analysis of CRD data (Excel).

Data	Tray	Fluid	Microwave	Vacuum	Freeze	Pulsed fluid	Pulsed microwave
			Reconstitution data: g H_2O/100 g				
Replicate 1	37	45	37	43	37	56	36
Replicate 2	48	35	19	39	50	45	43
Replicate 3	39	41	23	31	47	51	31

ANOVA: Single factor

Summary

Groups	Count	Sum	Average	Variance
Tray	3	124	41.33	34.33
Fluid	3	121	40.33	25.33
Microwave	3	79	26.33	89.33
Vacuum	3	113	37.67	37.33
Freeze	3	134	44.67	46.33
Pulsed fluid	3	152	50.67	30.33
Pulsed microwave	3	110	36.67	36.33

ANOVA

Source of variation	SS	df	MS	F	p-value	F-critical
Between groups	1020.00	6	170.00	3.98	0.02	2.85
Within groups	598.67	14	42.76			
Total	1618.67	20				

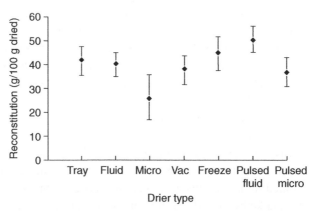

Fig. 9.1   Means ($\pm$sd) of reconstitution by drier type (Excel).

9.6   Chemical analysis applications

The foregoing accounts apply to any instrumental measurement system and the data produced. Analytical chemistry has received much more specialised treatment in these respects. As explained, this is due to the critical importance of the consequences of invalid analyses being performed. Thus, the chemical analyst is interested not only in the end result but also in the ***uncertainty*** of the estimation; some researchers state that unless a measure of uncertainty is included the results themselves are useless (Mullins 2003). This view could well apply to all scientific measures, but there are still occurrences of it not being adopted for chemical data.

Many investigations have taken place to examine error components and to quantify their contribution to uncertainty. Also, cost considerations are included in these studies as reducing uncertainty usually means additional analyses and hence cost in terms of time, resources and personnel. The interest here is in the balance between *gain in certainty*, against the *increased cost* to the laboratory (Lyn *et al.* 2002; FSA 2004a). Another unique aspect of studies in this topic is that the uncertainty is examined not only for location measure estimates such as the mean but also for that of the level of variability – thus, the uncertainty of the standard deviation is also of interest. ***Method proficiency testing*** is one aspect of this protocol that has been developed for some common standard methods with measures all focused on uncertainty in analytical chemistry. In addition to analysis coming under this latter umbrella, where analytes such as pesticides are determined at very low level, there are many proximate analyses and 'crude content' methods used for food. These may exhibit higher levels of uncertainty, but their results and in fact, those from any instrumental measure can be subjected to some of the calculations detailed below. Food analysis methods have received special attention via The ***Food Analysis Performance Assessment Scheme***. Patey (1994) described the initial stages and progress of this initiative – there was some improvement, but not for all analytes and all laboratories. A relatively simple check on performance for proficiency testing schemes is based on calculation of a form of z-score (Section 2.4.1):

$$z = (\text{Test result} - \text{reference value})/\text{sigma}$$

Sigma is a designated 'target value' for the standard deviation of the method data, based on realistic estimates. The larger the discrepancy (error) between the test and the reference, the larger the z-value. The calculation produces z in a standardised form – values *equal or less than 2* are required for the laboratories' result to be declared 'satisfactory'.

9.6.1   *Accuracy and bias in chemical analysis*

As stated above, accuracy of a chemical method is a measure of how close it is to the 'true' value. Variation from the true can occur due to error in the form of ***bias*** (Kane 1997). This circumstance can apply to a number of stages in the analysis (Table 9.5).

It is crucial that this source of error is quantified and removed, or at least accounted for in any analytical determination, although this is not always done (O'Donnell and Hibbert 2005). Bias can be calculated as the ***error of the mean*** (Section 3.4.3), and by the location of the range specified by a ***confidence interval***. The 'true value' is represented by ***reference samples*** or the nearest equivalent.

Table 9.5   Bias sources[a].

Source
Operator
Lab
Preparation
Run
Method

[a]These also affect precision.

Reference materials

Reference materials (RMs) are prepared samples of matter of the same or similar nature to the food material being analysed. They come in a variety of forms and can be made up by the analysts themselves. They cannot provide the exact true value as this is possessed by the test material itself. The assumption is, that if on analysis, the analyst obtains a result that is within the confidence interval for a reference, or within a 'close' region, then the result for the unknown sample can be considered accurate to a similar degree.

The quality of the result in this respect, therefore, depends on the quality of the reference material. Low-cost, 'laboratory-made' versions can be formulated from materials of known composition, or by use of a product or previous sample with identified levels. In the simplest form, this could be via a nutritionally labelled food product of similar nature to the test samples. Nutrition composition labels will give an approximate guide, but the analyst should expect at least ±10% variation. A similar way would be to look up or calculate the expected value based on chemical composition tables or databases. Again, a similar disparity can be anticipated (±10%), but in some cases, it can be much higher. Heinonen *et al.* (1997) found that analysed values for carotene and retinol compounds were 50–60% less than database figures. Materials analysed in this way can be frozen and used over time for stable analytes.

For higher levels of confidence, the analyst must contract analyses out to obtain valid figures from another laboratory or turn to certified reference materials (CRM). These are supplied by organisations such as the Laboratory of the Government Chemist and Office of Reference Materials (LGC, London). They come with a certificate, which details the content of the analyte, plus for most, an interval of uncertainty. The way in which this is expressed depends on various factors such as how many chemical analysis methods were involved in the determination. Thus, it may not be a confidence interval as such. The interval will vary in magnitude for different analytes and are typically of the order of ±1–5%. Others analytes may be quoted without an interval and these are included as 'indicative' levels.

For foods, the most available CRMs are for trace elements in the form of flour, vegetable tissue and animal liver and milk powder. Other certified analytes include moisture, pesticides and aflatoxins. Some CRMs are made up by 'spiking' a material, i.e. adding a known pure amount of the analyte (a technique that could also be employed by the analyst). Traceability of RMs relies on the facility of following standard RMs, through CRMs and ultimately to ***primary RMs*** (XiuRong 1995) that possess the highest attainable accuracy (and cost). More specific food RMs are increasing in scope with new 'wet matrices', which are more like the original food and with certified values for additives becoming available (Anonymous 1999).

One difficulty with RMs is the shelf life of such materials in that the analyte level may change with time. Stable forms of dried material are, therefore, common. Some analytes can be more unstable and sulphur dioxide is one of them. Levermore and McLean (2002) describe development of a stable sulphited food reference material based on a frozen sausage meat mix, which showed stable values for up to 8 months storage.

Forms of bias

Bias has been defined and explained in a general sense, but it can manifest itself in two specific ways. One is ***fixed bias*** where it remains constant over all analyte levels; secondly, it can change with the level as when ***relative bias*** exits, and both can be present (BSI 2000). Once identified, the *significance of bias* can be determined by use of a *1-sample t-test*, provided that a reference figure is available. Calculations of the necessary measures have been shown in Table 3.6 and Section 5.5. Some are used here to illustrate fixed bias (Table 9.6).

Table 9.6 Bias calculations (Fe mg/100 g) (Excel).

	Data	
Analyte	3.4	
Fe content mg/100 g	2.9	
	3.4	
	3.2	
	3.6	
Count (n)	5	
Mean	3.30	
Standard deviation	0.26	
Standard error (se)	0.12	
Accuracy:		
'True value'	3.9	
Error of mean (em)	−0.6	
%REM	−18.18	
$CI_{95.0\%z}$	0.26	
$t_{0.05,\ df} = n - 1$	2.776	
$CI_{95\%t}$	0.33	
CI lower	2.97	
CI upper	3.63	
Significance test:		
1 s t-test:		
t (t)	5.07	= ABS(em)/se
df	4	
p-value (two-tailed)	0.007	= T.DIST.2T^a(t,df)

^{a}TDIST(t,df,2) Excel 2003.

Fixed bias

Five end determinations were performed on the content of iron in barley. Is there evidence of bias (assume fixed bias)? The data were entered into the table with the formula of Tables 3.6 and 5.12.

It can be seen that bias is present as the test mean is below the true value and the confidence interval does not contain the true value. The 1-sample *t*-test reveals that this bias is significant. That is, there is evidence of *significant negative bias*.

The presence of fixed bias was assumed above, but to confirm this would require construction of a graph plotting a range of standard concentrations against the experimental values. *Fixed bias*, if present alone, will show (if large enough cf. random error) a *constant deviation* from the line and a *non-zero intercept* (see the example below).

Relative bias

This form of bias can be located by plotting standard concentrations against measured values as above, along with a 45° line (Calcutt and Boddy 1983). Relative bias will show a *varying deviation* from the line, and the slope of the points will not be equal to unity. If both are present, there will be a *varying deviation* from a *non-unity slope coefficient* plus a *non-zero intercept*.

Example 9.6-1 *Bias calculation and display for chemical analysis data*
Prior to a programme of mineral content analysis on foods, a check was made on possible bias in the atomic absorption spectroscopy methods. One mineral analyte was measured with a set of standards for absorbance, over a range of 10–100 μg/100 g concentration. Is there evidence of any form of bias?

Excel can illustrate this by constructing a scatter graph and performing regression on the standard concentrations against the test determinations. This was done for the data on the mineral content (Table 9.7).

The scatter graph (Fig. 9.2), a plot of observed versus fitted (predicted) concentrations, has two lines – one (dashed) is the *predicted concentration values* of the standards if *no bias* was present – this line has a intercept value at the origin and slope is equal to 1 (45°). The second line (solid) deviates from the first – the *intercept is non-zero* and the *slope is greater than 1*, indicating both forms of bias.

Note that the table has been truncated and values rounded off to improve readability, but all figures should be retained if they are required for later calculations using the regression equation. Fixed bias is given by the intercept – in this case, it is +2.18, which can be used as a *fixed bias correction factor*. Relative bias can be expressed by quoting the slope coefficient or expressing it as a percentage above 1, e.g. 1.13 or 13%. Correction factors for this form must be applied to individual test values, based on the appropriate bias for each level of concentration.

Assessing bias does depend on several factors including sample size and the variability of the instrumental method. Accepting that bias is not present, there are

Table 9.7 Regression analysis (fixed and relative bias (Excel).

Data

Standards conc. (μg/100 g)	Observed conc. (μg/100 g)
10	15
20	24
40	47
60	69
80	90
100	117

Summary output

Regression statistics

Multiple R	0.999
R^2	0.998
Adjusted R^2	0.998
Standard error	1.841
Observations	6

ANOVA

Source of variation	df	SS	MS	F	Significance F
Regression	1	7705.78	7705.78	2274.65	0.00
Residual	4	13.55	3.39		
Total	5	7719.33			
	Coefficients				
Intercept	2.18				
X variable 1	1.13				

still possible sources of error that can arise and these can be detected by a fuller analysis of the data from calibration standards. Error can affect the standards and consequently any calculation that depends on them.

9.6.2 Calibration studies

Certain methods of chemical analysis include determination of a series of reference or standard samples along with unknown sample. This allows

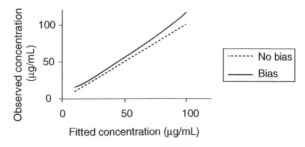

Fig. 9.2 Relative and fixed bias (Excel).

calibration of the analysis. As this involves more analysis time, it is particularly suited to auto-analyser systems rather than wet traditional methods. The procedure is performed routinely on each analysis occasion. In this way, slight variation in reagents, techniques, etc., will affect both unknowns and standards, thus reducing error. Calculation of unknown concentrations is achieved via statistical methods of correlation and regression (Chapter 6). Values are predicted by *interpolation* from the relationship between concentration and the particular measure used (e.g. absorbance).

The assumption is that there is a *linear relationship*. A **calibration graph** is drawn, then unknowns can be 'read-off' the graph, or calculated using the *regression equation*. This procedure, like any other end measurement, is subject to error.

How to check the accuracy of a calibration graph

A number of factors affect the calibration procedure in terms of accuracy:

- Linear relationship
- Background signal from reagents (blank)
- Range of standards is wide enough to include all likely unknowns
- Error may occur in both standard preparation and measurement, i.e. X and Y
- Y measure error can depend on concentration

Assuming adequate control over the standards (i.e. minimisation of error), inclusion of a blank, adequate range, etc., then the procedure for assessing the quality of the calibration is as given in Sections 6.3 and 6.4 using Excel, namely prepare a scatter graph, determine the coefficient r for the data, fit a regression line and perform a regression analysis. These procedures have been applied to the data of Table 9.8.

The scatter graph (Fig. 9.3) shows that some points are not precisely on the line, but there are no suggestions of non-linearity, which would lead to errors in r. The correlation coefficient is not provided, but is quickly obtained by taking the square root of R^2. Ideally, r should be unity (perfect correlation), but in practice it is likely to be less, but unless technique is very poor, r will be >0.9, as is the case in this example.

Regression analysis uses the **Data Analysis Toolpak** option in Excel (Section 6.4). For this set of data, the regression fit is high and the ANOVA is significant (Table 9.8).

The regression equation of Y (absorbance) on X (concentration) made up by the coefficients (the slope for X and the intercept) in the regression analysis is displayed on the graph, and it can be expressed as:

$$\text{Absorbance} = 0.0791 \times \text{concentration} + 0.017$$

Table 9.8 Regression analysis of calibration data (Excel).

Summary output			STD (mg/100 g)	ABS		
Regression statistics						
		D	0	0.02		
		A	1	0.08		
Multiple R	0.998	T	2	0.18		
R^2	0.996	A	4	0.35		
Adjusted R^2	0.995		6	0.47		
Standard error	0.021		8	0.68		
Observations	7		10	0.79		

ANOVA

	df	SS	MS	F	Significance F	
Regression	1	0.523	0.523	1153.04	0.000	
Residual	5	0.002	0.00046			
Total	6	0.526				

	Coefficients	Standard error	t-Stat	p-value	Lower 95%	Upper 95%
Intercept	0.017	0.013	1.300	0.250	−0.017	0.051
X variable 1	0.079	0.002	33.956	0.000	0.073	0.085

Residual output			Probability output	
Observation	Predicted Y	Residuals	Percentile	Y
1	0.017014	0.002986	7.142857	0.02
2	0.096075	−0.01608	21.42857	0.08
3	0.175137	0.004863	35.71429	0.18
4	0.333259	0.016741	50	0.35
5	0.491382	−0.02138	64.28571	0.47
6	0.649505	0.030495	78.57143	0.68
7	0.807628	−0.01763	92.85714	0.79

The equation can be used to predict Y (absorbance) values given fixed concentration values, but here calculated values of the *unknown concentrations* for a given absorbance are required. This can be obtained by rearranging the regression equation:

$$\text{Concentration} = (\text{absorbance} - 0.017)/0.0791$$

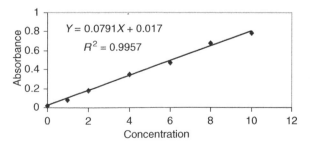

Fig. 9.3 Calibration data with regression line.

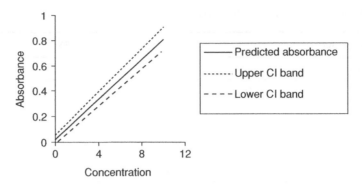

Fig. 9.4   Confidence bands for predicted *Y* (absorbance) (Excel).

Note that the blank (zero standard) value is incorporated into the slope value. Miller and Miller (1993) state that it is incorrect to subtract the blank from each reading (before plotting the graph) as the blank itself is subject to errors.

Errors can be calculated for the slope and intercept. This is done by considering the deviation from the 'true' values of the standards taken as being 'on the line' compared with the actual reading. These are the residuals (Section 6.4) and a measure of their variance can be obtained. This is then used to calculate the errors as confidence intervals. The Excel regression tool has performed this automatically (Table 9.8).

Excel gives predicted *Y* (absorbance) (fitted values) and the corresponding *residuals* (the difference between the predicted and actual absorbance) – so these can be used to calculate the CI for the line. This is not directly available in Excel, but they can be calculated using the CI values for each coefficient (Table 9.8). These should be substituted into the equation above to calculate the lower and upper values, e.g. the standard with 4 mg/100 g concentration has a predicted absorbance of 0.33, with lower CI $= 4 * 0.073 - 0.017 = 0.28$ and upper CI $= 4 * 0.085 + 0.051 = 0.39$. The confidence limits can be displayed as bands on either side of the line (Fig. 9.4).

Minitab can do this directly from the data. In the example, *CI limits* are clearly visible. They vary according to the position on the line (see below) and this means that any predicted absorbance will have uncertainty. While such *CI ranges* and bands are effective in illustrating the error in the standard line, these confidence bands give a CI for a particular predicted value of *Y* (in the above example, absorbance). This is not the main objective in use of a calibration line for chemical analysis, where prediction of unknown *X* values (concentration) is required.

Unknown sample concentrations

Concentration of the analyte in unknown samples is obtained by using the rearranged calculation. Unless the CIs for the standards are zero, there will be some error in the estimate of the unknown. Calculation of a confidence limit for each unknown is more complex. Miller and Miller (1993) and Calcutt and Boddy (1983)

Table 9.9   Confidence bands for unknown concentration (Excel).

	A	B	C	D	E
1		**STD (X)**			**ABS (Y)**	
2		0	D	0.02	
3		1	A	0.08	
4		2	T	0.18	
5		4	A	0.35	
6		6		0.47	
7		8		0.68	
8		10		0.79	
9			Centroid	0.37	$=$ AVERAGE(D2:D8)
10			Sum of squared deviation X	83.71	$=$ VAR.S^a(B2:B8) $*(n-1)$
11			Standard error(se)	0.021	From regression Table 9.8
12			Intercept	0.017	From regression Table 9.8
13			Slope	0.079	From regression Table 9.8
14			Unknown abs. (u_{abs})	0.7	As entered
15			n	7	As entered
16			Centroid$_{abs}$	0.367	$=$ Average all standard abs
17			Sum deviations x	83.71	$=$ Var(x values) $*(n-1)$
18			$t_{95\%}$	2.57	$=$ T.INV.2T^b(0.05, $n-2$)
19			Concentration	8.65	$=(u-$ intercept)/slope
20			CI $t_{95\%}$	0.80	$=$ Formula below^c
21		$=((t*\text{se})/\text{slope})*\text{SQRT}(1+1/n+(((u_{abs}-\text{centroid}_{abs})/\text{slope})\char`\^2)/\text{sum dev }x)$			

^{a} VAR() Excel 2003.
^{b} TINV() Excel 2003.
^{c} Calcutt and Boddy (1983).

give formulae which approximate, but few software packages perform these calculations. Excel can be programmed with the equation of Calcutt and Boddy (1983) to give the CI for X (concentrations) (Table 9.9). Most of the required values are available in the regression output of Table 9.8.

The formula has been applied to a single unknown. Values can then be calculated for a range of unknown absorbance, which allows the upper and lower points of the confidence bands to be calculated and plotted for the unknowns (Fig. 9.5).

In the plot, the limits vary, depending on the position on the line. This may seem unusual compared with the format of a single CI as described previously (Section 3.4.3), but it is a consequence of the values obtained from the regression line. Unknowns closest to the centroid of the line will have the narrowest interval contrasting with those further away. The regression line calculation is based on the centroid, defined as the overall mean of concentration and the signal (absorbance) for the standards (Miller and Miller 1993).

The *CI limits* can appear large when only single standards and sample are used (i.e. there are no true replicates). This is especially so considering the high

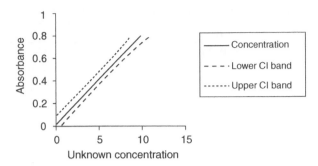

Fig. 9.5   Confidence bands (95%) for unknown concentration (Excel).

correlation coefficient obtained. Increasing replication will narrow the bands, as will increasing the range of the standards, or using an established measure of residual error based on a larger number of standards. Checking of assumptions in regression can be performed as described in Section 6.4.

9.6.3   Precision studies

Precision for repeated determination is measured by calculation of the sd, variance, %CV, etc. In addition, the width of a confidence interval for the estimated analyte content provides another view – the wider the interval, the lower the precision. CIs can make differences more obvious, especially when presented graphically as shown above. Like bias, precision can be relative and can depend on concentration. This can be revealed by comparing standard deviations of measures for different concentrations via tables or graphs (in a similar style to that used in Fig. 9.1).

Levels of precision depend on a number of factors such as the method of chemical analysis itself, as well as the effect of the technician, and the circumstances of the laboratory, etc. Values of 1–5 for %CV are attainable (Table 9.2). The significance of precision can be checked in a number of ways, but one method is to compare each set of replicates with the other(s) using a significance test, as for separate sample units. When relatively high precision is present, the test should not locate any significance, i.e. the data can be viewed as originating from the same population.

The definition of precision in its simplest form is as a single measure of variation, typically, the standard deviation. This can be applied to a wider range of circumstances, e.g. on the laboratory bench for two determinations or variation of repeated determinations over several days. On this basis, precision terms have been defined for specific conditions in analytical chemistry laboratories. These are the terms ***repeatability*** and ***reproducibility***.

Repeatability

This term refers to the variation within a single laboratory under prescribed conditions. This includes the BSI definition (BSI 2002) and the term is expressed

Table 9.10   Repeatability calculation data (Excel).

Data	Moisture Content (g/100 g)
Duplicate 1	71.5
Duplicate 2	70.9
mean	71.2
sd (pop.)	0.35
sd (unknown)	0.42
$t_{95\%}$, 1 df	12.71
Repeatability$_z$	0.97
Repeatability$_t$	7.62

as an interval in which repeated determinations should be located with a level of confidence, i.e. *a confidence interval of two determinations carried out under repeatability (within-laboratory) conditions.*

Example 9.6-2 Precision calculations for chemical analysis data
Data gathered during routine chemical analysis of moisture content in foods were examined for the level of precision. Mean values were in the range 70–72 g/100 g, and based on the databank, the population standard deviation was taken as 0.35 g/100 g. A duplicate measure was carried out under repeatability conditions. What is the precision of the data?

Table 9.10 shows the result of the duplicate moisture content determination. Assuming repeatability conditions as defined above, precision (repeatability) can be calculated in two ways.

As the population standard deviation (sigma (σ)) is known, then:

$$\text{Repeatability } 95\% = z_{95\%} \times \text{square root } (2) \times \sigma$$
$$= 1.96 \times \text{square root } (2) \times \sigma$$

This is essentially the confidence interval for a duplicate determination, i.e. $n = 2$. The z-value is the confidence level factor based on a normal distribution as explained in Section 3.4.3. In the example, repeatability has a value of 0.97%. Thus, duplicate determinations of moisture by the particular method in the same laboratory, same technician, reagents, etc., should differ by not more than 0.97%. The population sigma can be obtained from previous data as in the example, or by carrying out an initial larger set of determinations to give an improved estimate.

If sigma is estimated as the sample sd, then:

$$\text{Repeatability } 95\% = t_{95\%,\ 1\ \text{df}} \times \text{square root } (2) \times sd$$

The confidence level factor is based on the t-distribution. This results in a much higher value ($>7\%$), but it is also possible to use a t-value based on a larger earlier set to give more representative measure of repeatability, and a narrower interval. In practice, duplicate moistures by oven drying give %CVs of <1%, thus the former estimate of repeatability is not usual.

Some texts define repeatability as 'within laboratory' in a broader way in that it includes different operators, which is more realistic as it cannot be guaranteed that the same technician will analyse all incoming samples at one session, etc. Repeatability can also be obtained via the 'within variance' estimate in a two-way ANOVA (see below).

Reproducibility

Repeatability is a critically important measure for a laboratory, but there are many analytical laboratories and there is concern that results can vary depending on which laboratory is used. There are many examples of such discrepancies in the literature (e.g. Thompson 1994), with z-scores (as above) attaining values well above the ± 2 limit for some laboratories (± 10 in some instances). This has given rise to a further definition of precision. The term is expanded to cover variation between different laboratories – *reproducibility*. This is the variability where all aspects other than the material being analysed are different, i.e. analysis in different laboratories, hence different technicians, reagents, times etc. The definition is calculated in a similar manner to that for repeatability, with the inclusion of the 'different laboratory effect': *reproducibility is the magnitude of the interval for two determinations by any two laboratories.* The calculation reflects the wider source of variation by incorporating the variance of both *within- and between-laboratory sources*:

Reproducibility$_{95\%}$(population variance known):
$$= z_{95\%} \times \text{square root } (2) \times \text{square root (variance}_{\text{within}} + \text{variance}_{\text{between}})$$

And

Reproducibility$_{95\%}$(variance estimated):
$$= t_{95\%,1} \times \text{square root } (2) \times \text{square root (estimated variance}_{\text{within}} + \text{estimated variance}_{\text{between}})$$

The estimates in the second formula are conveniently obtained by ANOVA, which provides a 'pooled within variance' as well as the 'between variance'. This is shown for two laboratories – laboratory A compared with laboratory B (Table 9.11).

The variance of each laboratory is given and the average of this is the 'pooled within variance' – also shown as the 'within group mean square' (MS). The 'between variance' is the 'between groups MS'. These were used to calculate the reproducibility of these two laboratories. As seen, the reproducibility figures are higher than those of repeatability, especially when the variance estimates are based on two determinations (again, this can be improved using a known or an established more confident variance estimate).

Table 9.11   Estimation of within- and between-laboratory variance using ANOVA (Excel).

Data	Moisture content (g/100 g)				
	Laboratory A	Laboratory B			
	71.1	72.8			
	71.7	72.9			
Mean	71.4	72.85		Estimate	Known
sd (pop.)	0.18	0.09	Within variance	0.093	0.065
sd (unk.)	0.42	0.07	Between variance	2.10	1.5
$t_{95\%,\,1df}$	12.71	12.71			
Repeatability$_z$	0.50	0.25	$t_{95\%,\,1df}$	12.71	
Repeatability$_t$	7.62	1.27	Reproducibility$_z$	3.47	
			Reproducibility$_t$	26.62	

ANOVA: single factor
Summary

Groups	Count	Sum	Average	Variance
Laboratory A	2	142.8	71.4	0.18
Laboratory B	2	145.7	72.85	0.005

ANOVA

Source of variation	SS	df	MS	F	p-value	F-critical
Between groups	2.103	1	2.10	22.73	0.041	18.51
Within groups	0.185	2	0.093			
Total	2.288	3				

The above analysis is limited in that it cannot provide any indication of interaction of laboratories with different concentrations or levels of a particular analyte, e.g. a food with a higher or lower level of moisture. Inclusion of a second set of replicates from another food type allows interaction to be assessed (Table 9.12).

Here, reproducibility is determined via estimates of *within*, *between* and *interaction*, which are adjusted to take into account the number of determinations, etc. Reproducibility is still high when more than one concentration of the moisture content is considered, but the differing concentration does not appear to cause a significant effect. Such work is done in inter-laboratory proficiency-testing schemes, and large precision studies can involve ten or more laboratories examining a particular method at several levels of concentration.

9.6.4   Uncertainty

The ultimate way to gauge the quality of a result for a chemical analysis is by a composite measure and this is provided by **uncertainty**. It can be expressed in more than one way, but broadly, it is intended to provide a composite view of *both accuracy and precision e*ffects. It is defined in the Association of Public Analysts protocol publication (1986) in this manner:

$$U = 100 - \text{bias} \pm 2 \times \text{CV} \ (\%\text{CV of two determinations})$$

Table 9.12   Reproducibility with interaction (Excel).

Data	Moisture content (g/100 g)	
	Laboratory A	**Laboratory B**
s1	73.4	75.1
	72.7	74.2
s2	66.5	67.9
	68.0	67.2

Variance estimates:		
Within laboratory	0.51	= Within
Between laboratory	0.24	= columns – interaction/number of determinations per lab
Interaction	0.17	= interact – within/number in each cell
$t_{95\%,\,4df}$	2.78	df = within
Reproducibility	3.76	
ANOVA: two-factor with replication		

ANOVA

Source of variation	SS	df	MS	F	p-value	F-critical
Sample	83.21	1	83.21	164.76	0.0002	7.71
Columns	1.81	1	1.81	3.57	0.13	7.71
Interaction	0.84	1	0.84	1.67	0.26	7.71
Within	2.02	4	0.51			
Total	87.88	7				

Other definitions use variances (as sd) values for sources of variation, which cause uncertainty of the final measure (Miller and Miller 1999), where:

$$\text{Expanded uncertainty } (U) = \text{standard uncertainty } (u) \times \text{coverage factor}$$

Standard uncertainty is essentially an sd value and the coverage factor is 2 for 95% confidence, thus:

$$U_{95\%} = 2 \times \text{sd}$$

Recent work with designed experiments has used a factorial structure to estimate uncertainty, combined with robust forms of analysis. Hill and von Holst (2001) have evaluated sources of variation as 'sampling', 'preparation' and 'analytical' and these are summed to give the overall uncertainty measure. FSA (2004b) carried out a study of uncertainty in the food sampling process using quality control charts to monitor sampling precision.

A simple uncertainty experiment could involve drawing two samples from a batch in a random manner and in the same style as would be done for routine analysis, then performing an end determination on two subsamples of each sample unit. This can provide variance measures of 'between sample' (sampling uncertainty) and 'between replicates' (analytical uncertainty) using two-way ANOVA. The data and analysis for such an experiment are presented in Table 9.13. Although replication is present, Excel's 'ANOVA two-factor without replication' is

Table 9.13   Uncertainty calculation (Excel).

Data	Glucose content (g/100 mL)	
	s1	**s2**
Replicate 1	6.9	8.2
Replicate 2	6.5	8.7

ANOVA: two-factor without replication
Summary

	Count	Sum	Average	Variance	sd	Exp. U 95%
Replicate 1	2	15.1	7.55	0.84	0.919	1.838
Replicate 2	2	15.2	7.6	2.42	1.556	3.111
					Average U	2.47
s1	2	13.4	6.7	0.08	0.283	0.566
s2	2	16.9	8.45	0.13	0.354	0.707
					Average U	0.64

ANOVA

Source of variation	SS	df	MS	F	p-value	F-critical
Rows	0.003	1	0.003	0.012	0.93	161.45
Columns	3.06	1	3.06	15.12	0.16	161.45
Error	0.20	1	0.20			
Total	3.27	3				

used because both sample and replication itself are treated as possible contributing sources of variation. Absence of error (and uncertainty) would result in the two samples (s1 and s2) being the same in terms of analyte content, as would repeat determinations (replicate 1 and 2) on the same material. This, of course, is not the case. Examination of the summary table shows that the *sample means* differ giving relatively large variances (MS) for them (columns). Analytical variance is very low as seen by the closeness of the replicates for each sample. Uncertainty calculations have been added to the output. If we assume that the U calculation is to be based on two determinations then we have two measures each for both replicate and sample. Standard deviations were calculated from the variance values then the 'Expanded U' formula above was applied, then averaged. The $U_{95\%}$ value is much larger for the samples. Therefore, in this example, there is more uncertainty for the sampling stage than that caused by replication during analysis by the particular method employed.

As could be envisaged, the more factors that are included for uncertainty estimates, the more chemical analyses are required, and this involves more cost. As a result, there is considerable interest in examining the balance of cost against risks and benefits of uncertainty testing, ultimately in the form of optimisation experiments as described previously. Another possibility is the use of fractional factorials to reduce the number of sampling treatments (Jülicher *et al.* 1999).

9.7   Analysis of relationships

In many investigations, other measures are included along with instrumental methods and sensory assessments are a common example. Williams *et al.* (1988) give an account of this aspect in the context of food acceptance with examination of relationships between and chemical, physical and sensory data. Examples of relationship analysis have been presented in previous chapters (Sections 6.3 and 8.5.5). Additional examples of relating instrumental data to sensory and consumer data are covered in Chapters 10 and 12.

References

Abdullah, B. M. (2007) Properties of five canned luncheon meat formulations as affected by quality of raw materials. *International Journal of Food Science and Technology*, **42**, 30–35.

Anonymous (1999) Reference materials update. *VAM Bulletin*, **20**, 33.

AOAC (1999) *Official Methods of Analysis of the Association of Official Analytical Chemists*, 15th edn. Association of Official Analytical Chemists, Arlington, VA.

Association of Public Analysts (1986) *A Protocol for Analytical Quality Assurance in Public Analysts' Laboratories*. Association of Public Analysts, London.

BSI (2000) BS ISO 5725:1. *Accuracy (Trueness and Precision) of Measurement Methods and Results – Part 1: General Principles and Definitions*. British Standards Institute, London.

BSI (2002) BS ISO 5725:2. *Accuracy (Trueness and Precision) of Measurement Methods and Results – Part 2: Basic Method for the Determination of Repeatability and Reproducibility of a Standard Measurement Method*. British Standards Institute, London.

Calcutt, R. and Boddy, R. (1983) *Statistics for Analytical Chemists*. Chapman and Hall, London.

FSA (Food Standards Agency) (2004a) *Optimised Uncertainty at Minimum Cost to Achieve Fitness for Purpose in Food Analysis*. FSA Project No. E01034. Available at www.food.gov.uk (accessed 7 March 2013).

FSA (Food Standards Agency) (2004b) *Pilot Study of Routine Monitoring of Sampling Precision in the Sampling of Bulk Foods*. FSA Project No. E01049.

Gao, Y., Ju, X. and Jiang, H. (2006) Studies on inactivation of *Bacillus subtilis* spores by high hydrostatic pressure and heat using design of experiments. *Journal of Food Engineering*, **77**, 672–679.

Greenfield, H. and Southgate, D. A. T. (2003) *Food Composition Data*, 2nd edn. Food and Agriculture Organisation of the United Nations, Rome, pp. 149–162.

O'Donnell, G. E. and Hibbert, D. B. (2005) Treatment of bias in estimating measurement uncertainty. *Analyst*, **130**, 721–729.

Heinonen, M., Valsta, L., Anttolainen, M., Ovaskainen, M., Hyvönen, L. and Mutanen, M. (1997) Comparisons between analytes and calculated food composition data: carotenoids, retinoids, tocopherols, tocotrienols, fat, fatty acids and sterols. *Journal of Food Composition and Analysis*, **10**, 3–13.

Hill, A. R. C. and von Holst, C. (2001) A comparison of simple statistical methods for estimating analytical uncertainty, taking into account predicted frequency distributions. *Analyst*, **126**, 2044–2052.

Jülicher, B., Gowik, P. and Uhlig, S. (1999) A top-down in-house validation based approach for the investigation of the measurement uncertainty using fractional factorial experiments. *Analyst*, **124**, 537–545.

Kane, J. S. (1997) Analytical bias: the neglected component of measurement uncertainty. *Analyst*, **122**, 1283–1288.

Levermore, R. J. and McLean, B. D. (2002) *Development of a Stable Sulphited Food Reference Material, Campden and Chorleywood Food Research Assocaiton Group*. FSA Project code E01041. Available at www.food.gov.uk (accessed 7 March 2013).

Lyn, J. A., Ramsey, M. and Wood, R. (2002) Optimised uncertainty in food analysis: application and comparison between four contrasting 'analyte–commodity' combinations. *Analyst*, **127**, 1252–1260.

Martín-Diana, A. B., Rico, D., Fías, J. M., Barat, J. M., Henehan, G. T. M. and Barry-Ryan, C. (2007) Calcium for extending the shelf life of fresh whole and minimally processed fruits and vegetables: a review. *Trends in Food Science and Technology*, **18**, 210–218.

May, N. S. and Chappell, P. (2002) Finding critical variables for food heating. *International Journal of Food Science and Technology*, **37**, 503–515.

Miller, J. C. and Miller, J. N. (1993) *Statistics for Analytical Chemistry*, 3rd edn. Ellis Horwood, Chichester.

Miller, J. C. and Miller, J. N. (1999) *Statistics and Chemometrics for Analytical Chemistry*, 4th edn. Ellis Horwood, Chichester.

Moreira, M. D. R., Ponce, A. G., Del Valle, C. E. and Roura, S. I. (2006) Ascorbic acid retention, microbial growth, and sensory acceptability of lettuce leaves subjected to mild heat shocks. *Journal of Food Science*, **71**(2), S188–S192.

Mullins, E. (2003) *Statistics for the Quality Control Chemistry Laboratory*. Royal Society for Chemistry, Cambridge.

Nielsen, S. S. (2003) *Food Analysis*, 3rd edn. Kluwer Academic/Plenum Publishers, New York.

Patey, A. (1994) The food analysis performance assessment scheme. *VAM Bulletin*, **11**, 12–13.

Severini, C., Baiano, A., De Pilli, T., Carbone, B. F. and Derossi, A. (2005) Combined treatments of blanching and dehydration: study on potato cubes. *Journal of Food Engineering*, **68**, 289–296.

Therdthai, N., Zhou, W. and Adamczak, T. (2002) Optimisation of the temperature profile in bread baking. *Journal of Food Engineering*, **55**, 41–48.

Thompson, M. (1994) Proficiency testing in analytical laboratories – the international harmonised protocol. *VAM Bulletin*, **11**, 4–5.

Villavicencio, A. L. C. H., Araújo, M. M., Fanaro, G. B., Rela, P. R. and Mancini-Filho, J. (2007) Sensorial analysis evaluation in cereal bars preserved by ionizing radiation processing. *Radiation Physics and Chemistry*, **76**, 1875–1877.

Williams, A. A., Rogers, C. A. and Collins, A. J. (1988) Relating chemical/physical and sensory data in acceptance studies. *Food Quality and Preference*, **1**(1), 25–31.

XiuRong, P. (1995) A view on the traceability of certified values of chemical composition RMs. *VAM Bulletin*, **12**(reference material special), 18–19.

Yann, D., Didier, H. and Daniel, B. (2005) Utilisation of the experimental design methodology to reduce browning defects in hard cheeses technology. *Journal of Food Engineering*, **68**, 481–490.

Chapter 10
Food product formulation

10.1 Introduction

Product development is essential for food companies to remain viable, but the process of development is costly and failure rates are high. Many reasons have been postulated for this, but 'trail–and-error' methods and a general lack of an organised approach contribute.

The development process encompasses several stages (Brody and Lord 2000). For the laboratory worker, one of the tasks in making up a new product is that of *formulation*. At the simplest level, this involves combining ingredients together, processing, if necessary, and then examining the resulting product. The experience and knowledge of the food development technologist can be applied in trial-and-error experiments, but a much more efficient way is to apply *experimental design*. Evaluation of test formulations can be done by a variety of methods, but one or more of these measures will be used for the ultimate objective of optimising the ingredient balance in the product.

The purpose of this chapter is to consider this phase in the development process – that of the make-up of a formulation or prototype product for testing by the food development team. Mainly product aspects are covered, but similar techniques can be applied in process development studies.

10.2 Design application in food product development

The essentials of formulation studies are that ingredients are identified as factors and are combined at various levels to produce a set of prototypes. These can be assessed and the 'best one' identified to allow continuing work. From such

Statistical Methods for Food Science: Introductory Procedures for the Food Practitioner, Second Edition. John A. Bower.

Table 10.1 Design features in food formulation.

General design term	Formulation experiment term
Treatments	Formulations
Factors	Individual ingredients selected for study
Levels	The quantities, proportions, ratios or identities of selected ingredients
Responses	Instrumental, sensory, consumer measures on the formulations
Controls	Existing standard product or a competitors product

experiments, much information can be gained. At the commencement, there can be several unanswered key questions:

- Which ingredients have critical effects on quality, cost, storage stability, etc.?
- What are the optimum levels of ingredients for sensory acceptability, nutritional suitability combined with minimum costs?

All such features can be incorporated, with factorial designs providing a common route. While ingredients are the focus, other factors can influence the formulation, such as consumer wants and desires in any proposed product. An example could be that during initial market research, consumers expressed a desire for a wide range of flavours, but a low-sugar content in a new ice cream.

With small experiments, the data can be analysed by simple methods, but larger studies can employ predictive models and these can provide powerful decision aids in the development process. The principles of design deal with the various formulation details as shown in Table 10.1.

The ingredients of the proposed product are the factors with either quantities or types as the levels, i.e. quantitative and qualitative ingredients. Responses can be one or more. Usually, for sensory optimisation, a single key measure of quality is required, but a full sensory profile is possible. More objective responses, including chemical and physical measures, are possible. Table 10.2 indicates some designs used in formulation. Arteaga *et al.* (1994) provide a full review of designs for food formula optimisation. Selection of which ingredients and factors to manipulate come from a variety of sources including marketing data and category concept studies (Moskowitz 1994).

The simplest designs are based on factorial experiments, with one-way, two-way, etc. Section 7.4.1 has identified the disadvantages of doing one factor in

Table 10.2 Design types for formulation.

Design
One variable (factor)
Factorial
Fractional factorial
Composite
Response surface
Mixture

isolation to others, so ideally all factors (ingredients) should be examined together as they are to be in the final formulation. The exception might be where one key ingredient has as major effect and has little or no interaction with other ingredients. As the number of factors and levels increase, the size of such experiments can get very large and compromises have to be made. Reduction of experimentation can be achieved by certain designs such as *fractional factorials*. These can be used as a type of *screening process* to identify critical ingredients. Another route is to base selection on pre-knowledge of effects. Factorial experiments can give more information, but ultimately, the more advanced designs are required for optimisation.

The aim of all such studies is to reach a high standard of quality in the results as expounded upon by Joglekar and May (1987) in their article on *product excellence by experimental design*. More detail of some of the above designs is presented below, but as an initial example, a single ingredient and its quantitative variation are illustrated.

10.3   Single ingredient effects

With one ingredient, a relatively simple experiment can determine the formula content. This can be performed by making up a series of formulations of the product with increasing concentration of the ingredient, i.e. a *one-variable experiment*. These 'prototypes' can then be tested by sensory and instrumental methods. Results can be analysed and summarised in the form of a scatter diagram (Section 6.3). In formulation studies, the **independent variable** is the fixed level of the ingredient, on the X-axis, and on the Y-axis is the **dependent** response.

An example of a single ingredient manipulation is given below.

Example 10.3-1 *Display of single ingredient effect in formulation*
A product development exercise on new desserts included formulation of a custard-type product. Trial mixes (five) were made up with varying levels of a thickening agent (0.1–0.7 g/100 g). Subsamples were assessed for 'sensory consistency' (0–100 graphic line scale) by a panel of trained assessors and the panel means were calculated for each sample. Subsamples of the same formulation mixes were also presented to a small consumer panel who rated them for degree of liking (DOL) using a 9-point hedonic scale (1 = 'dislike extremely' to 9 = 'like extremely'). What do the data reveal about the relationships between thickening agent concentration and the sensory and hedonic measure?

For this analysis, the thickening agent concentration is assigned as the independent variable (X) and the sensory and hedonic as the dependent variables (Y). The data were plotted in scatter graphs (Figs 10.1 and 10.2).

The first graph forms an approximate straight line, and there appears to be direct relationship between the agent concentration and the sensory consistency.

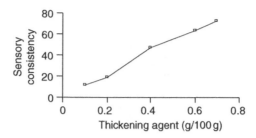

Fig. 10.1   Scatter graph of ingredient variation and sensory response (Excel).

For this type of response, the graph shows that increasing the stimulus will, over a limited range of quantities at least, show a straight-line relationship. In Fig. 10.1, this is a positive one – as the stimulus is increased, the sensory response increases. The graph cannot show the 'correct' or required level directly. This would require a decision on the level of 'sensory consistency' necessary for that product, based on a 'target value' from other data, e.g. a 'thick' product was considered desirable in a consumer concept study. If the formulator wants a higher consistency then obviously increasing the agent content will achieve this up to a limit, when the response may level off. Wider concentration ranges can be used to give a more comprehensive picture, but in these cases, the logarithms of the X and Y values could be required to produce a linear relationship, as the psychophysical relationship is non-linear over wider ranges. This applies to all effects of this nature when a sensory response is plotted against a stimulus.

Hough *et al.* (1997) gave some explanation of the use of Stevens law in such relationships, and then applied the principles in experiments to identify levels of sugar, cocoa and gum in sensory optimisation of a powdered milk chocolate formula.

If the response is non-sensory, i.e. instrumental, then a linear relationship is also likely, e.g. achieving a certain level of viscosity in a sauce product. In either case, the scientist has a rudimentary model of the effect, which can be used to determine a 'target' level of perceived sensory consistency (or instrumental viscosity) based on some criteria, such as consumer wants or technical, production requirements.

For the second graph (Fig. 10.2), the hedonic response appears to increases with the agent concentration up to a point and then to decrease. This is a

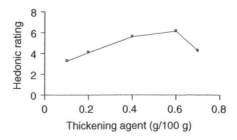

Fig. 10.2   Scatter graph of ingredient level and hedonic response (Excel).

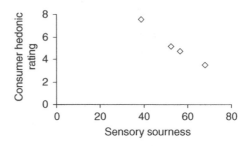

Fig. 10.3   Scatter plot of sensory stimulus and hedonic response (Excel).

non-linear response case, typically found with *affective* measures, and a characteristic 'inverted U-curve' can be seen if the stimulus covers a wide enough range.

The peak of the graph identifies the highest liking and a vertical line to the *X*-axis will provide the amount of ingredient to achieve this level of liking. Other graphs of this form are given by Moskowitz (1989) and in several other texts by the same author. A problem arises here. Most methodologies for formulation work are based on linear relationships and effects. Data as above show that hedonic responses may not be linear over wide ranges of acceptance or liking. This has ramifications for the design stage.

Accepting these limitations for now, the above results have the advantage of all simple experiments in that they are easy to interpret. Their scope can be extended to include stimuli in the form of 'sensory settings' as measured by a trained sensory panel and others based on instrumental readings such as chemical content, refractive index, viscosity, etc. Correlation features strongly in the analysis part and scatter graphs can illustrate many examples.

Hence, assuming that another ingredient, lemon juice, is studied in the dessert example, then a graph of sensory 'sourness' (as the independent) can be plotted against 'consumer hedonic' (Fig. 10.3).

This reveals that sensory sourness has a negative relationship with consumer hedonic rating over the range examined.

Data from instrumental measures can be combined, provided scales of measurement are similar or rescaled to be so (Fig. 10.4).

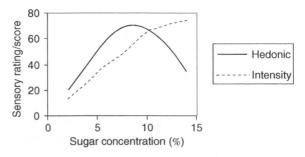

Fig. 10.4   Scatter plot of sensory and consumer responses against instrumental settings (Excel).

This graph displays both hedonic and sensory responses to increasing sugar content in a formulation. The hedonic relationship is not linear, but it provides the formulator with useful information on which level of sensory 'sweetness' is likely to be preferred by consumer, i.e. approximately a score of 55 on the intensity scale. If the consumer data are now linked to compositional data then a formulation figure can be ascertained – approximately 7% sugar. Relationships of this nature can be examined and compared for separate segments of the consumer population as an aid to marketing (Moskowitz 1994).

Ultimately, regression studies follow correlation and prediction via the regression equation can allow a simple evaluation of the ingredient level required for the desired level of the response. Non-linear responses can be accommodated by fitting an alternative equation. Using the above data in simple regression analysis and fitting a linear trend line (Section 6.4) produces:

Sensory consistency score $= 105 \times$ thickening agent concentration $+1 (R^2 = 0.9905)$

This prediction has high fit in that the R^2 value is very high; therefore, sensory consistency can be predicted with confidence within the limits of the experimental data.

Similar treatment (Section 6.4) to the hedonic response gives a very poor linear regression model in terms of predictive power. This is improved dramatically by use of a *quadratic polynomial* fit:

Linear fit:
Hedonic $= 2.7692 \times$ thickening agent concentration $+ 3.5923$
$(R^2 = 0.3599)$
Quadratic fit:
Hedonic $= -21.429 \times$ (thickening agent concentration)^2
$+ 19.912 \times$ thickening agent $+ 1.278 (R^2 = 0.8473)$

Using the latter equation, and substituting various values for X (the concentration of the thickening agent) soon gives a maximum value of Y (hedonic) in the region of 5.8–5.9 for a concentration of approximately 5% of the agent.

Of course, as pointed out in Chapter 7, 'OVAT' (one variable at a time) designs suffer from their inability to identify several effects at once and the interactions. In the examples above, the effect of one ingredient was measured. A second ingredient would then have to be measured, but the effect of both together is not possible without the use of the more powerful factorial experiment. Hence, the level of the first ingredient to achieve maximum or optimum effect may differ in the presence of the second. It could possibly be lower and thus be less costly for the product.

Experiments involving single factors are regularly reported in the literature, even though the formulations include several ingredients, but the variation is limited to one and others are kept constant. Other possibilities include variation of more than one ingredient, but the factor is simply assigned as the 'formulation'. This method was used by Ordóñez *et al.* (2001), who studied various properties of conventional and low-fat frankfurters. Fifteen ingredients were in the formulations, but nine of these were fixed. The other six were not varied systematically, but a limited number of variations were incorporated in four formulations after some preliminary experimentation. Conclusions from this type of study can only refer to the formulation effect as a whole rather than to specific ingredient effects. Another complication is that more than one response is likely to be taken into account. Overall liking is one, but there may be other constraints on the formulation such as fat content, cost, etc. (see below).

10.4   Two or more ingredients

The study of more than one ingredient in an efficient design requires a factorial experiment. This will allow assessment of all factors on all responses. A two-factor formulation experiment illustrates this.

Example 10.4-1 *Two-ingredient effect in formulation*
In development of bread products, two types of flour (white/low-fibre and high-fibre) and an improving agent were examined for their effects on quality. A simple experiment with the two flours and two different amounts of the improver was planned. Laboratory scale batches were made up and all other ingredients and process factors were kept constant. There were four treatments, carried out in random order. At the end of the bake process and cooling of the loaves, the response was assessed by loaf height as an index to volume (and quality). What observations and conclusions can be made about the experiment and the effect of flour and improver on loaf quality?

The factors are represented by the ingredients – the flour and the improver, and the levels by the flour type and the amount of improver, respectively (Table 10.3).
The improver is assigned as factor 1 and it is a *quantitative factor* and involves using two amounts of improver in the mix. Factor 2 is the flour type – it is a

Table 10.3   Factors and levels in a two-ingredient-formulation experiment.

Factor	Identity	Level
1	Improver	Low
		High
2	Flour	Low fibre (control)
		High fibre

qualitative factor as it specifies a variety of flour type (although quantities could be the same). The number of factors and levels decide the size of the experiment in terms of the number of treatments, i.e. the formulations (in this experiment, $2 \times 2 = 4$ treatments). As this is a small (preliminary) experiment, two levels were selected for each factor. At that point, there was an important choice – which levels? The levels in a factorial experiment are selected to 'trap' the optimum, if possible. They can be decided by experience or by guidelines from the supplier of the ingredient, etc. In this case, two quantitative levels were allocated for the improver within a suitable range (via the improver manufacturer's guidelines) and these are designated as 'high' and 'low'. For the flour type, a standard white flour (low-fibre) was used as a control, with a high fibre version as the second level.

The intention was to identify if any ingredient had a significant effect on response measure(s). This was assessed by loaf height. Again, there can be several other response measures for baked products, such as specific volume, crust colour, firmness, sensory, hedonic, etc.

10.4.1 Significance of effects (ingredients)

The significance of effects can be determined by ANOVA. The data from the experiment (Table 10.4) were entered into Excel and analysed using **ANOVA: Two-Factor Without Replication** in the Toolpak. The null hypothesis was that each ingredient has no effect on the loaf height when used at the levels in the experiment. The alternative was that they do.

Table 10.4 ANOVA table for factorial experiment on formulation (Excel).

Data	F2(1) control	F2(2) high fibre
F1(1) improver (low)	15.8	12.4
F1(2) improver (high)	17.6	13.7
ANOVA: two-factor without replication		

Summary	Count	Sum	Average	Variance
F1(1) low improver	2	28.20	14.10	5.78
F1(2) high improver	2	31.30	15.65	7.61
F2(1) low fibre	2	33.40	16.70	1.62
F2(2) high fibre	2	26.10	13.05	0.84

ANOVA

Source of variation	SS	df	MS	F	p-value	F-critical
Rows	2.40	1	2.40	38.44	0.10	161.45
Columns	13.32	1	13.32	213.16	0.04	161.45
Error	0.06	1	0.06			
Total	15.79	3				

Relative effects of different ingredients are readily identified (Table 10.4). Immediately, effect magnitude is indicated by the F-statistic level and significance by the p-value. The summary table takes the average of each factor level across the levels of the other factor. This can be confusing to interpret unless the identity of the factor levels is entered into the output as has been done. Thus, it can be seen that the factor level 'low improver' when used with low-fibre and high-fibre flour gives an average height of 14.1 cm. The lowest average height is found for the high-fibre flour used with low and high improver, and the highest with the low-fibre flour. The analysis of any significance of each factor is given by the F-ratio value and its p-value. The row and column effects represent factors 1 and 2, respectively (as per the layout of the data at the top of Table 10.4). The variation due to factor 1 (row) is not significant ($p > 0.05$), but factor 2 (column) is significant with $p < 0.05$. Thus, flour type has had a much more marked effect on loaf height, significantly so, compared with the improver. The low-fibre flour (control) produces loaves with a significantly larger height when used with a high level of improver.

Based on these preliminary results, incorporation of high-fibre flour types may cause problems, in that a well-risen loaf could be difficult to accomplish. Further examination of levels and other alternatives would need to be pursued. One piece of information was that a combination effect may be present – a higher loaf height was produced when the improver was at a higher level in the formulation. This point, and others within the experiment above, raises some important issues regarding formulation studies.

Size of experiment

With food formulations, a major drawback is that the number of ingredients, and hence treatments, can be large. The experiment above studied two ingredients, but there can be many more. Additionally, if process factors are also included then the size can be unwieldy. With foods, this leads to logistical difficulties in formulation trials and assessment procedures, especially if sensory work is included. For example, four ingredients (factors) at two levels each, generates 2^4 formulations ($2 \times 2 \times 2 \times 2 = 16$); including replication of all gives 32. This number of treatments would be difficult for a sensory panel to assess in one session if there are many attributes as with a descriptive profile. ***Blocking techniques*** could be used. Restricting the factors to two levels does reduce numbers of treatments, but has the disadvantage of relying on ***linear responses***. With some affective measures, linearity may be absent.

The experiment did not employ *replication*, thus it was unable to assess *interaction*, and it did not assess *batch-to-batch variation*, a measure of which is important in eventual larger experiments and pilot-scale work. Only ***main effects*** were assessed – any interaction between flour type and improver remains unknown. Inclusion of one full replication for each treatment doubles the size of the experiment.

Choice of levels and their format

The formulator has to decide on the magnitude (quantitative) or identity (qualitative) of levels for the factors. With the latter type of factor, there may be several alternatives, all potentially suitable. For quantitative factors, the problem is to ensure that the amounts used are realistic and viable for the product, i.e. they should not result in formulation 'failure' (such as a subsample that cannot be tasted or measured). If possible, the range of the levels should be outwith any possible non-linear regions for responses. This can be difficult in some cases, especially when several other ingredients are present and not all interactions can be detected until the formulations are made. Experience of staff can guide this, as well as use of simpler preliminary experiments with level manipulation.

Another aspect of quantitative level settings is the format that they take. In the example, they were specific amounts of improver in the recipe formulation, i.e. a **concentration level**. For such formulations, water is often varied to 'take up the slack' so that all treatments have the same weight. There are other formats that can be used, such as *ratios* and *mixture balances* – which is most suitable and can all such formats be used with factorial experiments? Lack of care when selecting amounts can lead to apparently different treatments being the same in terms of the ratios (*relative proportion*) of the ingredients. This can occur when the absolute weight in high and low levels is the same for two factors, e.g. levels of 25 g/25 g for low + low, and 50 g/50 g for high + high have the same ratio.

Inadequacy of factorials for formulation

Factorial experiments have many advantages, one being the relatively simple and clear interpretation of results. In the formulation case, they may be inadequate as indicated above in respect of quantitative level format. Factors levels are decided by the developer and it can be difficult to judge that the optimum will be contained between them. Level settings can be inefficient if too close together (Nakai 1981).

These issues indicate that more comprehensive design features require to be included.

How can the size of the experiment be reduced in food formulation?

Many of the above arguments have been expounded upon in Chapter 7 (Section 7.4.4). There are a number of possible solutions for formulation and optimisation experiments:

1. Pre-screen by preliminary experiments
2. Pre-screen by use of fractional or incomplete block designs

Using **screening experiments** can help reduce the number of treatments by identifying the factors (ingredients) that have a major effect on response(s). Preliminary experiments can identify combinations that are not suitable and possibly

Table 10.5 Effect parameters of a full three-factor factorial.

Treatment	Factor			Effect parameter
	A	**B**	**C**	
1	High	Low	Low	Main effect of factor A
2	Low	High	Low	Main effect of factor B
3	Low	Low	High	Main effect of factor C
4	High	High	Low	Interaction effect of AB
5	High	Low	High	Interaction effect of AC
6	Low	High	High	Interaction effect of BC
7	High	High	High	Interaction effect of ABC
8	Low	Low	Low	I ('identity element')

point to certain ingredients that have minimum effects, especially when knowledge of the product type is detailed. Screening designs can incorporate all ingredients, but instead of a full factorial with at least two levels for each ingredient, a fractional structure is used. There are a number of ways of achieving this, but the ***two-level fractional factorial*** is common. The fraction is constructed by removing treatment combinations with effects of lesser importance and of lower effect. In most circumstances, ***higher order interactions*** (above two factors) tend to have less effect. By removing treatments that assess these, the experiment becomes smaller.

Consider a three-factor experiment, A, B and C, at two levels. A full factorial without replication creates eight treatments (Table 10.5).

Such a design can be fractionated to give four treatments (a '$^1/_2$ fraction') by ***confounding*** some of the effect parameters.

Confounding of two or more effects means that they cannot be separated or partitioned in the analysis. Using the logic above, one obvious candidate would be the three-way interaction (ABC). However, in this example, there is less selection available to get a half replicate – thus all the two-way interactions and the three-way one can be confounded together along with the grand mean (Gacula 1993). Thus, the fraction will assess *main effects only* – A, B and C, along with *confounded effects* (Table 10.6).

This gives four treatments (the identity element I is not assessed as an effect). With more factors, it is possible to take larger fractions and obtain a design that fits the need of the experimenter. Ideally, main effects and all second-order interactions

Table 10.6 Half fraction of three-factor factorial.

Effect	Confounded with	Factor and levels		
		A	**B**	**C**
A	BC	1	0	0
B	AC	0	1	0
C	AB	0	0	1
I	ABC	1	1	1

should be assessed. Confounding should be such that main effects are confounded with three-way or higher interactions that are less likely to be significant. Designs are available for more factors and a one sixteenth fraction of an eight-factor experiment can be performed with 16 treatments, with no aliases for the main effects. Ellekjær *et al.* (1996) applied a fractional design to the study of seven factors influencing cheese processing and formulation. A full two-level seven-factor experiment would have required 128 runs of the process, but fractionation reduced this to a base size of 32 experiments.

Fractionation has other useful applications. Replication causes a major increase in treatment number. With a fractional factorial, it is possible to use the higher order interaction treatments as a substitute by extending the logic of non-significance for higher orders. Thus, they give a measure of error on the basis that they are not significant and any variation is simply 'background'; replication of the design is avoided.

The technique of blocking in experimental design is used in one form as a device to take account of experimental runs that cannot be performed easily within one session, as with an experiment performed over several days (blocks). Blocking serves its purpose, but the ***block effect***, which accrues, will usually not be of direct interest. In some larger experiments, the block effect can be confounded with the effects of interest. This can be alleviated by running a series of fractions over the times with the block effect confounded with higher order interactions. This latter application and the computations for construction and analysis of fractional factorial are explained by Gacula and Singh (1984). It is possible to avoid the tedium of construction of such designs by consulting published tables of designs or by use of specialist software. Unfortunately, analysis of fractional factorials cannot be done with the basic versions of Excel or Minitab, but a demonstration of the possibilities in screening for formulation studies is presented below.

10.5   Screening of many ingredients

An example to illustrate some of the characteristics and advantages of fractional designs is one involving screening of ingredients in a formulation study on a new 'autumn fruit' soft drink.

Example 10.5-1 *Screening experiment in formulation*
In an attempt to diversify their product base a food company initiated development of new fruit drink, in particular one using 'autumn fruits'. A consumer concept study identified certain requirements such as absence of additives, use of local fruits, a cloudy appearance and sugar acid balance as important. This information was used to decide on the general nature of the product and the ingredients. Four ingredients were selected – fructose, lemon juice, food dye (red) and mixed autumn fruit pulp. Two levels were used for each ingredient and these were based on common fruit drink features, e.g. 10% sugar, 5–40% fruit, etc. A fractional design was required and a series of prototype formulations were

prepared. Initial analysis was done by sensory methods. The formulations were made up in random order with coding and blinding to minimise bias. Coded subsamples from each batch were presented to a trained panel that assessed a number of sensory attribute responses using graphic lines scales (0–100) along with anchors and references. The samples were also assessed for DOL ('degree of liking' with 'dislike extremely' (0) to 'like extremely' (100)) by a consumer panel. What comments can be made about the design and what conclusions can be made from analysis of the fractional design data?

This experiment employed fractionation but some additional features were included. Two blocks were assigned that allowed inclusion of a centre point in each, and two existing retail products were incorporated. The final design comprised a $^1/_2$ *fractional factorial*, using **partial replication** by means of the two **centre points**. Centre points are achieved with levels set at mid-point. Eight formulations were generated, along with the two centre points and two commercial references, bringing the total number of samples assessed to 12 (Table 10.7).

The software package used to generate the design and perform the analyses was **Design-Expert (R)** (Stat-ease Inc., Minneapolis, USA). For each response, a **half-normal plot** (Section 3.5) was prepared and ANOVA was used to assess selected effects.

The design used has a number of features that affect its efficiency:

- Reduced number of treatments
- Centre points (replication and curvature check)
- Reference products

Centre points are particularly useful as they provide a measure of *pure error* (i.e. the centre point is replicated) and they add an element of '*three-level*' *structure* to the design that can allow detection of non-linear effects (curvature).

Table 10.7 Fractional factorial formulation experiment (Design-Expert (R)).

Formulation	Factor A: Fructose%	Factor B: Food dye%	Factor C: Autumn fruit pulp%	Factor D: Lemon juice%
1	15.00 *(hi)*	0.01 *(lo)*	10.00 *(lo)*	2.00 *(hi)*
2	15.00 *(hi)*	0.10 *(hi)*	10.00 *(lo)*	0.00 *(lo)*
3	3.50 *(lo)*	0.10 *(hi)*	40.00 *(hi)*	0.00 *(lo)*
4	3.50 *(lo)*	0.01 *(lo)*	40.00 *(hi)*	2.00 *(hi)*
5	9.25 *(cp)*	0.055 *(cp)*	25.00 *(cp)*	1.00 *(cp)*
6	3.50 *(lo)*	0.01 *(lo)*	10.00 *(lo)*	0.00 *(lo)*
7	15.00 *(hi)*	0.10 *(hi)*	40.00 *(hi)*	2.00 *(hi)*
8	15.00 *(hi)*	0.01 *(lo)*	40.00 *(hi)*	0.00 *(lo)*
9	3.50 *(lo)*	0.10 *(hi)*	10.00 *(lo)*	2.00 *(hi)*
10	9.25 *(cp)*	0.055 *(cp)*	25.00 *(cp)*	1.00 *(cp)*
11	Competitor 1			
12	Competitor 2			

hi, high level; lo, low level; cp, centre point level.

Table 10.8 Treatment means by attribute (Design-Expert(R)).

Response	Formulation treatment											
	1	**2**	**3**	**4**	**5**	**6**	**7**	**8**	**9**	**10**	**11**	**12**
Red colour	22.3	50.0	86.6	69.9	72.5	27.9	89.8	80.1	51.5	64.1	91.0	75.6
Cloudiness	6.5	67.2	87.9	55.0	64.3	25.1	90.0	65.1	79.8	71.6	14.6	12.5
DOL	48.9	55.1	23.3	17.9	47.2	8.6	53.2	47.0	15.1	56.1	3.9	53.4

The panel mean values for the sensory attributes and DOL were entered into the software and a selection of these is presented in Table 10.8.

High ranges are seen for the sensory attributes, but DOL has less in this respect. DOL scores did not attain high levels with this sample set, especially for one of the current products. There were a limited number of products with the necessary characteristics on the market at the time of the experiment. Based on the sensory results, these samples cannot be taken as having 'ideal' or 'brand leader' status.

10.5.1 Graphical analysis

Initial analysis comprised preparation of *half-normal plots* (similar to the normal plots in Section 3.5). These can reveal the relative magnitude of an effect by its distance from the line. If all factors have minimal effect then the sample set appears as if they all come from the same population and all points will be close to the line. The first response 'red colour' illustrates this convincingly (Fig. 10.5).

The ingredients that have relatively large effect on 'red colour' are the red dye (B) and autumn fruit pulp (C); this is logical as they are the only source of colour in the formula.

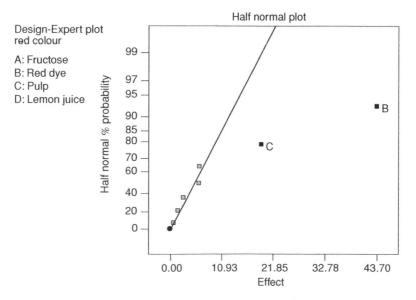

Fig. 10.5 Half normal plot of 'red colour' response (Design-Expert(R)).

Table 10.9   ANOVA for selected factorial model – 'red colour' (based on Design-Expert(R) analysis).

Source	Sum of squares	df	Mean square	F-value	Probability > F
Model	4575.98	2	2287.99	66.69	0.0002
Curvature	116.28	1	116.28	3.39	0.1250
Residual	171.55	5	34.31		
				R-squared	0.9639

				Standard t for H$_0$	
Factor	Coefficient estimate	df	Error	Coefficient = 0	Probability > |t|
B-food dye	21.85	1	2.07	10.55	0.0001
C-pulp	9.72	1	2.07	4.70	0.0054
Center point	8.52	1	4.63	1.84	0.1250

Final equation in terms of actual factors:
Red colour = +11.47+ 1.46 * Pulp + 216.11 * Food dye.

These two factors have been selected for further analysis (Table 10.9). The picture for 'cloudiness' is not quite so convincing (Fig. 10.6).

There is one possible (interactive) effect, but the coefficients for it are small so they were not included in the model. Hence, the fruit pulp has most effect with a contribution from the red food dye, although this needs exploration. The pulp contains compounds that clearly affect 'cloudiness'. In the case of the food dye effect, it may be a perception issue by the sensory panel, or simply stronger colour (i.e. more dye) appears darker and more cloudy. The final graph in the sequence is for the consumer measure (Fig. 10.7).

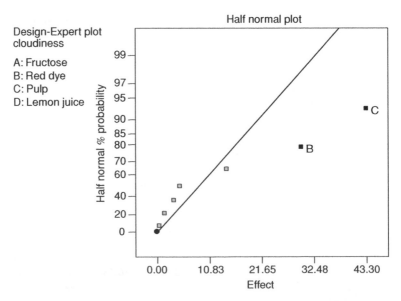

Fig. 10.6   Half normal plot of 'cloudiness' response (Design-Expert(R)).

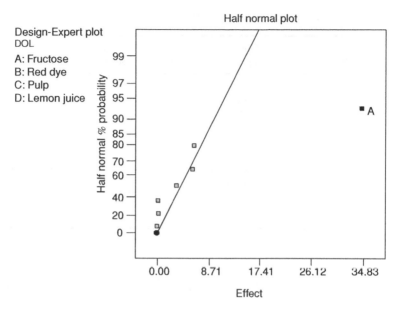

Fig. 10.7   Half normal plot of 'DOL' response (Design-Expert(R)).

Only the sugar ingredient appears to have any noticeable effect on DOL. There are other suggestions of effects as shown by other points offline, but their effects are minimal compared with that of fructose, for this particular experiment.

10.5.2 ANOVA

Analysis of variance was performed on the selected factors and responses.

Table 10.9 shows that the ANOVA model for 'red colour' is highly significant (p-value = 0.0002), but that curvature of this response is not. R^2 (Section 6.4) is very high showing that the selected ingredients explain more than 96% of the variation in colour. Individual effects are tested below and both are significant.

The ANOVA model is presented at the foot of the table and this can be used to predict the effect of manipulation of ingredients on the colour. Both ingredients have a positive effect on colour intensity (positive coefficient estimates).

A similar result was obtained for 'cloudiness' in that each factor was significant and positive in effect and there was absence of curvature. The analysis for 'DOL' was different in that significance was revealed for curvature (Table 10.10).

For this analysis, the ingredient factor (fructose) was highly significant, but because of the **curvaure effect** the model fitted (linear) would be inadequate. R^2 is high, but there are only three points available for assessment. At present, it indicates that 'DOL' will rise directly with fructose content. This may indeed be the case, but the curvature effect will cause a levelling off and eventually a decrease at higher concentrations of the sugar. Thus, a curved response model (quadratic) would be required for 'DOL', similar to that given in Section 10.3 (Fig. 10.2).

Much more analysis is possible with the above data before an adequate picture of the overall effects on sensory and liking are ascertained. Ultimately, one or more

Table 10.10   ANOVA for selected factorial model – 'DOL' (based on Design-Expert(R) analysis).

Source	Sum of squares	df	Mean square	*F*-value	Probability > *F*
Model	2425.56	1	2425.56	81.51	0.0001
Curvature	519.12	1	519.12	17.44	0.0058
Residual	178.55	6	29.76		
				R^2	0.9314

				Standard *t* for H_0	
Factor	Coefficient estimate	df	Error	Coefficient = 0	Probability > |*t*|
A-fructose	17.41	1	1.93	9.03	0.0001
Center point	18.01	1	4.31	4.18	0.0058

Final equation in terms of actual factors:
DOL = +5.63 + 3.03 * Fructose.

of the ingredients could be selected for a more detailed study as in a three-level factorial.

10.5.3   Three-level factorials

Increasing the number of levels for one or more factors will also increase the number of treatment formulations. Such designs would, therefore, be used when the number of factors has been narrowed down by screening, as above, and have been identified as critical ingredients in terms of the response effect. Three levels have some extra advantages in that they give more information on the linearity or not of the response. Of particular interest are affective responses, which are unlikely to be detected in two-level designs.

Visualisation of results

Up to this point, ingredient effects have been examined individually for the most part. Graphical displays appeal, but another difficulty arises – how can several ingredient effects be plotted together? This leads to plotting in more than two dimensions as in *response surface methods*. Peaks and troughs are shown for effects in the form of *contour maps*.

Response surface methods in food formulation

Response surface methodology (RSM) comprises optimisation procedures for the settings of factorial variables such that the response attains a desired maximum or minimum value. It is suitable for quantitative factors only. The response is in effect modelled by factorial techniques and ANOVA, but these are extended for more detailed modelling of the effects (Gacula 1993). RSM usually follows and it is based on the factorial study results (screening, then three-level factorial), and is a type of augmentation where extra treatments are added to focus the effects

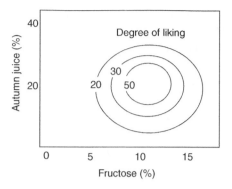

Fig. 10.8   Response surface plot.

and improve the predictive power of the model. The additional treatments are located within the factorial space (***centre point***) and outwith it (***star points***) – such a structure is known as a ***central composite design*** (Dziezak 1990). On analysis by regression, the enhanced model is produced and the equation can be used to plot the response surface. Contour plots reveal not only optimum levels for factors but also regions where a response shows the same magnitude. This can be used to visualise alternative products that use less of an expensive ingredient, etc. Figure 10.8 displays an idealised representation of a response surface from another experiment on the fruit drink formulation. Two ingredients (fructose and autumn fruit juice) had significant effects on DOL. The contours show points where DOL ratings are the same. They rise to a peak ('50' on the 0–100 scale) at approximately 10% sugar and 20% juice, i.e. the optimum level of the ingredients.

Mixture designs are a type of response surface design where the combination of all the quantitative factors must add up to 100%. All foods are mixtures and this would seem an ideal model for formulation. A difficulty arises in that in a mixture design each quantity can be present in amounts from 0% to 100%, which cannot apply to all food ingredients in some formulae, thus more complex analysis is required. The best application is with blending of liquids in beverages where the full span of percent content can be examined (e.g. Huor *et al.* 1980). The mixture model is used to plot a response surface.

The latter designs based on RSM are very efficient, but can some times run into the same difficulties as designs above, in terms of the number of treatments and the complexity of the outputs, which get more difficult to interpret with more than three factors.

10.6   Formulation by constraints

The formulation examples above have considered sensory responses to ingredient content balance, but there are other characteristics of importance:

- Cost
- Nutritional value, e.g. vitamin content

- Anti-nutritional value, e.g. heavy metal content or sodium content
- Legal chemical composition, e.g. minimum fat content
- Perceived quality, e.g. colour/flavour strength
- Functional quality, e.g. stabilising power
- Other properties critical to quality, e.g. pH, water activity

Factorial experiments can assess functional properties and sensory responses, etc., but some of the other factors cannot be left to chance, namely legal requirements of composition. Thus, in formulation of meat and dairy products, cognisance must be taken of the relevant legal minimum/maximums and arithmetical calculations are required to ensure that the raw materials can supply the necessary constituents. An initial formulation can be made up, examined for suitability and then modified to account for other possible factors. Cost is obviously important to the manufacturer and general eating quality, nutritional image, etc., are all now key facets of food products that can decide success on the open market. It is possible, using mathematical techniques, to rapidly manipulate and 'juggle' all these factors and other not so obvious ones, so that all desirable characteristics are maximised, or at least optimised, and all undesirable characteristics are minimised. The technique of *linear programming* has been used in food and other industries for many years for such purposes (Bender and Kramer 1983). Harper and Wanninger (1970) illustrate linear programming in its role as an optimisation tool. Arteaga *et al.* (1994) describe the application of the method in food formula calculations.

One of the main assumptions of this application is that a *linear relationship* exists. This may appear to be obvious, e.g. if 50 g of ingredient A provides 10 g protein to the product then 500 g will provide 100 g of protein – a graph of weight of ingredient A versus weight of protein provided would produce a straight line. Other factors may be linear or non-linear, namely cost – if it costs £20 for 1 kg of ingredient A then it could cost £200 for 10 kg, but the supplier could give discounts for large orders and charge £180 resulting in a non-linear relationship between ingredient purchase weight and cost. Similar examples may be found with large-scale production. In some cases, it may be possible to use a linear function of a non-linear relationship, e.g. possibly the logarithm of cost would be linear with purchase weight, etc. Most software, such as Excel, allows linear or non-linear assumptions to be imposed in problem solving.

With these techniques, it is possible to 'tailor make' a food product to suit a wide range of requirements covering cost, nutritional value, shelf life and legal composition, etc. An example of the possibilities is given in the paper by Strino (1984), which describes a procedure for 'nutritional synthesis' of a food product. Soden and Fletcher (1992) describe similar use in modifying diets to satisfy nutritional requirements and Ferguson *et al.* (2004) used the method for development and testing of dietary guidelines. Use in other formulation problems includes Nicklin (1979) on general use in formulation and Beausire *et al.* (1988), who developed an acceptability constraint for inclusion in linear programming analysis.

Problems of this type can be tackled with Excel's **Solver** tool (available in **Add-ins**). Several items of information are required for solution of the formulation problem:

- The type, cost, composition, functional properties, etc., of all ingredients to be included
- The limits or constraints to be imposed on the formulation
- The specific target to be achieved within the above constraints (typically, to minimise the cost)

Example 10.6-1 *Linear programming in formulation*

A product development team were assigned the task of formulating a new 'low-meat-vegetable' pie. The product was subject to constraints of levels of protein, fat and meat content. Cost was also important and had to be minimised. Initial guidelines required that the product be kept within the 'meat regulations' and that an equal weight of pastry was to be combined with the filling. Various ingredients were selected: meat, a fat plus seasoning mix, a vegetable mix, cereal and water. The problem was to solve the formula for the filling under the constraints of:

- Meat content at absolute legal minimum (10.5% × 2 = 21%)
- Not less than 10% protein
- An upper limit of 25% fat
- A batch weight of 100 kg
- A minimum cost
- Maximise vegetable content

The meat content was a general meat content level for pies with vegetable. The filling content was twice that of the eventual quantity, once the pastry was added. The team had access to compositional data for the ingredients. Provided with this brief how can the team implement linear programming for an initial trial formulation and what conclusions can be made on the procedure and result?

The procedure starts with set-up of an Excel spreadsheet. The initial values and their unit cost are entered for each ingredient (Table 10.11). The weight figures at this stage can be any, but conveniently addition to 100 makes rescaling easier. Compositional data are required for the ingredients and this is entered in a table at the foot of the sheet. These data are used to obtain the multiplication factors for the content of protein, fat, etc., in the product.

Next, the constraints are required. They consist mostly of compositional minimum and maximums and they relate to the characteristics required for a good quality product as judged at the time of initial formulation, based on the knowledge and judgment of the formulator. In Table 10.11, values have been entered for protein, meat content, fat and water.

At this point, formulae must be entered to calculate the content of each constituent based on the initial ingredient content. Thus, the actual protein content is

Table 10.11   Formulation sheet for use of Excel's Solver.

	A	B	C	D	E	F	G	H	I	J
1	Product name/code: Veg. pie filling (uncooked)									
2			Initial/final values							
3										
4	Added ingredients:					Product composition/content constraints (%)				
5	No.	Identity	Wt. (kg)	£/kg	No.	Identity	Actual	Min.	Max.	Detail:
6	1	Meat	20.00	4.00	1	Protein	7.40	10	100	Not less than 10%
7	2	Veg mix	20.00	2.00	2	Meat co.	34.03	21	21	= 21%
8	3	Cereal	20.00	1.00	3	Fat	20.00	00	25	Upper limit of 25%
9	4	Fat seas	20.00	1.50	4	Moisture	62.20	50	70	
10	5	Water	20.00	0.00	5					
11	6				6					
12						Meat Nitrogen:	0.58			
13		Total wt =	100.0	= SUM(C6:C10)						
14		Cost (£) =	170.00	= SUMPRODUCT(C6:C10,D6:D10)						
15	1 Protein = 0.18 * Meat + 0.07 * Veg mix + 0.12 * Cereal									
16	2 Meat content = Nitrogen/3.55 * 100 + 0.1 * Meat + 0.1 * Veg mix + 0.78 * Fat seas									
17	3 Fat = 0.1 * Meat + 0.1 * Veg mix + 0.02 * Cereal + 0.78 * Fat seas									
18	4 Moisture = 0.71 * Meat + 0.81 * Veg mix + 0.4 * Cereal + 0.19 * Fat seas + Water									
19	[Total Nitrogen = ((0.18 * Meat/6.25) + (0.07 * Veg mix/6.25) + (0.22 * Cereal/5.7))]									
20	Meat Nitrogen = (0.18 * Meat/6.25)									
21		Compositional data								
22			%Fat	%Protein	%Moisture	£/kg				
23		Meat	10	18	71	4.00				
24		Veg mix	10	7	81	2.00				
25		Cereal	2	12	40	1.00				
26		Fat seas	78	0.1	19	1.50				
27		Water	0	0	100	0.00				

calculated by adding together the protein contents of the ingredients, i.e. compositional data must be available. The meat content calculation is more complex and requires an intermediate calculation of nitrogen from the meat source.

The total nitrogen is made up by that of meat itself and other sources such as, in this case, vegetable and cereal protein. When working from chemical analysis data on an unknown meat product, total nitrogen is determined by the Kjeldahl or other method and further analyses or estimates are required for the amount of carbohydrate, etc., to enable the non-meat nitrogen to be calculated and subtracted from the total (analytical) nitrogen.

Calculations of this nature should be based on current legislation. Formulae can be obtained from the appropriate regulations or from texts such as *Pearson's Composition and Chemical Analysis of Foods* (Kirk and Sawyer 1991). For example, the following is one form of the calculation:

$$\%\text{Total meat} = \frac{\%\text{total nitrogen} - \%\text{non} - \text{meat nitrogen}}{\text{Nitrogen factor}(= 3.55 \text{ for beef})} \times 100 + \%\text{fat}$$

Here, the meat content value includes the fat content, which may not be the case with more recent regulations. However, when composition data are available, it allows direct calculation of the meat nitrogen alone and the contribution by non-meat sources is excluded from the calculation, i.e. meat content (%) = % nitrogen from meat source/3.55 × 100 + % fat.

These formulae are presented in the table in the form of named ingredients (using the identities in B6:B10), and the factors are derived from the compositional data at the foot of the table. One additional formula is entered in cell C13 that sums all ingredients.

Access to the solver is gained by **Data/Solver** (Excel 2010) or **Tools/Solver** (Excel 2003). On the solver menu, set cell C14 as the target to be minimised – edit if required; set the **By Changing Variable Cells** box (**By Changing Cells** in Excel 2003) to include the range of cells that contain the ingredient amounts (C6:C10).

Set the constraint cells – note that a minimum and a maximum require two constraints; enter the constraints box, select **Add** then enter the cell references in column G along with the minimums and maximums as defined in columns H and I; select **Change** if you require to edit. Initial constraints are:

C13 = 100 (total weight)
C6 > = 0 (meat)
C7 > = 10 (vegetable mix)
C8 > = 10 (cereal)
C9 > = 5 (fat and seasoning mix)

Some of the ingredient cells are set to be zero or over to ensure that there are no negative values and that certain minimum levels are used. Water (C10) could also be included, but it has minimum and maximum values in the constituents' section.

These latter settings ('Actual') are also included in the constraints box:

G6 > = 10	G6 < = 100 (protein)
G7 = 21 (meat content)	
G8 > = 0	G8 < =25 (fat)
G9 > = 50	G9 < = 70 (water)

For Excel 2010, choose **Simplex LP** (linear) in the **Select a Solving Method** box. Leave all other settings at default values. The menu can be left at this point (**Close**) and the file, which now includes the previously entered constraints, etc., can be saved. Select **Solver** again and this time run the program by selecting **Solve**. In Excel 2003, the solving method is chosen by clicking the **Options** button and selecting **Assume Linear Model** – this will speed up the solution.

If all goes well, and a solution is found, the file and the final values can be saved. If not, there may not be a solution, so check the constraints and the raw data entered earlier for errors and run again. The result should be obtained as shown in Table 10.12.

Table 10.12 shows the initial solution. The formula is possible within the specified constraints. A decision can be made on the cost and the actual figures produced. Often the initial values may not be acceptable but reformulation using the spreadsheet is simple. Manipulation of constraints can be explored. Many other 'what if' tests can be performed on the above example, e.g. changing the fat content maximum to see if a new 'low-fat' product is feasible.

Table 10.12   Solution of formulation (Excel's Solver).

	A	B	C	D	E	F	G	H	I	J
1	Product name/code: Veg. pie filling (uncooked)									
2			Initial/final values							
3										
4	Added ingredients:					Product composition/content constraints (%)				
5	No.	Identity	Weight (kg)	£/kg	No.	Identity	Actual	Min.	Max.	Detail:
6	1	Meat	12.92	4.00	1	Protein	10.00	10	100	Not less than 10%
7	2	Veg mix	10.00	2.00	2	Meat co.	21.00	21	21	= 21%
8	3	Cereal	58.12	1.00	3	Fat	12.58	00	25	Upper limit of 25%
9	4	Fat seas	11.70	1.50	4	Moisture	50.00	50	70	
10	5	Water	7.26	0.00	5					
11	6				6					
12						Meat Nitrogen:	0.37			
13		Total weight =	100.00							
14		Cost (£) =	146.35							

Once an initial solution 'on paper' is found for a formulation, it does not mean that it will be successful – it must be manufactured and assessed and usually adjustment of constraints is required before further testing. The mathematical techniques provide a speedy route to the formula on paper, but the food scientist/technologist must still employ other skills and knowledge to produce the article that is to compete in the modern food market.

References

Arteaga, G. E., Li-Chan, E., Vazquez-Arteaga, M. C. and Nakai, S. (1994) Systematic experimental designs for product formula optimization. *Trends in Food Science and Technology*, **5**, 243–254.

Beausire, R. L. W., Norback, J. P. and Maurer, A. J. (1988) Development of an acceptability constraint for a linear programming model in food formulation. *Journal of Sensory Studies*, **3**(2), 137–149.

Bender, F. E. and Kramer, A. (1983) Linear programming and its implementation, Chapter 7. In *Computer-Aided Techniques in Food Technology* by I. Saguy (ed.). Marcel Dekker, New York.

Brody, A. L. and Lord, J. B. (2000) *Developing New Food Products for a Changing Marketplace*. Technomic Publishing, Lancaster, PA.

Dziezak, D. (1990) Taking the gamble out of product development. *Food Technology*, **44**, 110–117.

Ellekjær, M. R., Ilsing, M. A. and Næs, T. (1996) A case study of the use of experimental design and multivariate analysis in product development. *Food Quality and Preference*, **7**(1), 29–36.

Ferguson, E. L., Darmon, N., Briend, A. and Premachandra, I. M. (2004) Food-based dietary guidelines can be developed and tested using linear programming analysis. *Journal of Nutrition*, **134**, 951–957.

Gacula, M. C. (1993) *Design and Analysis of Sensory Optimization*. Food & Nutrition Press, Trumbull, CT, pp. 29–34.

Gacula, M. C. and Singh, J. (1984) *Statistical Methods in Food and Consumer Research*. Academic Press, Orlando, IL.

Harper, J. M. and Wanninger, L. A., Jr. (1970) Process modelling and optimization 3. Optimization design. *Food Technology*, **24**, 590–595.

Hough, G., Sanchez, R., Barbieri, T. and Martinez, E. (1997) Sensory optimization of a powdered milk chocolate formula. *Food Quality and Preference*, **8**(3), 213–221.

Huor, S. S., Ahmed, E. M., Rao, P. V. and Cornell, J. A. (1980) Formulation and sensory evaluation of a fruit punch containing watermelon juice. *Journal of Food Science*, **45**, 809–813.

Joglekar, A. M. and May, A. T. (1987) Product excellence by experimental design. In *Food Product Development From Concept to the Marketplace* by E. Graf and I. S. Saguy (eds). AVI, Van Nostrsand Reinhold, New York, pp. 211–229.

Kirk, R. S. and Sawyer, R. (1991) *Pearson's Composition and Chemical Analysis of Foods*, 9th edn. Longman Scientific and Technical, Harlow.

Moskowitz, H. R. (1989) Sensory segmentation and the simultaneous optimization of products and concepts for development and marketing of new foods. In *Food Acceptability* by D. M. H. Thomson (ed.). Elsevier Applied Science, Barking Essex, pp. 311–326.

Moskowitz, H. R. (1994) *Foods Concepts and Products*. Food & Nutrition Press, Trumbull, CT, pp. 233–292.

Nakai, S. (1981) Comparison of optimization techniques for application to food product and process development. *Journal of Food Science*, **49**, 1143–1148.

Nicklin, S. H. (1979) The use of linear programming in food product formulations. *Food Technology in New Zealand*, **14**, 2–7.

Ordóñez, M., Rovira, J. and Jaime, I. (2001) The relationship between the composition and texture of conventional and low-fat frankfurters. *International Journal of Food Science and Technology*, **36**, 749–758.

Soden, P. M. and Fletcher, L. R. (1992) Modifying diets to satisfy nutritional requirements using linear programming. *British Journal of Nutrition*, **68**(3), 565–572.

Strino, E. (June 1984) The computerised planning of food products. *IFST Proceedings*, **17**(2), 79–91.

Chapter 11
Statistical quality control

11.1 Introduction

Consumers rely on food products being reasonably constant in terms of quality.
They also expect the standard of quality to be high. Quality covers all aspects of
a food product's nature, including sensory, nutritional, microbiological, chemical
and physical as well as safety and legal. Thus, consumers expect safe, nutritious and
tasty foods, which are storage stable and which adhere to legislative requirements
in respect of chemical content and weight or volume, etc. To achieve these goals,
food manufacturers can employ statistical quality assurance programmes, but they,
in turn, expect high quality levels and relatively constant quality specifications in
the raw materials that they receive.

Quality control procedures include periodic sampling of all food materials from
the time they arrive at the factory to the time they exit as the finished product.
Statistical techniques provide a means to monitor levels of quality at one or more
critical points in procedures, hence the term *statistical quality control* (SQC). This
can be extended to *statistical process control* (SPC) because, in addition to the
food itself, the efficiency of plant machinery is also monitored as it affects several
aspects of the products make-up and weight, etc. In fact, any pertinent process can
be monitored in this respect, such as data generated by an analytical laboratory
(Mullins 2003).

An appreciation of statistical variability and population distribution is invaluable
in these applications. Once again, acknowledgement of the fact that 'all measures
vary' (Chapters 2 and 3) lays a solid foundation in the understanding of statistical
QC. This variation must be taken into account, but the food manufacturer cannot
allow below-standard food products to pass through the QC 'net'. That is, it is
not sufficient that the average, central measure meets the standard. The spread of
the population distribution, which applies to the measurand, must be above the
standard by an amount that ensures that below-standard occurrences are negligible
or within specified limits.

Statistical Methods for Food Science: Introductory Procedures for the Food Practitioner,
Second Edition. John A. Bower.
© 2013 John Wiley & Sons, Ltd. Published 2013 by John Wiley & Sons, Ltd.

Table 11.1   Some applications of statistics in food quality and process control.

Application	Function
Control chart (variable)	
Monitor over time, during production of products or data from any ongoing procedure	
Average chart	Change in mean value of small samples
sd Chart	Change in standard deviation of small samples
Range chart	Change in range of small samples
Attribute charts	
Monitors change in defectives of small samples or units	
p Chart	Change in percentage of any defectives in sample
np Chart	Change in number of specific defective in sample
c Chart	Change in number of defectives in single unit
Acceptance sampling	
Monitor raw material on delivery	
Variable measures	
Attribute measures	

11.2   Types of statistical quality control

Statistical applications for quality control come under two main types. Firstly, there are procedures that monitor trends in various quality measures over time (Table 11.1).

These comprise a variety of ***control charts***. They display a quality measure against time, either as a variable or as an attribute category. This measure is obtained from a sample drawn from the production line, QC laboratory, chemical analysis or sensory laboratory, or any process at periods according to a sampling plan.

A second common application is that of ***acceptance sampling*** procedures. They are used for a different purpose – that of monitoring the quality of incoming raw materials and for examining the number of defective articles in consignments of goods. Acceptance sampling procedures are of major consequence because they decide actions in distinctly positive or negative manner, as 'accept' or 'do not accept'. They can be applied to the 'presence or absence' type of defect or to variable measures, e.g. 'any soluble solids content below 60% is classified as a defect'.

11.2.1   Types of end determination measure in SQC

All levels of measurement are possible in QC data from nominal occurrences up to continuous ratio scales. Rapid measures of assessment are used for control within production, where adjustment can be made as the process is 'live'. In other applications, such as analytical methods where analyses take longer, data may be collected on a daily basis, etc. Monitoring of microbial growth could come into this circumstance, although rapid biochemical methods such as that in the study by Hayes *et al.* (1997) have been applied in SQC analysis.

Data originating in QC can be plotted to provide a graphical display of ongoing variability or as records of compliance/non-compliance. In SQC, the term *variable* is used for data that are usually ratio level of measurement and continuous, and *attribute* is applied to nominal data measures such as 'accept', 'not accept'.

Thus, **variable control charts** monitor measures such as net weight, volume, solubility, hook overlap percentage and colony counts. Attribute control charts deal with counts of defective items. Similarly, acceptance sampling can deal with occurrence of defects or faults in material supplies based on simple tallies or on variable measures as explained above.

11.3   Sampling procedures

Sampling for quality control requires a full appreciation of the nature and function of the sample selection process (Section 2.3). The type of sample taken depends on the variability of the population and whether or not identifiable subgroups, which have uneven variability, exist. All forms of sampling are possible in QC work, but they are usually of a *random nature* (simple random, stratified, etc.). The common problems arise when deciding on *sample size*, but also on the *frequency* and *location* of sampling points in a multi-stage production process. Other contributing factors include which form the end determination measure takes, as this has implications for time and resources.

11.3.1   Sample size and frequency

A feature of sampling in QC is the intermittent gathering of samples during production. The population in these circumstances is viewed as *infinite*, if the same materials and equipment are in operation. Samples can be drawn randomly from this population, or randomly from identified strata such as different production lines, fillers, ovens, etc. Samples are therefore not of the single 'one-off' type, but are 'small and frequent' in the control chart situation. In the acceptance sampling procedure, there are more instances of an identifiable finite population (e.g. a consignment of freshly harvested peas). This would be sampled once according to a plan and a decision made, and possibly followed by a second sample selection.

11.3.2   Sampling point location

Location of where to take samples in a production process situation can be ascertained by various methods. It depends on such factors as the previous history of quality variation. It can be guided by the knowledge of experienced staff, but ultimately, it may have to be located by a systematic identification of critical points by implementation of a hazard analysis and critical control point system (HACCP; FAO/WHO Codex Alimentarius Commission 1997).

Many factors influence the location in respect of the effect on quality. Decisions can be based on:

- The criticality of any fault in quality which could develop at that point
- Variability of quality at the point (crucial information)
- The cost of sampling at that point (finance, time delay, etc.)

This applies across the whole spectrum of food product manufacture from raw material delivery to final goods production. HACCP schemes are crucial for all aspects of quality and provide an ideal layout of the stages where sample gathering can be prioritised.

Some of these points can be illustrated by acceptance sampling of a raw material delivery. The manufacturer requires that the consignment be of an acceptable quality before it can be permitted entry. Based on a sample, an 'accept/reject' decision is made. The sample size and location depend on the previous history of the supply and the form in which it is delivered – on pallets, or in a tanker, open lorry, or silo, etc.

Sampling can be based on every pallet, or a random selection from a larger bulk – it depends on the likely variability and the criticality of making an error (e.g. in accepting a poor load). A history of top quality and many successive 'accept' decisions may allow relaxed sampling (every second load – or at random intervals) to be used.

Similar decisions could apply to a *variable measure* on a metric scale, e.g. the viscosity of a particular component in a product. If this property had to be within tight specifications (i.e. it was critical to the quality of the final product) and it was known that the viscosity had high variability then sampling would be more frequent and more comprehensive to ensure less chance of a below-specification batch being accepted.

Valuable aids to sampling in QC are provided by use of existing tables of plans drawn up in QC texts and by means of analysis of sampling plans by an *operating characteristic (OC) curve*. This method allows identification of a region within a sampling scheme where the sample result is acceptable and another region where it is unacceptable. OC curves can be prepared for both acceptance sampling and control chart use.

An important principle in sampling and analysis is that of the *central limit theorem* (CLT) (Section 2.4.1). In QC, it may be that many small samples are taken over time. The distribution of the average values on such a series will be approximately normal even if the parent population is not. This allows distributional knowledge to be applied to the variation occurring during data generation with time in QC. Further details on sampling are dealt with below.

11.4   Control charts

This technique is used by the food manufacturer to ensure that a standard of quality is maintained. The *control chart* requires that an ongoing output of data is occurring and samples are drawn at intervals. A common application for the *variable control chart* is net weight control, but other quality aspects can also be monitored, such as volume, physical dimensions, colour, chemical contents, etc. In these situations, the measure is the *mean value*, but in other charts, it can be the range or *standard deviation*, and *attribute charts* display *incidence* of defectives. The account below deals mostly with the variable control chart where the summary

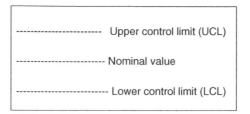

Fig. 11.1   Basic structure of a control chart.

measure is the mean. The statistical abbreviation for the sample mean (an estimate of the population mean) is $\bar{x}$ and the chart is known as the ***x-bar chart***.

11.4.1   The x-bar control chart

A control chart shows a live display of how a process is varying based on a summary measure. At the centre of the chart, a line marks the expected summary value, referred to as the ***nominal value*** (not to be confused with nominal level), which in the case of the *x*-bar chart is the mean. Some variation is expected due to random causes. This is demarcated by a ***normal distribution*** spread. To display the spread, ***limit lines*** are marked on the chart (Fig. 11.1).

The limit line positions are based on the population distribution. The normal distribution or other distribution provides a model of the expected variability. As explained above, the normal distribution will apply to the population of sample averages even if the parent population is not (according to the CLT). Deviation from the projected population shape can indicate sources of unusual variation. Thus, *excessive drift* from the central region can be detected. There will be some variation from the centre and this is deemed as random, uncontrolled or ***unassignable error***. This variation cannot be countered and it originates from machine sources such as play in bearings, slight fluctuation in electrical supply or slight variation in food materials, etc. This contrasts with *large variation,* which will show up more obviously as ***assignable error***. These assignable sources of variation occur due to *gross error* by machines or operators, e.g. incorrect readings or biased reading of analogue devices, a sudden failure of a machine component or large variation in raw material nature.

Chart construction

Chart construction begins by obtaining a summary measure from a sample, which is then plotted on the control chart. This can be based on a value obtained from previous data or by doing an initial run with 30–50 samples. Sample size for the online situation is typically five individual units every 30 min from the product line. The expected variability (via the distribution) allows demarcation of the ***limit lines*** on the chart. These are positioned above and below the central region. Limit lines can be of a number of types (inner and outer) that are drawn horizontally along with the centre line (Fig. 11.1).

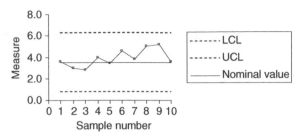

Fig. 11.2   Control chart showing within-limit condition (Excel).

In this way, it can be envisaged that the distribution is turned through 90° with the tails forming the upper and lower limits. Several limits lines can be positioned, but the basic chart can show a process that is 'in control', i.e. within the expected limits of variability (Fig. 11.2).

Example 11.4-1 *'x-bar' control chart construction and monitoring*

A manufacturer sets up a canning line for a new liquid product. While net content was important, initial concern centred on achieving a headspace of 3.5 cm. Product consistency and filling machines were set to achieve this latter level, but a preliminary data set shows variation that was possibly excessive. Samples (ten in total) of five cans were drawn off every 25 min, opened and the headspace (cm) measured (values as expressed as nearest integer to simplify the data). How can the manufacturer construct a control chart from this data and what are the conclusions from the display?

An *x*-bar chart was constructed from the data as displayed above (Fig. 11.2).

To construct the chart using Excel, the small set of data comprising ten samples was entered into the spreadsheet. The means and standard deviations were calculated for each (Table 11.2).

Table 11.2   Data for construction of *x*-bar control chart (Excel).

Sample	Unit 1	Unit 2	Unit 3	Unit 4	Unit 5	Mean	Nominal	sd	LCL	UCL
1	2	6	4	1	5	3.6	3.5	2.07	0.79	6.21
2	1	7	2	4	1	3.0	3.5	2.55	0.79	6.21
3	2	4	1	5	2	2.8	3.5	1.64	0.79	6.21
4	5	6	4	2	3	4.0	3.5	1.58	0.79	6.21
5	1	2	2	7	5	3.4	3.5	2.51	0.79	6.21
6	4	1	7	4	7	4.6	3.5	2.51	0.79	6.21
7	2	5	5	6	1	3.8	3.5	2.17	0.79	6.21
8	3	6	6	5	5	5.0	3.5	1.22	0.79	6.21
9	5	7	8	3	3	5.2	3.5	2.28	0.79	6.21
10	6	4	2	2	4	3.6	3.5	1.67	0.79	6.21
							Overall sd = 2.02			
							x-bar sd = 0.90			

Mean and sd are calculated using the AVERAGE( ) and STDEV.S( ) functions (STDEV( ) Excel 2003). Two further measures were calculated. Firstly, the overall sd, which is obtained by taking the average of the column of individual sds above. Secondly, the **x-bar sd**, which is the standard deviation of the average values, and which uses the formula for the standard error calculation (see Table 3.6):

$$x\text{-bar sd} = \text{overall sd}/\sqrt{5}$$
$$= 2.02/2.24 = 0.90$$

The final calculations were the ones for the limits lines. They can be obtained in two ways:

- ± Sigma method
- ± Probability method

Where are the limits set?

Action and warning limits

The positioning and number of limit lines depends on the variability of the process and other factors such as the type of chart. Referring to the normal distribution (Fig. 2.11), it can be seen that for the mean value, approximately 95% of the measures lie within ± 2 population standard deviations (σ) and 99.7% lie within $\pm 3\sigma$. Thus, the natural variation of the process will result in some values reaching and exceeding $\pm 2\sigma$ and some isolated occurrences of some reaching $\pm 3\sigma$. There is very low probability (approximately 0.3%) of some values exceeding the $\pm 3\sigma$ region, and they form an indicator not only of very rare natural occurrences but also of possible non-natural *out-of-limit* ('ool') or *out-of-control* values. Control charts based on the mean value have limit lines set at the $\pm 3\sigma$ level as outer limits (the *action limits*) and some have inner limit lines (*warning limits*) set at $\pm 2\sigma$. Thus, a convenient and simple way of calculating the position of the limits lines is based on using ± 2 and 3σ, which, for the sampling distribution, is the x-bar sd ($\sigma_{\bar{x}}$) value.

The above '*sigma* method' is approximate with respect to the probabilities and limits can set by using *exact probabilities*. These are calculated for a given probability using the *z-value* (portion of the normal distribution; Section 2.4.1) and the standard error (x-bar *sigma*) in a similar manner to that given for a confidence interval (Section 3.4.3). The procedure requires specification of a risk probability, e.g. manufacturers may specify that they wish to control out-of-limit event detection to exactly 1%. This value is divided by 2 to give the risk for the

Table 11.3   Probability of out-of-limit work at various levels of risk.

Probability	Out-of-limit risk	z	Location x-bar ±	Approximate *sigma*
(0.05) 5%	1 in 20	1.96	1.96 × se	±2
(0.01) 1%	1 in 100	2.58	2.58 × se	±2.6
(0.002) 0.2%	2 in 1 000	3.09	3.09 × se	±3
(0.001) 0.1%	1 in 1 000	3.29	3.29 × se	±3.3

se, standard error.

lower and upper limit. Assuming that the *x*-bar *sigma* value above is pertinent, then:

$$
\begin{aligned}
&\text{Risk of 'ool'} = 1\%, (0.01), \sigma_{\bar{x}} = 0.9 \\
&\text{Lower risk} = 0.5\%, (0.005) \\
&\text{Upper risk} = 100 - \text{lower} = 99.5\%, (0.995) \\
&z_{(\text{lower risk})} = z_{0.5\%} = \text{NORM.S.INV}^a\,(0.005) = -2.58 \\
&z_{(\text{upper risk})} = z_{99.5\%} = \text{NORM.S.INV}\,(0.995) = +2.58
\end{aligned}
$$

[a]NORMSINV( ) Excel 2003.

Thus, limit lines would be set at $\pm 2.58 \times \sigma_{\bar{x}}$. Similar values can be calculated for various levels of risk (Table 11.3). To achieve a similar setting to $\pm 3\sigma$ would require a risk of 0.2%. Apart from the data shown in Table 11.3, *z*-values can be obtained for any probability using Excel's NORM.S.INV( ) function.

In a similar manner, probabilities can be calculated for a given *z*, using the NORM.S.DIST( ) function, and this can be used to calculate the exact probabilities of the ± 2 and 3σ method. Thus, for $z = 2$ and 3 what is the probability of 'ool' work?

$$
\begin{aligned}
&z = 2, \text{Probability} = 1 - \text{NORM.S.DIST}^a(z,1) = 0.02275\ (2.3\%),\ \text{two-tailed} = 4.6\% \\
&z = 3, \text{Probability} = 1 - \text{NORM.S.DIST}(z,1) = 0.00135\ (0.14\%),\ \text{two-tailed} = 0.28\%
\end{aligned}
$$

[a]NORMSDIST(z) for Excel 2003.

As seen, these are very close to the 5% and 0.3% values given above, so the systems are very similar. The method used depends on the region. Practitioners in the United Kingdom and Europe tend to use the probability limits – in the United States, limits are based on *sigma* (Montgomery 2001). The probability methods represent the risk of committing the Type I and II errors in use of the control charts. Thus, a probability limit of ± 0.001 where 0.2% of work is 'ool' in the population constitutes a risk of committing a Type 1 error of 0.002 (0.2%), i.e. approximately 2 out of 1000 will cause a false alarm. Calculation of the Type II error is more complicated, but it is reduced by narrowing the limits.

For the example data in Table 11.2, the '*sigma* method' was used and upper and lower limit lines were calculated as:

Upper control limit (UCL) = nominal value $+ 3 \times \sigma_{\bar{x}} = 3.5 + 3 \times 0.90 = 6.21$
Lower control limit (LCL) = nominal value $- 3 \times \sigma_{\bar{x}} = 3.5 - 3 \times 0.90 = 0.79$

A nominal value of 3.5 was allocated in this case, but it could have been estimated by the overall mean of all the values. These data were selected in the spreadsheet and the three lines were plotted in a line graph (select non-contiguous columns using **Crtl-click**). Figure 11.2 shows that the process is 'in control' in that all plotted points are within the action lines, although five consecutive points are above the centre line and this would require vigilance if the trend were to continue (see below). However, the magnitude of the headspace variation is large, with values ranging from 1 to 8 cm for individual cans. This is not satisfactory for several reasons, including net weight and expansion effects, etc. It could be that there are assignable causes for this and a tighter specification for headspace is required.

Thus, the usefulness of the chart depends on the estimate of *sigma* values. The process must be in control (i.e. random variation only) when these are measured as above, otherwise the variation could be *overestimated*. Such a situation would allow out-of-control material to pass through unnoticed. Alternatively, the variability must reflect the likely variation of the process as used under all possible intended conditions. If not, the variability could be *underestimated* and 'false alarms' would be likely.

Limits set at ± 2 and 3σ will have sample averages exceeding the line at approximately 5% and 0.3% of the time, respectively. Any exceeding $\pm 3\sigma$ are very unlikely to be due to chance. They are viewed as possible 'out-of-limit' points and action may be taken to identify the cause. Figure 11.3 shows the above process with an out-of-control situation developing.

Control chart operation allows a number of ways of deciding on possible out-of-control occurrences. These are based on certain patterns displayed by the points, such as 'any point outside an action limit' and 'seven points above or below the centre line' in sequence'. Fuller details are given in texts such as Hubbard (1996) and Montgomery (2001). In Fig. 11.3, there are five consecutive points above the

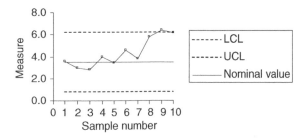

Fig. 11.3 Control chart showing points within limit and drift out-of-control (Excel).

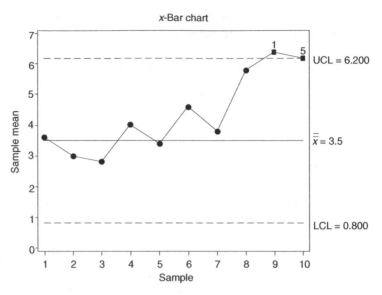

Fig. 11.4   *x*-bar chart with out-of-control markers (Minitab).

centre line, and two over the limit line, so this is a warning that the process is 'out-of-control' (see below).

Minitab and MegaStat provide a selection of control charting facilities, which can produce charts with the original data (i.e. without the calculation required for the Excel example). Minitab also has the ability to carry out up to eight checks on the points. Figures 11.4 and 11.5 show the analysis of the data above in Minitab's *x*-bar and range chart facility.

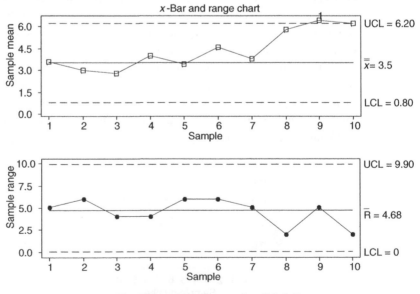

Fig. 11.5   *x*-bar and range chart (Minitab).

The *x*-bar chart has detected two types of 'out-of-control' points – one point beyond the upper limit line and 'two out of three' in the upper region or beyond.

The second plot includes a range chart, which indicates that although the mean values are out-of-control, the variability is remaining within the limits (again, as stated above, this situation is based on the initial data, which may have assignable causes contributing to variation).

The control limits for *x*-bar charts can also be constructed by use of tables of factors (available in texts on quality control), which give an improved estimate of the population standard deviation. A common version uses the range (easier to measure in pre-electronic calculator days) as the starting point for estimation.

Control charts based on the mean are used extensively to monitor 'non-production' processes in food science, such as chemical analysis data and monitoring of sensory panel training. In the latter case, a median-based chart is recommended by Gacula (2003) as training data are subject to distortion by skew and outlying points. Median values for each panellist can be plotted with the panel median as the centre line. Other possibilities are plotting of individual replicate scores along with the average to illustrate consistency of panel and panellists, and comparing several attributes on one chart.

Specification limits

In addition to the limits above, at least one other limit level can be used: *specifications*. These limits specify a standard quality level, which is set by the food manufacturer for the particular measure, and takes the form 'within ± specified values', e.g. pH 6 ± 0.3. The **specification limits** usually require to be inside the $\pm 3\sigma$ action limits above, otherwise the process will not be able to perform 'in spec'. If they were set near $\pm 2\sigma$ then 2.5% of 'below-standard' work would get through the process, as would 2.5% 'above standard'. In these circumstances, the manufacturer risks producing below-standard products, which in some cases would be infringing legal requirements. Specification limits nearer to $\pm 3\sigma$ are more achievable. In the theoretical normal distribution, where the population is infinite, the tails extend to infinity, but beyond $\pm 3\sigma$ only a very small proportion (0.3%) of the population would occur – 3 in every 1000. The acceptance of this low occurrence by the manufacturer depends on several factors such as the implications in terms of legal issues, consumer complaints and cost. For higher level of control, *sigma* requires to be reduced so that specification limits can be brought inside a wider setting. If action limits were set at $\pm 6\sigma$ and specifications were just inside then the occurrence of out-of-specification work would reduce to 2 in a billion (Joglekar 2003). Another point is that if a measure such as **mean net weight** is being controlled then the main specification is the lower limit. Net weights above this are all legal, but if excessive, result in extra costs (see below).

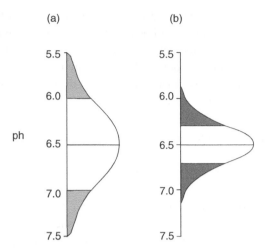

Fig. 11.6   Original distribution (a) and sampling distribution (b) of the same population (Excel).

11.4.2   *Sampling for control charts*

The detection of out-of-limit work depends on the sampling procedure. In the control chart circumstance, samples are continuously drawn and both *sample size* and *sampling frequency* are important. Individual unit sampling is convenient and rapid, but the distribution will be broader, i.e. the original distribution will apply (Fig. 11.6a). Taking a single unit every half hour will monitor the process, but there is a higher probability of missing points outwith 2 and 3σ, even when sampling units on a more regular basis. Such procedures are uneconomical of sampling resources and knowledge of the CLT provides a solution.

Plotting values based on a *summary measure* of several units gives a narrower peak and tighter limits (Fig. 11.6b). These limits allow a much more economical detection of out-of-limit work. Shifts in the plotted summary values are much more sensitive to wider deviation and the chances of detecting out-of-limit work have improved. Both Figs 11.6a and 11.6b are from the same population, but Fig. 11.6b has a narrower peak. Thus, for variable control chart use, typical sampling parameters are a small number of units over a short period (*five units every half hour*).

The procedure of taking small samples on a regular basis means that the plot is of the variation of the **sampling distribution**, which has less variation than the original distribution. This does not mean that the distribution has changed, but detection has increased. If a process has a 1% of work below the required standard then this will apply to both the original and the sampling distribution. The only way to reduce this figure is to reduce *sigma* as discussed above, i.e. the limit lines are set by *sigma* – the smaller *sigma* is, the tighter the lines.

The 'x-bar' control chart in use

A more specific examination of the *x*-bar chart in use is given by the circumstance of control of minimum net weight in packaged food products. The problem facing

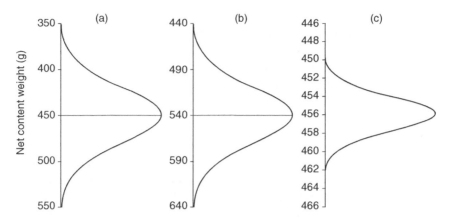

Fig. 11.7   Sampling distributions at different nominal target settings and *sigma* values (Excel). (a) 450 g, 30 g; (b) 540 g, 30 g; (c) 456 g, 2 g for nominal and sigma values, respectively.

the packer is that a minimum is specified legally, but not a maximum. The consumer expects the minimum and legal enforcement backs this view. Food packing and filling machines cannot deliver an exact weight repeatedly. The weight is subject to random error giving variation around any set weight. Systematic error can be eliminated by calibration of the machine and by adjusting the setting. The weight delivered will follow a normal distribution. If such a machine is set to a nominal target of 450 g (the pack declared minimum weight) then approximately 50% of the packs will be above the minimum and approximately 50% will be below and will infringe the legal requirements (Fig. 11.7a).

Resetting the machine to a higher nominal value will rectify this, but what should the setting be? Most of the values in a normal distribution will lie within $\pm 3\sigma$, so assuming $\sigma_{\bar{x}}$ of the sampling population is 30 g, the nominal weight can be set at $450 + 90$ g $= 540$ g (Fig. 11.7b). Now only 0.13% of packs are under weight, but 99.87% of packs are over! In essence, the manufacturer is 'giving away' this product material to the consumer. This, of course, is not economically viable and there may be other implications in that the containers used for the pack may be unable to hold some of the higher weights, etc. The amount of *giveaway* depends on the number of packs and the magnitude of $\sigma_{\bar{x}}$. The extreme example just given would result in a very large giveaway of approximately 90 kg per 1000 packs, a small number being as much as 180 g overweight. To overcome this problem, the packer must lower the nominal weight and reduce variability (σ) to a much tighter spread. This can be done by examining the fill or packing machine control to optimise it and to ensure that ingredients are controlled in respect of variation in any property that would affect product weight and final pack weight. This depends on the product and could be particle size, viscosity of coating of sauces or new batch-to-batch variation in raw material flow, etc. Machine suitability can be examined and replacement may be required, but this is much more expensive. Control charts can help here by examining variability of ingredient properties on the final pack weight.

Assume that nominal weight is adjusted to 456 g and by extreme efficiency, $\sigma_{\bar{x}}$ is reduced to 2 g, then giveaway is reduced to a fraction of the original weight per 1000 packs, and the upper limit is only 12 g overweight (Fig. 11.7c). Once the setting is considered acceptable, the process can now be monitored by control charts of the average x-bar and range or sd type.

11.4.3   Compliance issues

How do food manufacturers ensure net weight compliance?

Food manufacturers are subject to monitoring of their goods by trading standards, etc. These enforcement agencies employ sampling schemes for testing of products in respect of net weight, meat content, additive level, etc. The detailed nature of the sampling and testing scheme depends again on the governmental agencies involved in different parts of the world, but certain features are common. Essentially, a sample is taken and tested. A certain amount of leeway is given in some systems for a level of failure, e.g. for net weights in the United Kingdom, a 450 g weight class can be up to a maximum of 3% under. In the United States, stringency of control can vary and one scheme permits 0% fail on a screening check, but another allows around 4%. Exact figures also depend on the weight class, possible moisture loss and sampling error may be taken into account (Hubbard 1996).

Overall results are expressed as sample average compared with the label claim. Usually, the sample average must be equal to or greater than the claim, but the exact 'rule' depends on the specific weight class and any corrections. Thus, some systems may allow the sample average to be greater than or equal to label claim minus two standard deviations. According to a survey conducted by Grigg *et al.* (1998), some food manufacturers do not follow codes of practice for net weight control or indeed implement SPC due to its cost and complexity. They ensure compliance by check weighing and by over-filling on the basis that the costs will balance out. This view could be considered blinkered, as there are other advantages in implementation of the system, especially with less complex forms (Grigg and Walls 1999).

What can manufacturers do to ensure control?

Attempting to control a quality parameter such as net weight by sampling at the end of production can be wasteful and expensive – such 'after the fact control' may mean discarding material or re-packing. The manufacturer needs control charts for immediate reaction to drifts. It is also possible to calculate the risks and costs involved in any prescribed system.

There are two aspects to this, in that the manufacturer can ensure a *high level of compliance*, thereby minimising the risk, but at the same time may have higher 'giveaway', i.e. a higher 'over-fill' cost. The converse is to cut this cost, but increase the *non-compliance risk*. The calculations for these parameters use the *sigma* value and knowledge of the maximum variation that any enforcement system allows. Enforcement rules are complex and some simplification is used with the example calculation below. Grigg *et al.* (1998) and Hubbard (1996) give more comprehensive details for the UK and US systems, respectively.

Cost of over-fill and risk

To calculate this cost, the first requirement is the enforcement agencies' maximum allowable variation from the label claim. Assume that this **tolerable negative error** (TNE) is 3%. This can be subtracted from the declared weight to give the *minimum allowable weight* (MAW) for a 450 g pack:

> 3% of 450 g maximum = approximately 12 g
> Minimum weight allowable = 450 g – 12 g = 438 g

Secondly, the $\sigma_{\bar{x}}$ value is required; assume that this is equal to 9 g.

If the nominal weight (centre line) is set to the declared label weight then the proportion of under-fill/over-fill can be expressed as the area of the normal curve below the 'MAW' value. This involves calculation of the area above the 'MAW' figure, then subtracting this from unity (the total area):

> Area below MAW = 1 – area above MAW

These data and formulae were entered into Excel (Table 11.4).

To calculate the cost of over-fill requires calculating an average amount per pack, and then multiplying this by the number of packs over a selected period. An accurate assessment of this would require integration of the values above the nominal, but it can be approximated by summing the number of *sigma* increments above this weight. It is known that the area on either side the centre of the distribution occupies 50% of the total. In turn, the total is divided up into regions by the sigma values (Fig. 2.11) as 68% up to $\pm 1\sigma$, 28% $\pm$ between 1 and 2σ and 4% $\pm$ between 2 and 3σ (approximate percentages). These values correspond to the same percentage figures for the upper portion of the area, i.e. 68% of the upper region is *up to* $+1\sigma$, i.e. the values range from 0 g above nominal weight to 1σ above (on average, 0.5σ). Thus, most of the overweight packs are approximately 0.5σ above and the least are overweight by approximately 2.5σ. The number of packs is given by the %area:

	+0.5s	+1.5s	+2.5s
Above nominal weight	= 68	28	4

The overweights are calculated and summed to give the total 'giveaway' weight (Table 11.4). A cost estimate can be calculated using an arbitrary cost of 1 monetary unit per kilogram of product. The results show that 50% of the packs are going out over-filled. This corresponds to an approximation of 4 kg, costing 4 monetary units per day, assuming that 1000 packs are a daily output.

Table 11.4   Calculation of over- and under-fill cost and risk (Excel).

	A	B	C
1	Declared weight (NOM)	450	
2	*sigma x*-bar (sxb)	9	
3			
4	TNE (%)	3	
5	Minimum allow weight (MAW)	436.5	$= NOM - 0.03 * NOM$
6			
7			
8	*Sigma*s NOM to MAW	1.5	$= (NOM - MAW)/sxb$
9	Area above MAW	0.933	$= NORM.S.DIST(B8,1)$[a]
10	Area below MAW	0.067	$= 1 - B9$
11	Total area over-fill	0.500	$=$ half of total area (1)
12	Total area under-fill	0.067	$=$ area below MAW
13			
14	Total units	1000	$=$ per day or other
15			
16	Units above NOM	500	$= B14 * 0.5$
17	Giveaway	3870	$= (0.68 * B16 * 0.5 * sxb) + (0.28 * B16 * 1.5 * sxb)$ $+ (0.04 * B16 * 2.5 * sxb)$
18	Cost per kilogram	1.00	$=$ any monetary unit per kg
19	Cost of 'giveaway'	3.87	$=$ giveaway $*$ unit cost
20			
21	Units below MAW	67	$= B14 * B12$
22	Risk of detection:		
23	Random single pack	0.067	$= 1 - BINOM.DIST$[b]$(0,1,B12,TRUE)$
24	Random ten packs	0.499	$= 1 - BINOM.DIST(0,10,B12,TRUE)$

[a]NORMSDIST(B8) for Excel 2003.
[b]BINOMDIST ( ) for Excel 2003.

Non-compliance risk

Balanced against the cost are the consequences of non-compliance and the associated risk. This can be obtained by assessing the probability of a below compliance pack turning up in a random sample by enforcement officers. Based on a specified number of failed packs in a drawn sample, if a sample of one is taken randomly from the whole population of the above packing process then the risk is the same as the area below MAW (0.067% or 6.7% in the table). This figure can be confirmed by use of the binomial formula (Section 8.4.3):

$1 - BINOM.DIST$[a] (Number of successes $- 1$, number of trials, 0.067, TRUE)
---

[a]BINOMDIST ( ) for Excel 2003. (Binomial use assumes large population).

This gives the probability of drawing one or more packs that fail (where the number of 'fails' is taken to be 'successes'). The risk will increase with sample size, and for the above process there is an approximate 50% chance that at least one fail pack will be picked up when a sample of ten is drawn (a high risk).

Some enforcement systems specify an average test weight with some leeway below. In this case, the sampling distribution is used to calculate the standard deviation of the sample for a particular sample size. As explained above, this will be smaller depending on the number of units. The z-score (Section 2.4.1) for the level below the label weight can be calculated, then the using approximation of values, the probability can be ascertained:

Mean = 450 g, sigma = 9 g, sample size = 10, leeway = 1.5 × se
se (standard error) = 9/square root (10) ≈ 3
$z = ((450 - 1.5 \text{ x se}) - 450)/3 \approx (445 - 450)/3 = -5/3 = -1.67$
Probability = NORM.S.DIST[a]$(-1.6.7, 1) = 0.048 = 4.8\%$ (approximately 5 in 100)

[a]NORMSDIST(−1.67) Excel 2003.

If, on the other hand, the enforcement gives no leeway then the sample average must be equal or greater than the label weight:

$$z = (450 - 450)/3 = 0/3 = 0$$

The probability of a z-score of zero is 0.5 (50%), which again is high.

Are these risks acceptable?

The above risks are high is some cases and even the lower ones would probably be considered too high. To reduce them requires a reduction in variation (*sigma*). The whole process, including the process line, packing machines and ingredients would have to be examined to study the effect on sds at each stage, which could be expensive. The reader can experiment with different values for *sigma* and cost, etc., in Table 11.4, which allows visualisation of the possibilities. One change is to calculate the risk probability of other requirements. For instance, the UK system specifies that there must be a very low occurrence (1 in 10 000) of pack weight below 2 × TNE (6% in the example above). The probability of this can be obtained by changing %TNE to 6% – the probability is 0.0013 which is equivalent to '13 in 10 000', so the system does not comply.

11.4.4　Other variable control charts

Other types of variables control charts include the sd chart and the range chart; these are constructed in a similar manner to the average chart. The range chart is usually plotted along with the x-bar as it can detect changes in the level of variation, which can be cancelled out by the x-bar system of sample averaging (see Fig. 11.5).

11.4.5　Attribute charts

This type of chart monitors data of a nominal nature, although results on the chart are plotted as counts or percentages. The structure of the chart is essentially the same as that for the variable type, but the measure of interest is one of three types (Table 11.1). *p Charts* plot percentage defects for lots of fixed size and sample. The limits are based on the parameters of the binomial distribution (Section 5.1.2). These *percentage defective* or fraction defective charts are designed for measures of one or more defects in the sample where any of the defined defect being present will signify a 'defective' proportion in the sample. This can be expressed as:

> % Defective = number of defective units/Number of units in sample

This is similar to the binomial average value. The acceptable %defective levels or starting point for control can be obtained from existing records, management directives or by an initial run of 20–30 samples to obtain an estimate. Control limits are set at the ±standard deviation for the sample as:

> Control limits for p chart = % defective $\pm 3 \times$ square root $((p(1 - p)/n))$
> (assuming a 3σ limit)

The p chart classes a unit as nonconforming when any defect from a specified set is detected. If individual defects are of interest then a 'number of defects' chart or **np chart** can be used. The nature of the defect is defined, e.g. 'excessive chocolate coating'. Here, the standard value is given by the average number of defectives established previously or from an initial trial. This value is taken as being equal to the binomial average of np, from which p can be calculated. The limits are then calculated as before using the values of np, p and n:

> Number of defective $(np) = d$, $p = d/n$

A third type of attribute chart is the **c chart**. This chart plots the number of defects in a *sample of one*. This is applicable for units produced in large amounts, such as large commercial cans of foods. This is modelled by the **Poisson distribution**, which is similar to the binomial except events occur less frequently.

11.5 Acceptance sampling

In acceptance sampling, materials that arrive at the food factory are sampled and checked for compliance to one or more criteria. This procedure is critically dependent on the nature of the sample selection and sample size. Measures performed on the sample material are usually in the form of 'number of defective items', e.g. checking for defectives in a newly arrived batch of raw material vegetables. The manufacturer will accept the batch provided that on sampling a random selection of items (e.g. ten from each crate), no more than maximum number (e.g. two) are defective in terms of damage and bruising.

An analysis of the ability of any sampling plan can be done by construction of an **operating characteristic (OC) curve**. Construction of the OC curve requires specification of the *number of defectives acceptable* as a proportion. The probability of acceptance for all possible proportions of defectives is given by use of the binomial distribution (Cohen 1988). The exact probability of accepting the batch is calculated for a range of percent defectives.

Example 11.5-2 *Construction of an operating characteristic curve for acceptance sampling*

A quality control food technologist is asked to prepare an OC curve for a sampling plan used to monitor incoming raw materials that are causing concern. These arrive daily in large batch loads of 10–20 kg. The directive is that the maximum acceptable percentage of defective units is to be 20%. A member of the QC staff will require to be allocated to do the examination and as staff resources are under pressure sampling must be kept to a minimum. How can the technologist construct the OC curve and what conclusions come from it?

Table 11.5 OC data (Possible results with $c = 2$, $n = 10$, $p = 0$ to 1).

Actual proportion of defects	Actual number of defectives in 10	Accept probability	Reject probability (%)
0.0	0	1.00	0
0.1	1	0.93	7
0.2	2	0.68	32
0.3	3	0.38	62
0.4	4	0.17	83
0.5	5	0.05	95
0.6	6	0.01	99
0.7	7	0.00	100
0.8	8	0.00	100
0.9	9	0.00	100
1.0	10	0.00	100

c, maximum number of defectives acceptable; p, actual portion of defectives; n, sample size.

The starting point is to decide on an initial sample size based on previous experience, knowledge of sampling procedure, likely variability and time considerations. Based on existing data, assume that a sample of ten is chosen. When there is indeed 0.2 (20%) defective in the whole batch, the probability of finding *at most* two is:

Probability = BINOM.DIST[a](number defective, sample size, proportion defective, 1) = BINOM.DIST(2, 10, 0.2, 1) = 0.68 = (68%)

[a]BINOMDIST ( ) for Excel 2003.

The chance of locating up to 2 is 68% – 'no more than 2' can be stipulated as the maximum figure for acceptance, signified by c. As the real percent defective increases, the chance of getting up to 2 in a random sample of 10 (n) decreases (Table 11.5). The table calculates values for the setting of $c = 2$ and $n = 10$ for a range of actual proportions (p) of defects from 0 to 1.

These values are used to construct the OC curve using Excel (Fig. 11.8).

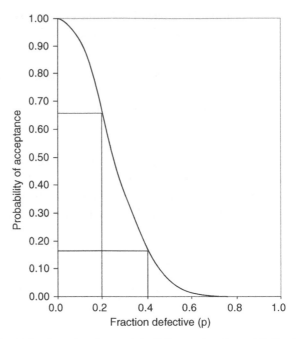

Fig. 11.8   Operating characteristic (OC) curve ($c = 2$, $n = 10$) (Excel).

The technologist can study the OC curve to make assessments of the sampling plan's potential. The plan is defined by n and the maximum acceptable proportion of defectives c, so each OC curve is unique. Changing n or c produces a different curve.

The curve can be used to determine the chances of acceptance for a given fraction of defectives. Thus, as above, if there are 0.2 (20%) defectives then based on the OC curve the chance of acceptance is 0.68 (68%). This means that there is a 32% chance that a 'true' batch (i.e. it in fact does not have more than 0.2 defectives) can be rejected. The supplier would no doubt be unhappy about such a sampling plan, when he or she has, in good faith, supplied goods to a specified standard, based on supply control measures. The other side of the coin is that if the level of defectives is higher, e.g. 40%, then there is a 17% chance that the manufacturer will accept the batch. Consequently, a load with 40% defective goods would 'slip through' the sampling examination. These two viewpoints represent the risks taken by the supplier (the seller or vendee) and the receiver (the buyer or vendor), respectively, in using such a plan. The technologist may feel that the risks are too high and will therefore have to examine ways to reduce them.

How can the risks be reduced?

The ideal sampling plan would be one with a sharper cut-off between the acceptance and rejection regions. This can be achieved by increasing the sample size. Figure 11.9 shows this for $n = 50$. Now the batches with 20% defective will be

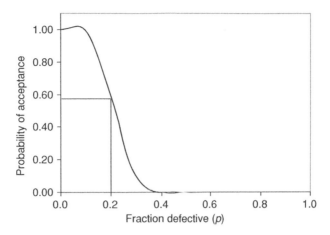

Fig. 11.9 Operating characteristic (OC) curve for sample size of 50 ($c = 2$) (Excel).

accepted at a slightly lower level (58%), but the plan is much more effective at 40% defective where acceptance is less than 1%. The disadvantage is that the receiver has to carry out more inspection and this may exceed staff availability. In other cases, large sample size can use up more products if testing is by destructive methods.

With further increases in sample size, the OC curve eventually has a vertical drop off – any 'good' batch will be accepted and any 'bad' one will fail. This is only achievable with 100% sampling with consequent expense in terms of resources.

Based on the above OC curves and analysis, the technologist may decide that a compromise between the limitations of the small sample size and the extra time required for the larger sample is the solution, and would go on construct curves for sample sizes 20–40, etc.

These examples have been for a single sampling plan. There are refinements with *double* and *sequential* plans where actions are taken (i.e. more sampling), when certain conditions occur.

The receiver (vendor) can also specify limits of quality as an ***acceptable quality level*** (AQL) and a ***rejectable quality level*** (RQL). Thus, a vendor could specify an AQL of 5% and a RQL of 30%. The OC curve can be marked with these regions corresponding to 'acceptable' and 'unacceptable' areas – in between, there is an area of *indecision* where good batches can be failed and bad batches can get through. The OC curve and sampling plan should attempt to minimise the indecision area (Fig. 11.10).

AQL has no risk with the plan, but the RQL does entail a 10% risk. The indecision area is large, and batches with 10% and 20% defective still have over 50% chance of being accepted. This can be reduced by further inspection or more samples as in the case of double and sequential sampling or by tighter specifications for the quality levels. OC curves can be used for other sampling circumstances, such as control charts, where the effect of changing sample size can be displayed.

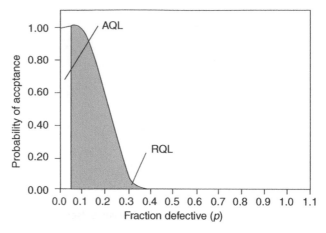

Fig. 11.10   Operating characteristic (OC) curve with acceptable quality level (AQL) and rejectable quality level (RQL) settings ($n = 50$, $c = 2$) (Excel).

References

Cohen, S. S. (1988) *Practical Statistics*. Edward Arnold, London, pp. 162–175.

FAO/WHO Codex Alimentarius Commission (1997) Hazard Analysis and Critical Control Point (HACCP) system and guidelines for its application. Annex to CAC/RCP 1–1969, Rev. 3 (1997) In *Food Hygiene Basic Texts*. Food and Agriculture Organization of the United Nations, World Health Organization, Rome.

Gacula, M. C., Jr. (2003) Language development in descriptive analysis and the formation of sensory concepts. In *Viewpoints and Controversies in Sensory Science and Consumer Research* by H. R. Moskowitz, A. M. Muñoz and M. C. Gacula, Jr. (eds). Blackwell Publishing Ltd., Oxford, pp. 313–336.

Grigg, N. P., Daly, J. and Stewart, M. (1998) Case study: The use of statistical process control in food packing. *Food Control*, **9**(5), 289–287.

Grigg, N. P. and Walls, L. (1999) The use of statistical process control in food packing. *British Food Journal*, **101**(10), 763–784.

Hayes, G. D., Scallan, A. J. and Wong, J. H. F. (1997) Appling statistical process control to monitor hazard analysis critical control point hygiene data. *Food Control*, **8**(4), 173–176.

Hubbard, M. R. (1996) *Statistical Quality Control for the Food Industry*, 2nd edn. Chapman & Hall, New York.

Joglekar, A. (2003) *Statistical Methods for Six Sigma*. John Wiley & Sons, Inc., Hoboken.

Montgomery, D. C. (2001) *Introduction to Statistical Quality Control*, 4th edn. John Wily & Sons, Inc., New York.

Mullins, E. (2003) *Statistics for the Quality Control Chemistry Laboratory*. Royal Society for Chemistry, Cambridge.

Chapter 12
Multivariate applications

12.1 Introduction

The foregoing chapters have given many examples of the form that data can take in science and the study of food. The focus in these instances has been with univariate and bivariate data. Beyond this, food data are becoming increasing multivariate in nature and response measures can comprise several univariate types. A typical large study can involve 10–20 responses including sensory and instrumental determinations. Ultimately, the purpose of such data gathering and experiments is to ascertain which, if any, of these variates have significant effects on, or relationships with, other responses such as consumer acceptability, purchase intention, or storage stability of products.

In this chapter, the methods illustrated require more sophisticated software that is less generally available, although most examples are analysed using SPSS (SPSS Software (IBM®/SPSS®)), which is available as a student version (e.g. with the text by Malhotra and Peterson (2006)). More of a descriptive style is taken with the analysis without the calculation detail given in the previous chapters. The essentials are covered with the intention of giving a brief guide to understanding and interpreting some of the main methods. More advanced accounts and details of methodology abound, and the reader is directed to several texts.

12.2 Multivariate methods and their characteristics

How do multivariate methods differ from univariate?
When studying many variables, it is important to get a clear picture of the function of the analysis and the role played by variables and objects under study. In the univariate case, independent and dependent variables have been identified. Here, there is one dependent variable. For the multivariate situation, the number of both independent and dependent variables can increase. Each method can be classified into groups according to the number and nature of the dependent (Hair *et al.*

Statistical Methods for Food Science: Introductory Procedures for the Food Practitioner, Second Edition. John A. Bower.
© 2013 John Wiley & Sons, Ltd. Published 2013 by John Wiley & Sons, Ltd.

Table 12.1   Some multivariate methods.

Multivariate method	Independents	Dependents	Function
Dependence methods			
Multivariate ANOVA	One or more	Two or more	Significance of effects
Multiple regression	Two or more	One	Significance of prediction
Discriminant analysis	Two or more	One	Significance of discrimination
Conjoint analysis	Two or more	One	Significance of effects
Partial least squares	Two or more	One	Prediction from latent variables
Interdependence methods			
Factor analysis, principal component analysis	–	–	Data reduction
Correspondence analysis	–	–	Nominal data optimal scaling
Cluster analysis	–	–	Grouping by distance
Preference mapping	–	–	Preference dimensions by PCA
Procrustes analysis	–	–	Consensus of different configurations

1998), but essentially, two main groups arise – those that involve *independent and dependent variables*, and those that look for *interdependence* (Table 12.1).

It must be stressed that these are just some of the multivariate techniques. Many others are in use, others are emerging continuously and existing methods are being refined and extended. Some explanation of a selection of methods follows with some graphical outputs of analysis. NB: SPSS outputs have a lot of information, some of which can be ignored. The essential parts are given in the examples.

12.3   Multivariate modes

Some methods are extensions of the univariate circumstance. Previous chapters have dealt with ANOVA, correlation and regression, and these have multivariate counterparts. They are ideal stepping stones to start the explanations where readers have some familiarity with terms and stages and a brief account of these is given to introduce the subject.

Perhaps the most daunting aspect of multivariate analysis is the use of *perceptual mapping graphs* that display results in more than two dimensions. Several methods are characterised by comparing objects and variables based on dimensionality in hyperspace. These comparisons are often done on *distance measures* of which there are a number of possibilities. A simple distance measure between two objects or for a variable can be obtained by subtraction. Representation of the distance among objects based on 20 measures requires display on several dimensions. Of all such methods, ***principal component analysis*** is arguably the most prevalent in the literature. It provides an ideal route for introduction to the topic and more detail is given.

Assumptions and checks are complex, but often multivariate normal distributions are assumed and that groups being compared have homogeneity of variability.

Checking of assumptions can involve detailed analysis and this is referred to in some examples.

12.3.1 Multiple regression

The ***multiple regression*** (MR) method is similar to simple regression (Section 6.4) in that there is one metric *dependent variable*, but *two or more independent* metric variables involved (strictly speaking, MR is not a multivariate method as it has a single dependent variable). The relationship between these and the dependent is established. On analysis, prediction of the dependent is possibly based on several measures, and thus, the model can be more effective (Pike 1986). Linearity is assumed in the basic MR model. Each independent is analysed for its contribution to variation in the dependent. Any that contribute little can be removed, and the model reassessed for the 'fit' to improve predictive ability. MR independents can be assessed for significance by ANOVA.

A processing experiment was conducted to study the effect of temperature and two beater settings on the viscosity of a product. The results were analysed by MR (Table 12.2).

Table 12.2 Multiple regression (SPSS).

Model summary

Model	R	R square	Adjusted R square	Standard error of the estimate
1	0.886[a]	0.786	0.754	106.78345

[a] Predictors: (Constant), beat_time, beat_speed, temp.

ANOVA[a]

Model		Sum of squares	df	Mean square	F	Significance
1	Regression	836895.2	3	278965.071	24.465	0.000[b]
	Residual	228054.1	20	11402.706		
	Total	1064949	23			

[a] Dependent Variable: viscosity.
[b] Predictors: (Constant), beat_time, beat_speed, temp.

Coefficients[a]

Model		Unstandardised coefficients		Standardised coefficients		
		B	Standard error	Beta	t	Significance
1	(Constant)	−209.774	184.035		−1.140	0.268
	temp	14.437	2.794	0.685	5.167	0.000
	beat_speed	−2.384	1.029	−0.279	−2.316	0.031
	beat_time	2.068	5.145	0.048	0.402	0.692

[a] Dependent variable: viscosity.

The regression model is significant in effect (p-value < 0.05) and the *goodness of fit* of the model is given by adjusted R square which is high. Thus, the independents explain 75.4% of the variation in viscosity. Two of the independent variables, temperature and beating speed have had significant effects (p-value of t is <0.05), one in a positive manner and one negative. Beating time appears to have much less of an effect and has smaller coefficients. The analysis can be redone with 'beating time' removed, which gives a slight improvement of 'adjusted R square' (76.4%). Both Minitab and Megastat can carry out the above analysis and produce the same result. Excel itself can perform MR, although the **Data Analysis Toolpak** does not indicate this. The procedure described for simple regression (Section 6.4.2) is followed with the data in columns. The independent variables must be in contiguous columns and entered as a block in **Input X̲ Range** (in this case of three columns). Excel output is similar to that above.

Swan and Howie (1984) used MR to model whiskey volatiles to identify compounds related to 'peaty' odour. Whiskies with high- and low-peaty character (sensory) were analysed by gas chromatography, and five phenols and seven unknown compounds were selected by the regression procedure.

MR can also be used with survey data where consumer demographics and status in respect of attitudes and behaviour can be analysed for relationship with another variable scored or rated on a scale. In these regression analyses, nominal responses are coded as *dummy variables* with values of 0 and 1. This corresponds to the 'absence' and 'presence' of a characteristic, respectively, and is applied to demographics such as gender, 'old'/'young' measures, or the nature of food-type purchase such as a 'low fat'/'conventional fat' preference, etc.

12.3.2   *Multivariate analysis of variance*

Multivariate analysis of variance (MANOVA) can assess two or more independent variables for any significance of effects on two or more metric dependents. It allows a joint analysis of each dependent rather than performing several univariate tests, thus avoiding multiple testing risks (Section 5.5.2). More information is gained by this procedure than would be the case with a series of univariate analysis. MANOVA can be applied to assess product sample differences when there are several responses to be compared. For instance, in sensory profiling where a number of attributes in the region of 10–20 are scored on intensities scales. Rather than carry out 20 individual univariate ANOVAs, a single MANOVA can achieve the same purpose.

12.3.3   *Principal component analysis*

Principal component analysis (PCA) is a form of factor analysis. This method has received extensive application in food science because of its ability to reduce large data matrices to more readable formats. It also provides a useful platform on which to base understanding of the multivariate concept (Piggot and Sharman 1986; Risvik 1996).

Table 12.3   Structure of data matrix for multivariate analysis.

Object	Response 1	Response 2	Response 3
1	. . .	. . .	. . .
2	. . .	. . .	. . .
3	. . .	. . .	. . .
4	. . .	. . .	. . .
. . .	. . .	. . .	. . .

Data for PCA typically originate from experiments that include several food objects and several measures (Table 12.3).

A summary table of mean values can be calculated in a univariate manner. This may show which objects have differing levels of the response, e.g. product 2 has high-sensory 'meatiness' and low-salt content compared with product 4. This becomes more difficult as the number of objects and variables increases. Additionally, the 'whole picture' of why one object dominates others or why a prototype has the desired balance of characteristics is difficult to see. PCA essentially reduces the number of variables to a smaller number of 'key' *composite variables*, which account for the majority of the variation present.

The procedure is based on the fact that when there are many measures on a particular object then some of these are likely to be correlated. Variables that are inter-correlated can 'represent' one another. For instance, if variables 1, 2, 3 and 4 are highly correlated with variable 5, then they will all change as variable 5 changes. A composite variable derived from these could reduce these five variables to one. In PCA analysis, this could constitute the first component. A second component (uncorrelated with the first) can then be derived to examine more variation. The analysis is often based on derivation of measures from a ***correlation matrix***, which displays the inter-correlation of all the variables. In this way, PCA reduces the data and it is referred to as a ***data reduction method***. It makes it possible to obtain a simpler composite view of a large number of measures made on the series of individual objects.

The objects can be a group of people (subjects), a selection of food products or a list of new food concept ideas, etc. The measures or variables can be made directly on the objects or obtained indirectly by questionnaire survey or by observation, etc. They can be analytical instrumental measures such as a meat content or mechanical firmness, sensory intensity measures such as crispness, consumer preference, perceptual data like 'appropriateness for use' of food types, or they can be attitudinal in nature – 'agreement' or 'disagreement' with new food policies, etc. Usually, the measures are on a scale that is at least ordinal in nature, but it is also possible to analyse 'yes/no', 'agree/disagree' dichotomous responses in survey data, provided the response is formatted as the count of one of the categories.

PCA looks for a smaller number of underlying factors that explain most of the variation exhibited by the larger number of measures made on the objects. This is based on the degree of correlation between the variables and there must be some correlation for the method to work. The number of measures taken ranges from 5 to 50 or more and PCA can reduce these to three factors, which may explain approximately 70% of the variation in the data. The factors are assumed to represent *latent information*, which is not observable directly via the original measures taken (Risvik 1996). This should allow increased understanding of the nature and relationships of the objects and variables.

The success of the method depends a lot on the experimenter's knowledge of the discipline where it is applied, e.g. food science, consumer studies or food studies and on the choice of objects and measures. The choice should be based on researched information or generated by focus group style discussion, etc. PCA does not make many assumptions about the data, although there are several tests for the appropriateness of its application. Ideally, a ratio of at least 5:1 of subjects to variables for 'general perceptual'-type studies and 2:1 of objects to variables when looking at specific products, etc. This guideline is often infringed in sensory science when PCA is applied to averaged data on a small number of food products, examined on many variables in the form of sensory attributes. No significance testing is involved in the basic form of PCA and it is an exploratory tool, which can also be used on pilot data to identify important features for future experiments.

Example 12.3-1 *Principal component analysis on descriptive analysis data*
A part-trained sensory panel assessed 24 samples of a type of baked fruit biscuit product on nine sensory variables. These included aspects of appearance ('yellowness', 'filling visibility'), texture ('chewiness', 'hardness', 'cohesion') and flavour ('toasted', 'sweet', 'fruity' and 'sour'). Product samples were coded as uppercase characters for display. Some samples were replicates, although the panellists were unaware of this and four sessions were held under full experimental design protocol, with no replicates in the same session. A PCA analysis was performed. What observations and conclusions can be made from the stages in the analysis?

The panel means were used for the analysis with the replicates retained as separate samples. The correlation matrix was prepared first (Table 12.4). It shows a variety of levels of correlation are present ranging from approximately −0.7 to +0.8 with coefficients above magnitude 0.42 being significant.

The matrix was analysed by PCA. The *components* are the factors extracted by analysis and three components explain most of the variation in the data (Table 12.5). Component 1 explains almost 40% and by three components, almost 80% is explained.

The component matrix in the table gives the *loading* of each variable on the components. A high magnitude (near to +1 or −1) for loading means that the variable is highly correlated with that factor, but >0.3 can be enough for importance. Thus, 'fruity' and 'sour' flavours load strongly on component 1, and 'cohesion' loads strongly and negatively on component 2, etc. The loadings across

Table 12.4 Correlation matrix of sensory variables (SPSS).

		a_yellow	a_fill	t_hardness	t_chewy	t_cohesion	f_toast	f_sweet	f_fruity	f_sour
a_yellow	Pearson Correlation	1	−0.540[b]	−0.082	−0.010	−0.370	0.472[a]	−0.188	−0.176	0.019
	p-value		0.006	0.702	0.963	0.075	0.020	0.380	0.411	0.929
a_fill	Pearson Correlation	−0.540[b]	1	−0.118	0.160	0.154	−0.321	−0.119	0.515[a]	0.364
	p-value	0.006		0.582	0.454	0.471	0.126	0.581	0.010	0.080
t_hardness	Pearson Correlation	−0.082	−0.118	1	−0.707[b]	−0.129	0.433[a]	−0.290	−0.601[b]	−0.573[b]
	p-value	0.702	0.582		0.000	0.547	0.035	0.170	0.002	0.003
t_chewy	Pearson Correlation	−0.010	0.160	−0.707[b]	1	−0.065	−0.280	0.137	0.463[a]	0.563[b]
	p-value	0.963	0.454	0.000		0.762	0.186	0.524	0.023	0.004
t_cohesion	Pearson Correlation	−0.370	0.154	−0.129	−0.065	1	−0.260	0.493[a]	−0.027	−0.240
	p-value	0.075	0.471	0.547	0.762		0.219	0.014	0.900	0.258
f_toast	Pearson Correlation	0.472[a]	−0.321	0.433[a]	−0.280	−0.260	1	−0.421[a]	−0.583[b]	−0.511[a]
	p-value	0.020	0.126	0.035	0.186	0.219		0.041	0.003	0.011
f_sweet	Pearson Correlation	−0.188	−0.119	−0.290	0.137	0.493[a]	−0.421[a]	1	0.159	−0.046
	p-value	0.380	0.581	0.170	0.524	0.014	0.041		0.458	0.831
f_fruity	Pearson Correlation	−0.176	0.515[a]	−0.601[b]	0.463[a]	−0.027	−0.583[b]	0.159	1	0.807[b]
	p-value	0.411	0.010	0.002	0.023	0.900	0.003	0.458		0.000
f_sour	Pearson Correlation	0.019	0.364	−0.573[b]	0.563[b]	−0.240	−0.511[a]	−0.046	0.807[b]	1
	p-value	0.929	0.080	0.003	0.004	0.258	0.011	0.831	0.000	

[a] Correlation is significant at the 0.05 level (two-tailed).
[b] Correlation is significant at the 0.01 level (two-tailed).

Table 12.5   Loadings of components and sensory variables, and variance explained (SPSS).

	Component matrix[a]		
	Component		
	1	2	3
a_yellow	−0.331	0.700	0.446
a_fill	0.530	−0.228	−0.687
t_hardness	−0.762	−0.229	−0.429
t_chewy	0.678	0.341	0.255
t_cohesion	0.156	−0.787	0.278
f_toast	−0.761	0.336	−0.015
f_sweet	0.318	−0.540	0.636
f_fruity	0.878	0.160	−0.151
f_sour	0.803	0.444	−0.141

[a]Three components extracted.
Extraction method: principal component analysis.

	Total variance explained		
	Initial eigenvalues		
Component	Total	% of variance	Cumulative %
1	3.553	39.475	39.475
2	1.956	21.734	61.209
3	1.445	16.051	77.260

Extraction method: principal component analysis.

the components provide the coordinates for the ***component plot*** (or *loadings plot*). Figure 12.1 shows components 1 and 2.

The plot shows a spread of the loadings with 'toasted', well separated from 'fruity', 'sour' and 'chewy' attributes on *dimension 1* (component 1). Component 2 shows a separation linked to high sweetness and a more cohesive structure from stronger yellow (more baked) appearance. In terms of *latent information*, it appears that fruity and sour flavours are associated with a chewy texture, possibly linked to a more obvious visible presence of fruit. There is a definite textural dimension for component 1, with a marked separation of hard from chewy textures. Cohesion is related to sweetness, which may reflect formulation contents in terms of sugars, syrup, etc.

The component plot does not show the products and information on these is obtained via calculation of the *object (product) scores*. An o***bject score*** is a composite measure calculated from the factor weights and the original variables. Scores describe the properties of the products or samples examined (or the views of the individual subjects in survey data PCA analysis). The ***score plot*** is similar to the component graph and shows a scatter diagram of the product scores against the factors (Fig. 12.2). It is now possible to locate the individual products and to examine their position in the graph along with component plot.

Objects positioned in a similar location to variable loadings are associated with higher measures for that variable. Objects with no associated loadings will have lower values for the measures. In the plot, there are several products in a large group

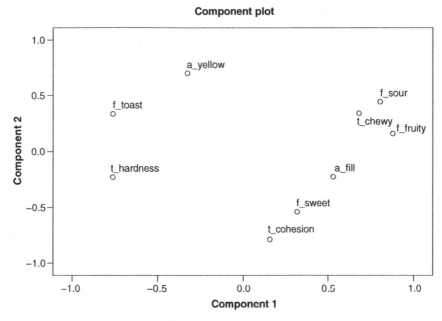

Fig. 12.1   Component plot of sensory variables (SPSS).

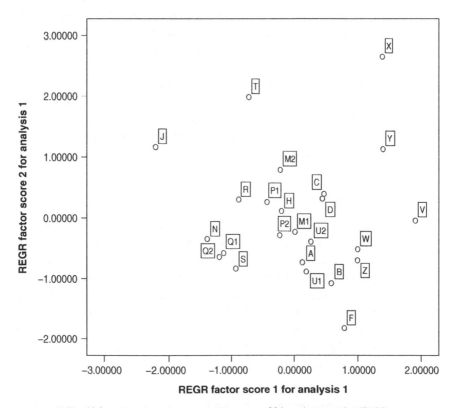

Fig. 12.2   Plot of regression (REGR) scores of 24 product samples (SPSS).

just off-centre. In addition, some individual products are scattered 'away from the pack', although not excessively so.[1] There is limited representation of hard-textured, toasted flavour products (J), and those with marked baked appearance (T). Three products have strong fruity, sour and chewy characteristics (X, Y and V). A larger but more separated group of products is associated with high sweetness and cohesion. A small tighter group of products (Q, N and S) is similar to these latter ones, but is located neared to harder texture. Note that the relative position of replicate samples for products P, M, U and Q provide a measure of the panel's consistency in measurement. This is 'good' in the case of product Q, but less so for others, especially M. The reasons for lack of a close grouping of replicates can also depend on the confusability of the sensory attributes as well as poor panel consistency.

The analysis has provided some new information concerning the characteristics of the biscuit products. Some appear to be very similar, but the positioning should be examined in other identified components (in the example, components 2 and 3 can be plotted together) before dismissing an object as having 'no characteristics'. Another possibility is that during vocabulary generation a key variable of importance was missed.

Much more information can be gleaned from the plots and analysis above, especially with additional data. Assuming that they are already on the market (competitor products) then information on ingredient content and sales figures, consumer liking, etc., can be linked to the PCA results. This can be used to indicate gaps in the market for future product development. Depending on further analysis, it seems that many of these products are similar and do not stand out. There is limited number of products with high-fruit content – if these have higher market shares then there may be room for more of this type. There are few products with hard textures and baked colour and flavour. Depending on sales and consumer liking, more possibilities may lie with this type of product. Some examples of linking the PCA results to preference data are illustrated below (Section 12.4).

12.3.4   Cluster analysis

Cluster analysis is a classification method, which forms subgroups of similar objects based on assessment of the distance between measured variables (Jacobson and Gunderson 1986). It is also an exploration method that searches with no prior definition. Examples of groups that could be formed are food products with similar characteristics, which could be measured not only on virtually any response variable including sensory, instrumental, storage stabilities but also on measures of prices, flavour variety, etc. Consumer demographics and attitudes with respect to food preference and purchase frequency, etc., can also be used to form clusters of people with similar likes and dislikes and levels of purchase frequency.

---

[1] Objects positioned at large distances from all others may be classed as ***outliers*** and they can distort the plots making interpretation difficult. Removal is one solution, but essentially objects should be part of a set of similar products and there should be representation of more than one of each type.

Virtually, any form of object can be classified on any measure. Many food and beverage groups have been examined by cluster analysis as a classification aid, including alcoholic beverages and cheeses, but it has also been applied to food-related materials as in the case of microbial strains, etc. In cluster analysis, where several disparate measures are possible on the objects, *standardisation* is required, otherwise measures on large scales with higher values than others would tend to dominate.

Using the profile data above, cluster analysis produced a ***dendrogram diagram*** (Fig. 12.3) that shows clustering of the objects (biscuit products) based on the sensory characteristics. In a sense, this is a test of the data because if the identity of the objects were known then it would be expected that clusters of similar products would form (assuming that the sensory measures have some discrimination ability).

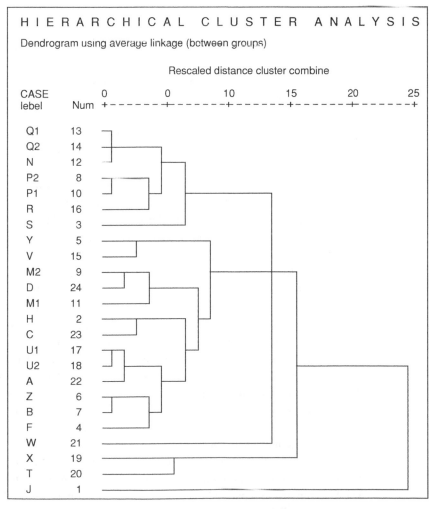

Fig. 12.3   Cluster analysis dendrogram (SPSS).

Encouragingly, for the panel, replicates are grouped within the same clusters. There are two larger clusters, two smaller ones and product J stands on its own – quite clearly, it cannot be classified along with any other product on the basis of the sensory measures. Generally, the grouping of the products is similar to that of the PCA score plot, although there are exceptions. An instance is with products X and W that are clustered together, but which are well separated on dimension 2 of the score plot above. These two products are similar in ingredient make-up, but the PCA appears to have picked up a feature (based on the sensory data) that separates them.

The usefulness of the cluster analysis will of course be more so when classification groups are unknown. One such application of this form was used for free choice profile data in sensory work to examine possible grouping of food objects based on the free choice terms (Baxter *et al.* 2001).

12.3.5   Correspondence analysis

Analysis of nominal data is limited by conventional methods due to the lower level of measurement (Section 2.2.3). Association of two nominal variable can be done using crosstabulation tables, etc., but interpretation can be cumbersome when several categories are involved in each variable, and when the number of variables increases. There are a number of multivariate methods that can be used for such circumstances. *Correspondence analysis* (CA) is one of a number of *optimal scaling* methods that can be applied in this way (SPSS 1990). Objects and measures can be explored together where differences are based on association of category (nominal data) terms.

The data must be nominal in nature or be converted to nominal. The methods produce *metric quantifications* from qualitative data, from which identification of order in categories can be obtained. These scales are 'optimised' by the method to explain variation. Two or more variables with many categories can be analysed together and perceptual mapping displays aid interpretation.

CA can be used for the simplest case with two variables. In the graphical output, a CA *biplot* is displayed where objects and variables can be viewed together (SPSS 1990). This can be envisaged as a type of 'graphic crosstabulation' that can be studied to locate possible association of objects and measure categories. An example biplot (Fig. 12.4) shows the association of sample apple juices with hedonic category as an exploration of preference data. The hedonic ratings were gathered using the nine-point scale (degree of liking; DOL), so the data were declared as nominal with the anchors allocated as categories. This may appear to lose information as the hedonic scale would be likely ordinal and possibly interval in nature, but this study was exploring the data to address this point, and the lack of differentiation of the products by conventional ANOVA.

The plot shows the visual proximity of products with scale anchors, which can reveal more about the preference structure and the balance of 'DOL' categories. The first dimension separated the products in a similar manner to ANOVA in that product 2 is 'disliked' and others are 'liked'. This latter group (p1 and p3–p6)

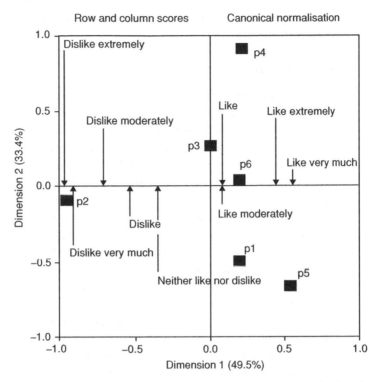

Row and column scores   Canonical normalisation

Fig. 12.4   Correspondence analysis biplot of products and hedonic categories (based on SPSS analysis).

was not distinguished by ANOVA. Dimension 2 reveals some new information in that the liked products are spread out and two (p1, p5) have some association with the neutral category ('neither like nor dislike'). This may indicate segmentation in consumer preferences, which could be useful for marketing.

The biplot can be further analysed by marking orthogonal projections ('right-angled lines') onto various parts of the graph. One set of interest for this study was a series of lines from the hedonic categories onto dimension 1. This reveals the order of use of the categories. In this case, they appear to be roughly in order as per the original hedonic scale, but not exactly so. The 'dislike' categories progress almost as intended, but the 'like' categories are more confused. This could indicate that consumers were less discriminating with that region of the scale and instead of using four points of liking, only two were used.

The type of analysis above lends itself to other situations where there may be concerns about the manner in which a scale was used or when there are doubts about the assumed linear relationships in correlation studies, or for relationship studies with mixed levels of measurement. There are forms of the analysis for multiple categorical variables (*homogeneity analysis*) and *non-linear principal component analysis* where there are more than two variables and mixed levels of measurement. McEwan and Schlich (1991/1992) describe the use of CA for sensory analysis, but it has also appeared as a technique for comparing relationships of predominant 'taste' and texture descriptors with hedonic categories between obese and lean subjects (Cox *et al.* 1999).

12.3.6   Conjoint analysis

During retail food purchase, consumers are presented with a choice from a variety of product 'conditions' in the form of variations in price, composition, packaging, flavour, quality, etc. Such situations are studied in food marketing research as part of product development. The product (or commodity) conditions are attributes that are assumed to be 'weighed up' for their value before a choice is made. Such a procedure can be set up in a factorial design as part of a *conjoint analysis* study (Hair *et al.* 1998).

The consumer can be presented with all the possible combination of the attributes in real situation, e.g. in a food store, and observations can be made of the choices. This has a major advantage in terms of being very valid in many respects, but there are several difficulties with this approach. One is that to test all the attributes adequately several levels for each would be required. If there are several attributes then the number of unique combinations becomes large, namely with five types of packaging, five flavours and five price levels, a full factorial would generate 125 treatments – an excessive amount to display in a store for one product group and for a consumer to evaluate. Thus, the store display approach is better left to a reduced number of treatments after screening of all the factors.

There are still logistical problems in displaying many treatments of real food products even in a food mall or hall test display. A convenient solution is to use photographs of all the various combinations, or simply cards or sections in questionnaires to display a printout of each set. Newer approaches employ computer consoles with full colour, sound and video to produce atmosphere to make the choice as realistic as possible. One difficulty still remains in that a large number of choices may still be unrealistic, especially if the consumer is an unpaid volunteer. Such design difficulties can be circumvented by use of some of the techniques discussed in Section 7.4.4. One solution is to use *incomplete block designs* for the treatment delivery and present a smaller number of combinations to consumers, such as 12 treatments per consumer. Another method is to use a *fractional factorial* structure to reduce the number of treatments.

The output of Table 12.6 shows analysis of a conjoint study of this latter type, with the response of purchase intent to varying price and label information. Both these attributes were significant ($p < 0.001$) in effect on 'purchase intention', and that the price outweighed the label in decision-making. The above type of approach is common in food conjoint analysis applications. Moskowitz and Silcher (2006) reviewed the method and its extensive application in food research and Moskowitz *et al.* (2006) highlighted its importance in the development of consumer research.

12.3.7   Discriminant analysis

Discriminant analysis (DA) examines a single non-metric dependent variable and several metric independents for significance of discrimination (Powers and Ware

Table 12.6   Conjoint analysis of price and label style effect (SPSS).

Averaged Importance	Utility	Factor		
		Price		
|68.92 |	0.7042	|----	1
	−0.2667	--|	2
	0.2875	|--	3
	−0.5708	---|	4
	−0.2333	-|	5
31.08|	0.1000	Label		
	0.1000	|-	1
	−0.3083	--|	2
	0.1625	|-	3
	−0.0083	|	4
	0.2667	|--	5
	3.0167	Constant		

Pearson's $R = 0.707$ Significance $= 0.0000$

1986). A set of groups or classes is *pre-defined* for the dependent and the analysis looks for independent measures that discriminate. This is unlike cluster analysis, which has no pre-defined groups.

Again, the applications are unlimited using multiple sensory and instrumental measures on food products to ascertain variables that can be used to 'tell the difference' between types. Such information is useful in identification procedures and for characterising why certain products are successful, whereas others are not. Another application is in sensory profiling methods where vocabulary generation methods can result in extensive lists. DA can be used to select attributes that are more efficient at differentiating product samples (Powers and Ware 1986).

This latter application is shown below with the above sensory data. DA was used to select the variables that distinguished most amongst the products (Table 12.7).

At the start of the analysis, all the sensory variables were entered. Discrimination can be based on a number of criteria, and in this case, it was the ability to demonstrate a significant difference between samples. As the analysis progressed, the attribute with the most discriminatory ability was entered first and then others followed up to a limit beyond which any remaining were not sufficiently able to discriminate. At the end, four (at the foot of the table) of the original nine were identified as the most effective at discriminating the product set. If an excessively large set of discriminators were available then such analysis provides one way of pruning the attribute list.

12.3.8   *Partial least squares regression*

Partial least squares (PLS) regression has elements of both inter-relationships and prediction. It searches for relationships between two or more sets of data.

Table 12.7   Discriminant analysis of sensory attributes (SPSS).

	Tests of equality of group means				
	Wilks' Lambda	*F*	df1	df2	Significance
a_yellow	0.005	40.306	19	4	0.001
a_fill	0.010	21.605	19	4	0.004
t_hardness	0.018	11.532	19	4	0.014
t_chewy	0.018	11.740	19	4	0.014
t_cohesion	0.070	2.805	19	4	0.164
f_toast	0.026	7.874	19	4	0.029
f_sweet	0.060	3.308	19	4	0.127
f_fruity	0.002	111.667	19	4	0.000
f_sour	0.026	7.993	19	4	0.028

	Variables in the analysis			
Step		Tolerance	*F* to Remove	Wilks' Lambda
1	f_fruity	1.000	111.667	
2	f_fruity	0.933	86.985	0.005
	a_yellow	0.933	31.401	0.002
3	f_fruity	0.244	134.684	0.000
	a_yellow	0.896	20.555	0.000
	t_hardness	0.257	13.627	0.000
4	f_fruity	0.031	495.915	0.000
	a_yellow	0.042	176.261	0.000
	t_hardness	0.029	55.654	0.000
	t_cohesion	0.039	14.857	0.000

The method differs from MR where the independent variables are analysed in their original form for the variation in the dependents. In PLS, the independents are analysed for a smaller number of latent variables. This analysis also includes variation in the dependent. The reduced set is used to search for relationships in both dependents and independents. One major advantage is the large number of independents that can be used, as eventually they are reduced to a few. MR cannot do this unless sample size is increased. Additionally, PLS gives less error with small sample sizes and gives prediction for the dependent, unlike another multivariate technique, canonical correlation, which is purely correlative (Martens and Martens 1986). Piggot *et al.* (1993) used PLS to predict flavour intensity in blackcurrant concentrates. Predictors were based on gas chromatographic peak area data, and these gave high positive correlations and high predictive power with sensory intensity.

12.3.9   Preference mapping

Internal preference mapping is basically a PCA treatment of consumer hedonic data and product samples (Greenhoff and MacFie 1994; McEwan 1996). The difference is that instead of a number of variable measures on each sample, each individual consumer is viewed as a measure. For instance, an investigation with 6 product samples and 100 consumer assessors would be viewed as a '6 objects plus 100 variables' study for an internal preference analysis. *External preference mapping* includes both sensory and consumer data in a more involved analysis based on regression of sensory profile measures as predictors of the consumer response. An example of internal preference mapping is illustrated below for a consumer panel test.

Example 12.3-2 *Internal preference mapping on hedonic data*
A consumer panel of 75 participants rated eight prototypes of a new high-fibre, low-fat product (p1–p8) using a nine-point hedonic scale. The data were subjected to internal preference analysis. What observations and conclusions can be made from the map?

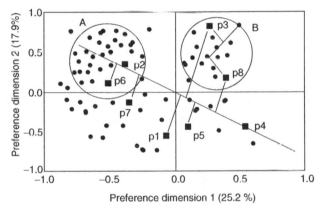

Fig. 12.5 Internal preference map of consumer hedonic rating of high-fibre, low-fat prototype products (based on SPSS analysis).

Internal preference analysis produced a map (Fig. 12.5). The prototypes (solid squares) are well separated, but most consumers (solid circles) are grouped around four of the products with much less for the others. Groups can be identified and demarcated for individual product preference by use of orthogonal lines drawn from the group through the origin. The line drawn from the largest group of consumers (A) shows that the group appeared to like prototype p6 the most, followed closely by p2 and p7, and they liked prototype 4 the least. Another smaller group (B) shows preference for the more isolated products p3 and p8.

Based on the analysis, prototypes p2 and p6 have most potential. These would have to be examined for their individual characteristics via other data, e.g. sensory and instrumental (see Section 12.4).

12.3.10   *Procrustes analysis*

Generalised Procrustes analysis (GPA) is a method that has been used in food studies largely for sensory profiling data, particularly free choice profiling. This latter technique attempts to avoid lengthy training by allowing assessors to generate their own individual attribute list for a product set. The difficulty comes when analysing the resulting data, in that each assessor's data are different in terms of attributes, even though the scale and products are the same. GPA can analyse such data by using a technique that stretches and rotates each individual matrix so that a *consensus configuration* is achieved (Dijksterhuis and Gower 1991/1992).

The method can also be applied to standard profiling data given that error sources such as assessor effect and assessor–product interaction can be present. This is particularly so when training time is limited and assessor understanding of attributes definitions and use of scales may be lacking (Dijksterhuis 1996).

12.4   Relationship of consumer preference with sensory measures

As an example of how some of the above multivariate data and methods can be combined, another example of a preference map is given below. This displays results from a consumer study using a selection of the biscuit products analysed above by PCA. A subset of nine of the products was used, selected because they exhibited a range of sensory quality and they showed a degree of separation in the PCA score plot. A group of 50 consumers assessed these using a nine-point hedonic scale. The ratings were analysed to give the internal preference map (Fig. 12.6).

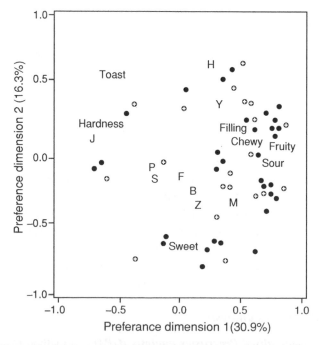

Fig. 12.6   Internal preference maps of hedonic data (consumer panel) with significantly correlated attributes (sensory panel) (based on SPSS analysis).

The preference dimensions derived from the analysis were examined for any correlation with the sensory attribute measures for these nine products. Any attributes that were significantly correlated (*correlation coefficient r* = ±0.7) with preference dimensions were used to plot the sensory data onto the preference map. In this way, attributes most important in liking can be identified.

There are some distinct groups of consumers, but most form around the fruity, surface-coated, chewy-textured biscuits with an obvious layer of fruit material – products H, Y and M. Several products are grouped in the lower half of the plot, although near to the origin. This usually means that the samples are not showing marked characteristics, but the position is also near the 'sweet' sensory attribute, which could point to a 'sweet tooth' group.

Consumer demographics and other characteristics can be incorporated into the preference map displays. In Fig. 12.6, the solid circles indicate consumers who were 'regular to frequent' purchasers of such products and the clear circles were 'infrequent to rare'. The small group of consumers, who possibly prefer sweet tastes, appear to have more incidence of 'frequent purchaser', and the main group shows this effect to a lesser degree.

References

Baxter, I. A., Dijksterhuis, G. B. and Bower, J. A. (2001) Converting free-choice data to consensus to explore the influence of demographic variables on consumer perceptions. Oral presentation abstract at *The 4th Pangborn Sensory Science Symposium*, 22–26 July 2001, Dijon, France.

Cox, D. N., Perry, L., Moore, P. B., Vallis, L. and Mela, D. J. (1999) Sensory and hedonic associations with macronutrient and energy intakes of lean and obese consumers. *International Journal of Obesity*, **23**, 403–410.

Dijksterhuis, G. (1996) Procrustes analysis is sensory research. In *Multivariate Analysis of Data in Sensory Science* by T. Næs and E. Risvik (eds). Elsevier Science BV, Amsterdam, pp. 185–220.

Dijksterhuis, G. B. and Gower, J. C. (1991/1992) The interpretation of GPA analyses and allied methods. *Food Quality and Preference*, **3**, 67–87.

Greenhoff, K. and Macfie, H. J. H. (1994) Preference mapping in practice. In *Measurement of Food Preference* by H. J. H. MacFie and D. M. H. Thompson (eds). Blackie Academic and Professional, London, pp. 137–166.

Hair, J. F., Jr., Anderson, R. E., Tatham, R. L. and Black, W. C. (1998) *Multivariate Data Analysis*, 4th edn. Prentice Hall International, Upper Saddle River, NJ.

Jacobson, T. and Gunderson, R. W. (1986) Applied cluster analysis. In *Statistical Procedures in Food Research* by J. R. Piggot (ed.). Elsevier Applied Science, London, pp. 361–408.

Malhotra, N. K. and Peterson, M. (2006) *Basic Marketing Research*, 2nd edn. International Edition, Pearson Education, Upper Saddle River, NJ.

Martens, M. and Martens, H. (1986) Partial least squares regression. In *Statistical Procedures in Food Research* by J. R. Piggot (ed.). Elsevier Applied Science, London, pp. 293–360.

McEwan, J. (1996) Preference mapping for product optimization. In *Multivariate Analysis of Data in Sensory Science* by T. Næs and E. Risvik (eds). Elsevier Science BV, Amsterdam, pp. 71–102.

McEwan, J. and Schlich, P. (1991/1992) Correspondence analysis in sensory evaluation. *Food Quality and Preference*, **3**, 23–26.

Moskowitz, H. R., Gofman, A., Beckley, J. and Ashman, H. (2006) Founding a new science: mind genomics. *Journal of Sensory Studies*, **21**, 266–307.

Moskowitz, H. R. and Silcher, M. (2006) The applications of conjoint analysis and their possible uses in Sensometrics. *Food Quality and Preference*, **17**, 145–165.

Piggot, J. R., Paterson, A. and Clyne, J. (1993) Prediction of flavour intensity of blackcurrant (*Ribes Nigrum* L.) drinks from compositional data on fruit concentrates by partial least squares regression. *International Journal of Food Science and Technology*, **28**, 629–637.

Piggot, J. R. and Sharman, K. (1986) Methods to aid interpretation of multidimensional data. In *Statistical Procedures in Food Research* by J. R. Piggot (ed.). Elsevier Applied Science, London, pp. 181–232.

Pike, D. (1986) A practical approach to regression. In *Statistical Procedures in Food Research* by J. R. Piggot (ed.). Elsevier Applied Science, London, pp. 61–100.

Powers, J. J. and Ware, G. O. (1986) Discrimination analysis. In *Statistical Procedures in Food Research* by J. R. Piggot (ed.). Elsevier Applied Science, London, pp. 125–180.

Risvik, E. (1996) Understanding latent phenomena. In *Multivariate Analysis of Data in Sensory Science* by T. Næs and E. Risvik (eds). Elsevier Science BV, Amsterdam, pp. 5–36.

SPSS (1990) *SPSS Categories*, SPSS, Chicago, IL.

Swan, J. S. and Howie, D. (1984) Correlation of sensory and analytical data in flavour studies into Scotch Whiskey. In *Proceedings of the Alko Symposium on Flavour Research of Alcoholic Beverages*, Helsinki, 1984, L. Hykanen and P. Leytonen (eds). Foundation for Biotechnical and Industrial Fermentation Research, **3**, 291–300.

Index

Food Science and Technology

GENERAL FOOD SCIENCE & TECHNOLOGY, ENGINEERING AND PROCESSING

Organic Production and Food Quality: A Down to Earth Analysis	Blair	9780813812175
Handbook of Vegetables and Vegetable Processing	Sinha	9780813815411
Nonthermal Processing Technologies for Food	Zhang	9780813816685
Thermal Procesing of Foods: Control and Automation	Sandeep	9780813810072
Innovative Food Processing Technologies	Knoerzer	9780813817545
Handbook of Lean Manufacturing in the Food Industry	Dudbridge	9781405183673
Intelligent Agrifood Networks and Chains	Bourlakis	9781405182997
Practical Food Rheology	Norton	9781405199780
Food Flavour Technology, 2nd edition	Taylor	9781405185431
Food Mixing: Principles and Applications	Cullen	9781405177542
Confectionery and Chocolate Engineering	Mohos	9781405194709
Industrial Chocolate Manufacture and Use, 4th edition	Beckett	9781405139496
Chocolate Science and Technology	Afoakwa	9781405199063
Essentials of Thermal Processing	Tucker	9781405190589
Calorimetry in Food Processing: Analysis and Design of Food Systems	Kaletunç	9780813814834
Fruit and Vegetable Phytochemicals	de la Rosa	9780813803203
Water Properties in Food, Health, Pharma and Biological Systems	Reid	9780813812731
Food Science and Technology (textbook)	Campbell-Platt	9780632064212
IFIS Dictionary of Food Science and Technology, 2nd edition	IFIS	9781405187404
Drying Technologies in Food Processing	Chen	9781405157636
Biotechnology in Flavor Production	Havkin-Frenkel	9781405156493
Frozen Food Science and Technology	Evans	9781405154789
Sustainability in the Food Industry	Baldwin	9780813808468
Kosher Food Production, 2nd edition	Blech	9780813820934

FUNCTIONAL FOODS, NUTRACEUTICALS & HEALTH

Functional Foods, Nutraceuticals and Degenerative Disease Prevention	Paliyath	9780813824536
Nondigestible Carbohydrates and Digestive Health	Paeschke	9780813817620
Bioactive Proteins and Peptides as Functional Foods and Nutraceuticals	Mine	9780813813110
Probiotics and Health Claims	Kneifel	9781405194914
Functional Food Product Development	Smith	9781405178761
Nutraceuticals, Glycemic Health and Type 2 Diabetes	Pasupuleti	9780813829333
Nutrigenomics and Proteomics in Health and Disease	Mine	9780813800332
Prebiotics and Probiotics Handbook, 2nd edition	Jardine	9781905224524
Whey Processing, Functionality and Health Benefits	Onwulata	9780813809038
Weight Control and Slimming Ingredients in Food Technology	Cho	9780813813233

INGREDIENTS

Hydrocolloids in Food Processing	Laaman	9780813820767
Natural Food Flavors and Colorants	Attokaran	9780813821108
Handbook of Vanilla Science and Technology	Havkin-Frenkel	9781405193252
Enzymes in Food Technology, 2nd edition	Whitehurst	9781405183666
Food Stabilisers, Thickeners and Gelling Agents	Imeson	9781405132671
Glucose Syrups – Technology and Applications	Hull	9781405175562
Dictionary of Flavors, 2nd edition	De Rovira	9780813821351
Vegetable Oils in Food Technology, 2nd edition	Gunstone	9781444332681
Oils and Fats in the Food Industry	Gunstone	9781405171212
Fish Oils	Rossell	9781905224630
Food Colours Handbook	Emerton	9781905224449
Sweeteners Handbook	Wilson	9781905224425
Sweeteners and Sugar Alternatives in Food Technology	Mitchell	9781405134347

FOOD SAFETY, QUALITY AND MICROBIOLOGY

Food Safety for the 21st Century	Wallace	9781405189118
The Microbiology of Safe Food, 2nd edition	Forsythe	9781405140058
Analysis of Endocrine Disrupting Compounds in Food	Nollet	9780813818160
Microbial Safety of Fresh Produce	Fan	9780813804163
Biotechnology of Lactic Acid Bacteria: Novel Applications	Mozzi	9780813815831
HACCP and ISO 22000 – Application to Foods of Animal Origin	Arvanitoyannis	9781405153669
Food Microbiology: An Introduction, 2nd edition	Montville	9781405189132
Management of Food Allergens	Coutts	9781405167581
Campylobacter	Bell	9781405156288
Bioactive Compounds in Foods	Gilbert	9781405158756
Color Atlas of Postharvest Quality of Fruits and Vegetables	Nunes	9780813817521
Microbiological Safety of Food in Health Care Settings	Lund	9781405122207
Food Biodeterioration and Preservation	Tucker	9781405154178
Phycotoxins	Botana	9780813827001
Advances in Food Diagnostics	Nollet	9780813822211
Advances in Thermal and Non-Thermal Food Preservation	Tewari	9780813829685

For further details and ordering information, please visit www.wiley.com/go/food

Food Science and Technology from Wiley-Blackwell

SENSORY SCIENCE, CONSUMER RESEARCH & NEW PRODUCT DEVELOPMENT

Sensory Evaluation: A Practical Handbook	Kemp	9781405162104
Statistical Methods for Food Science	Bower	9781405167642
Concept Research in Food Product Design and Development	Moskowitz	9780813824246
Sensory and Consumer Research in Food Product Design and Development	Moskowitz	9780813816326
Sensory Discrimination Tests and Measurements	Bi	9780813811116
Accelerating New Food Product Design and Development	Beckley	9780813808093
Handbook of Organic and Fair Trade Food Marketing	Wright	9781405150583
Multivariate and Probabilistic Analyses of Sensory Science Problems	Meullenet	9780813801780

FOOD LAWS & REGULATIONS

The BRC Global Standard for Food Safety: A Guide to a Successful Audit	Kill	9781405157964
Food Labeling Compliance Review, 4th edition	Summers	9780813821818
Guide to Food Laws and Regulations	Curtis	9780813819464
Regulation of Functional Foods and Nutraceuticals	Hasler	9780813811772

DAIRY FOODS

Dairy Ingredients for Food Processing	Chandan	9780813817460
Processed Cheeses and Analogues	Tamime	9781405186421
Technology of Cheesemaking, 2nd edition	Law	9781405182980
Dairy Fats and Related Products	Tamime	9781405150903
Bioactive Components in Milk and Dairy Products	Park	9780813819822
Milk Processing and Quality Management	Tamime	9781405145305
Dairy Powders and Concentrated Products	Tamime	9781405157643
Cleaning-in-Place: Dairy, Food and Beverage Operations	Tamime	9781405155038
Advanced Dairy Science and Technology	Britz	9781405136181
Dairy Processing and Quality Assurance	Chandan	9780813827568
Structure of Dairy Products	Tamime	9781405129756
Brined Cheeses	Tamime	9781405124607
Fermented Milks	Tamime	9780632064588
Manufacturing Yogurt and Fermented Milks	Chandan	9780813823041
Handbook of Milk of Non-Bovine Mammals	Park	9780813820514
Probiotic Dairy Products	Tamime	9781405121248

SEAFOOD, MEAT AND POULTRY

Handbook of Seafood Quality, Safety and Health Applications	Alasalvar	9781405180702
Fish Canning Handbook	Bratt	9781405180993
Fish Processing – Sustainability and New Opportunities	Hall	9781405190473
Fishery Products: Quality, safety and authenticity	Rehbein	9781405141628
Thermal Processing for Ready-to-Eat Meat Products	Knipe	9780813801483
Handbook of Meat Processing	Toldra	9780813821825
Handbook of Meat, Poultry and Seafood Quality	Nollet	9780813824468

BAKERY & CEREALS

Whole Grains and Health	Marquart	9780813807775
Gluten-Free Food Science and Technology	Gallagher	9781405159159
Baked Products – Science, Technology and Practice	Cauvain	9781405127028
Bakery Products: Science and Technology	Hui	9780813801872
Bakery Food Manufacture and Quality, 2nd edition	Cauvain	9781405176132

BEVERAGES & FERMENTED FOODS/BEVERAGES

Technology of Bottled Water, 3rd edition	Dege	9781405199322
Wine Flavour Chemistry, 2nd edition	Bakker	9781444330427
Wine Quality: Tasting and Selection	Grainger	9781405113663
Beverage Industry Microfiltration	Starbard	9780813812717
Handbook of Fermented Meat and Poultry	Toldra	9780813814773
Microbiology and Technology of Fermented Foods	Hutkins	9780813800189
Carbonated Soft Drinks	Steen	9781405134354
Brewing Yeast and Fermentation	Boulton	9781405152686
Food, Fermentation and Micro-organisms	Bamforth	9780632059874
Wine Production	Grainger	9781405113656
Chemistry and Technology of Soft Drinks and Fruit Juices, 2nd edition	Ashurst	9781405122863

PACKAGING

Food and Beverage Packaging Technology, 2nd edition	Coles	9781405189101
Food Packaging Engineering	Morris	9780813814797
Modified Atmosphere Packaging for Fresh-Cut Fruits and Vegetables	Brody	9780813812748
Packaging Research in Food Product Design and Development	Moskowitz	9780813812229
Packaging for Nonthermal Processing of Food	Han	9780813819440
Packaging Closures and Sealing Systems	Theobald	9781841273372
Modified Atmospheric Processing and Packaging of Fish	Otwell	9780813807683
Paper and Paperboard Packaging Technology	Kirwan	9781405125031

For further details and ordering information, please visit www.wiley.com/go/food

Printed and bound by CPI Group (UK) Ltd, Croydon, CR0 4YY

27/10/2024

14580311-0001